The Changing Water Utility

The Changing Water Utility

Creative Approaches to Effectiveness and Efficiency

Garret P. Westerhoff
Diana Gale
Paul D. Reiter
Scott A. Haskins
Jerome B. Gilbert

John B. Mannion, Editor

American Water Works Association
Dedicated to Safe Drinking Water

American Water Works Association
6666 West Quincy Avenue
Denver CO 80235

Production Manager and Senior Technical Editor: Mindy Burke
Cover design: Scott Nakauchi-Hawn
Production Editor and book design: Alan Livingston
Editorial and production services: WestWords, Inc.

Library of Congress Cataloging-in-Publication Data applied for

ISBN: 0-89867-944-3

Table of Contents

Acknowledgments

This book began with all the fits and starts of a teenager learning to drive a standard transmission, finally roaring away with the satisfying screech of accomplishment. In Phase One, Diana Gale, while discussing the many challenges in facing restructuring and demands for increased efficiency at Seattle Water Department, simply said, "We ought to write a book." The idea floated for several weeks while we were consumed by busy schedules.

We then got together in Seattle after a long day of working on Seattle's DBO project and moved almost offhandedly to Phase Two. We bounced the idea around and became more excited about the changes that were taking place in the water industry at a rate never before encountered. We agreed that many water utility managers were looking for a logical explanation of these changes and for a framework to help them learn from the rapidly evolving experience of others. Of particular concern was the swift rise of privatization in all its forms and guises.

As we talked to other utility managers and analyzed the work in which we were all involved, the concepts for this book flowed naturally. As it turned out, defining the structure of the book was the easy phase. Fitting the development of text into our schedules in a timely way became our albatross. Without the encouragement and support of numerous wonderful people who helped us to make this book a reality, the albatross would have been the victor. In the end, advice and assistance from leaders in dozens of utilities made it possible to assemble a work representative of the best North American experience in progressive refocusing of water utilities.

Our friend and colleague Jack Mannion, the former executive director of the American Water Works Association (AWWA), deserves special mention. He brought clarity, common sense, and a calm to the flurry of draft manuscripts that were being transmitted nonstop between Seattle, California, and New York from his den in the Colorado Rocky Mountain foothills. Functioning as our editor, Jack was patient and uniquely able to bring the contributions of the five of us into an integrated text.

The energy and leadership of Gary Westerhoff must be recognized here, for he was the driving force and chief organizer of the project from the outset. The book would not have come into being without him. Particular thanks to his employer, Malcolm Pirnie, Inc., for its generous support of this project.

Mary Ellen Hradek, an executive assistant to Gary Westerhoff at Malcolm Pirnie, Inc., also deserves special mention. Mary Ellen functioned as the central point for word processing and electronic transmissions. We thank her for her special skills and tireless effort in coping with our never-ending revisions to the manuscript.

We recognize and appreciate the special financial support of the Electric Power Research Institute (EPRI), through its Community Environmental Center in St. Louis, Mo., as well as AWWA.

We are indebted to numerous individuals who contributed materials and reviews in the development of many of the chapters:

Raul Torres (chapters 15, 16, and 17), Malcolm Pirnie, Inc.

Paul Eisenhardt (chapters 15, 16, 17, and 19), Malcolm Pirnie, Inc.

Tom Arn (chapters 6 and 9), Malcolm Pirnie, Inc.

Paul Busch (chapter 20), Malcolm Pirnie, Inc.

Kathi Mestayer (chapter 16), Malcolm Pirnie, Inc.

Dave Harris (chapter 9), Malcolm Pirnie, Inc.

Jack Jacobs (chapter 10), EMA, Inc.

Linda Paralez (chapters 12 and 13), Rose Enterprises

We also recognize the contribution of our research assistants Paul Fleming and Julia French of Seattle Public Utilities, and Naomi Starobin and Barb Luck of EMA, Inc.

We benefited greatly from a number of thoughtful and insightful individuals who reviewed and commented on drafts of our work. In particular we wish to thank:

- L.D. McMullen, general manager, Des Moines Water Works

- Diane Van de Hei, executive director, Association of Metropolitan Water Agencies

- Patrick Cairo, director, Technology and Research Division, Suez-Lyonnaise des Eaux, Paris

- David Rager, general manager, Cincinnati Water Works

Very special thanks and appreciation go to Peter W. Dobrolski, chief engineer, Department of Public Works and Engineering for the City of Houston to Frank Mangravite, Public Works Management, Inc., and to James S. McInerney, President and CEO of BHC Company in Bridgeport, Conn., for their peer review of the entire manuscript.

Finally, we could not have found better advocates and supporters than Jack Hoffbuhr, executive director of the American Water Works Association and Mead Noss, AWWA's manager of business and product development. They both saw the urgent need for our work and expedited its publishing. Monica Joda-Baruth, Mindy Burke, Scott Nakauchi-Hawn, and other AWWA employees cooperated with enthusiasm.

We are grateful to all of these individuals. Without them this book would not have become a reality. Although it required the help of many to bring this project to fruition, we reserve to ourselves the responsibility for any shortcomings the book may have. Our fundamental and heartfelt hope is that *The Changing Water Utility: Creative Approaches To Effectiveness And Efficiency* will help its readers advance the cause of world-class service and competitive management among the water utilities of North America.

Garret Westerhoff
Diana Gale
Paul Reiter
Scott Haskins
Jerry Gilbert

Contributors

Garret P. Westerhoff, P.E., D.E.E., is executive vice president of Malcolm Pirnie, Inc. and also serves as its Director of Water. He has more than 30 years of professional engineering and planning experience in the drinking water field and is a specialist in water quality, water treatment, creating more efficient water utilities, and privatization consulting services. An honorary member of AWWA, Mr. Westerhoff has authored more than 100 papers in the drinking water field and has contributed chapters to several water treatment textbooks. He has lectured throughout the United States on the subjects of creating more forward-looking water utilities and competitive ways to run water systems.

Diana Gale, Ph.D., is the managing director of Seattle Public Utilities and has 20 years of experience in city management. She has been Director of the Legislative Department, Director of Solid Waste Utility, City Budget Director, Superintendent of Water, and now manages all aspects of the operation of water, sewage, drainage, solid waste, and engineering services for the City of Seattle. As a city department director, Ms. Gale has a strong record of utility management, work process redesign, strategic management, resource management, financial analysis, and program development. Ms. Gale has a bachelor's degree from Wellesley College, and a master's and a doctorate in urban planning from the University of Washington.

Paul D. Reiter is Director of Strategic Policy for Seattle Public Utilities and has more than 20 years of experience in utility management, quantitative and economic analysis (focusing on integrated resource planning) strategic policy development, and public finance. As Director of Strategic Policy, he is responsible for strategic planning and policy development, regional issues, and special projects. Mr. Reiter has a master's degree in economics from the University of Oregon, where he specialized in urban economics and public finance. Mr. Reiter is a former adviser to the mayor of Seattle on utility issues and was also a private consultant in utility economics, with a decade of project work in complex and innovative approaches to the least-cost supply and conservation development for electrical and solid waste utilities.

Scott A. Haskins is the Resource Branch Management Executive with Seattle Public Utilities. He has more than 20 years of professional water utility management experience with Seattle Public Utilities. He has covered a range of assignments including administration and finance, contracting, budgeting, data processing, personnel, operations planning and scheduling, water resource management, water quality, project development and overall utility management functions. Mr. Haskins is co-directing the city and the consultant teams in developing and implementing the 120 mgd Tolt Water Filtration Project through its design-build-operate contract. He is chair of the West Coast Water Utilities Benchmarking Group and a participant in American Water Works Association Research Foundation studies on benchmarking, conservation, and customer satisfaction.

Jerome B. Gilbert, P. E., is an independent consulting engineer and has had more than 40 years of professional engineering and utility management experience both in the United States and abroad. He consults with public and private agencies on water supply matters including: water marketing, privatization (including design-build-operate contracting), planning, regulatory compliance, and economic analysis. Mr. Gilbert is the former general manager and chief engineer of both small and large urban water agencies, and has served as president of American Water Works Association and chair of the AWWA Research Foundation. Most recently, he has been a member of independent panels reviewing watershed protection programs and treatment issues, and is chair of the Scientific and Technical Council of the International Water Supply Association.

Foreword

The arrival of the third millennium will be accompanied by fundamental changes in the way the world does business. The astounding speed and capabilities of computer and communications technologies constitute but one dimension of this phenomenon. Equally influential will be the shift in organizational values and attitudes, in corporate culture and public expectations. The spur of competition is driving institutions, public and private, toward a more responsive, cost-effective, customer-oriented way of operating. Indeed, the tax-paying, rate-paying, consuming public has become and will continue to be the prime driver of change throughout the industrialized world. No sector of the economy will remain unaffected by these forces, not even the last bastion of conservative silent service, public water supply.

This book provides a comprehensive understanding of public and private approaches to creating more efficient public water utilities. The term "water utilities" is intended to mean drinking water systems, although nearly all of the techniques and approaches discussed apply equally to wastewater and to combined water and wastewater systems.

The authors have written this book in response to a national and global need for water operations that protect public health in a reliable, effective, and efficient manner. Its intent is to focus on the challenges that water utilities face in satisfying these needs and to help water utility managers understand and thoughtfully implement necessary changes by using proven and developing tools.

Water utilities are among the most conservative organizations in our society. This is not a criticism, but rather a fact that reflects the fundamental need to assure the public of its health and safety. This conservatism has expressed itself in many ways, sometimes to the advantage of consumers and sometimes not. Even the many systems that seem to meet all consumer demands for service and for a safe and reliable water supply could improve the way they do so. A number of utilities have stepped forward and begun the climb toward a higher standard of performance. Their successes will surely encourage their colleagues.

At the end of the twentieth century there is a new factor in the water equation. It is the insistence that government (i.e., providers of public services) should do more for less. This is in apparent conflict with the tradition in the drinking water industry of building systems that will survive for generations to deliver increasing quantities of potable water to expanding populations. The apparent dilemma is complicated by the fact that water systems are confronted by a population growth that will produce 100 million more people during the next 50 years in the United States.

We live in an era of remarkable change. Changes are taking place on all fronts and affect all aspects of modern life: in science, industry, society, the role of the individual in society, and even in the type and extent of medical care and preventive public health protection. These changes have significant impacts on the ability of water and wastewater systems to meet their service goals which, in principle, remain essentially unchanged: to assure the best possible public health and economic welfare of society at the least long-term cost. However, rather than making water service more difficult, the changes offer unparalleled opportunities to do a better job.

Utilities in growing numbers are considering new ways of doing business, some of which are commonly used in other fields and in other parts of the world. This new openness comes at a time when the electronic world offers significant and increasing opportunities for efficiency and labor savings. The possibilities for changing the way utilities function are almost limitless, with various options covering a broad spectrum from public management to private ownership choices. These changes come under a variety of labels: "re-engineering," "improving competitiveness," "privatization," and others, and they open the doors to new ways of doing business that can have long-term rewards. However, they can also create uncertainty and insecurity for those who manage and operate utilities and for all the stakeholders involved. If not done in a thoughtful manner, the anticipated efficiencies will produce unsatisfactory results.

It is in this context that this book is written. The basic underlying premises of the authors are:

- Customers have the right to expect a high level of efficient service and a reliable water supply that is protective of public health.

- Public utilities using a public work force can provide effective service, but they have an opportunity to be more efficient by adopting new tools and management approaches.

- There is a competitive and competent private market capacity available to assist public utilities to become more efficient.

- The effectiveness of public utility use of these opportunities depends on a partnership relationship built on open communication, mutual trust, and respect. The best way to achieve this is through a thoughtful planning and contractor selection process.

- The final decisions in each community as to how local water service should be structured, managed, and operated belong to the people of that community, who deserve the thoughtful advice and judgment of their elected officials and water professionals.

Of course there are people who firmly believe that water service ought to be owned and operated by local government. Others believe that water service can be more efficiently delivered by the private sector. The authors of this book do not approach the issue from either of these viewpoints. We believe that local water service should be structured and operated in the most appropriate, effective way possible in each community. We believe that public health and safety issues are paramount and that the long-term interests of the people must be well provided for. We believe that water operations must be stringently cost-effective and fully reliable as to service and quality. New technologies and management approaches must be adopted to achieve such a high level of performance.

The matter of system size merits a special comment here. Although this book is aimed chiefly at what we in North America call "larger systems"—those serving more than 50,000—its message is equally applicable and urgent for smaller systems. In almost every chapter we offer comments addressed to medium or smaller utilities to aid in applying our recommendation on those scales. Because of the economics of scale, smaller systems face more daunting obstacles to progress than larger systems, as a general rule. This is the reason smaller systems in the United States have been more apt to privatize or to outsource their operations. Smaller communities can make good use of this book in reviewing their operations and considering how to best provide quality service to their customers. At the end of the day, that is the very challenge confronting all systems, regardless of size.

The authors happily acknowledge the progressive innovations and leadership of a growing number of water utilities in the United States and Canada. They are trying new approaches in every aspect of utility operations and management. Clearly the industry is beginning to move in a new direction, often cutting new paths and forging new tools. We are all learning, and the authors thank the many forward-looking leaders from whom they have learned much of what is presented in this book. In fact, we see this book as an early contribution to a movement newly underway, a movement that has a long and open-ended future.

We must also ackowledge our colleagues in the United Kingdom and other countries in the European community, in Japan, Australia, South Africa, and elsewhere, for their bold leadership in exploring new ways to organize and manage water operations. Their aggressive applications of new technologies serve as a model we can emulate.

Those who own, manage, and operate water systems and who aspire to high quality performance will, we fervently hope, find this book to be a sound and practical guide for the times.

Introduction

Understanding the structure of this book will help the reader. The authors do not expect anyone to read it cover to cover in a few sittings. (If you do that, dear reader, please let us hear from you!) There are four parts to the book and three appendices. Part 1 is intended for all readers; it provides background and perspective important to the other parts, as well as an overview of approaches to utility performance improvement. The three following parts are a handbook of discrete but related topics. Readers are encouraged to read the part descriptions that follow, as a guide to understanding what to expect.

Part 1: A Framework for Improving Performance. This part begins with a brief review of the history of publicly owned and operated water supply systems in the United States to bring into perspective the present municipal model and its primary objective of protecting public health and safety. It then discusses the imperative to change from this model to one that is more appropriate for the twenty-first century. It then outlines a framework for analyzing present performance, identifying key opportunities to maintain effectiveness while improving efficiency, and the development of a strategy for action. These chapters are intended as an overview for managers and other readers. Greater detail is provided in later chapters.

Part 2: Enhancing Performance in Key Functional Areas. The more significant targets for improved efficiency are the treatment works, system maintenance and rehabilitation, customer service, information and control systems, and energy. Each of these areas is discussed in separate chapters in this part.

The discussion on improved efficiency at the treatment works is orientated around self-improvement using primarily public works staff with selective outsourcing.

Part 3: Tools for Improvement. This part presents a number of tools that a manager can utilize to improve efficiency, both within the key functional areas presented in part 2 and also in other functional areas in the water utility. Chapters 11–14 treat in greater depth what was sketched in chapters 3–5 as an overview.

Part 4: Implementation Issues. This final part discusses four issues that emerge during many of the activities for enhancing performance. These include political concerns, legal and financial issues, conflict of interest, and regulatory agency considerations.

Appendices. The first appendix consists of the synopsis of selected case studies to provide illustration and reference to utilities that have undertaken activities described in the text.

The final two appendices are comprised of model documents for contracting outservices, mainly the operation and maintenance of the treatment works.

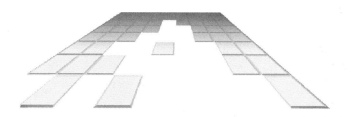

A Framework for Improving Performance

Foreword to Part 1:
A Framework For Improving Performance

The first two chapters provide the fundamental rationale for every chapter in the book. There would be no compelling reason, for example, to engage in a penetrating assessment of your utility's performance (chapters 3 and 12) or to consider managing a competition to award a contract to operate and maintain your water system (chapter 16) if life and the world were the same as they were 20 years ago. The changes in utility management and philosophy proposed in this book make sense only in light of the changes going on in the world around us, among the people we serve and who work for us, and among all the institutions of society—public and private.

Chapters 1 and 2 establish the authors' thoughts about the forces that are overturning previous assumptions in business and government, forces that are already visibly at work in the water industry. Understanding the forces of change helps us to cope with them effectively and to do a better job of positioning our organizations in society.

Chapters 3, 4, and 5 provide an overview of the process being used, in one form or another, by many organizations, including water utilities, to assess their effectiveness and to undertake a comprehensive program of improvement. This process calls for all those who have a stake in the utility to understand why change is necessary and to commit themselves to the demanding task of thinking, planning, and acting their way to new and higher levels of performance and service. The process of assessment, planning, and change is so detailed and comprehensive that we begin by outlining it in broad strokes in the book's first part, and then by going into greater detail in later chapters.

Utility managers and senior staff, public officials, and other key players in the industry are the audience for whom part 1 was especially written. The leaders in our field, locally and nationally, need to think and talk seriously about the future of public water service. This book, and particularly this part of the book, reflects the thinking going on today among our leadership and, we hope, makes a contribution to this important dialogue.

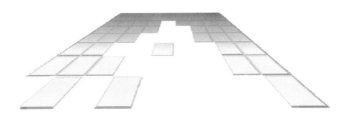

CHAPTER ONE

Imperative For Change

I'll give you the story in a nutshell. Bill had been working in the water utility 32 years, 14 as general manager. It all seemed to end in a one hour meeting with his mayor and city manager. They told Bill that they had been talking to a contract operator of water systems. The offer seemed to be too good for the city to turn down: $6 million up front, an annual payment, new equipment, and no new investment by the city. Costs would be cut and the staff reduced by 20 percent or more. What did this mean for Bill?

Bill asked a lot of questions, but his bosses were determined. They talked about the fine job he had done but also about the tight city budget and bonding limits and other difficulties they faced. The change was a good deal for the city and he should cooperate. They made it clear that if he didn't get with the program he would have to leave.

He walked out of that office a shaken man. He had done the job their way all his life, never rocked the boat, and taken the heat on rate increases and other customer gripes. He never dreamt they would support major staff reductions, automation of a big new plant, fancy monitoring and information systems, and other major changes. Now he had to go back to the office and break the news to the staff. How could he support such a radical new program?

In the weeks and months that followed, Bill still couldn't understand why it all happened. In his mind, he hadn't done anything wrong.

Utility managers can't pick up a magazine today or go to a meeting without facing demands for change. Some have anticipated the change and they are doing something about it in their organizations. Others are still dubious and suspect there is more than a hint of exaggeration and hype in the air. Of course, everyone wants to improve and do a better job, but talk of drastic changes and organizational turnarounds is over-reacting and could even lead to serious mistakes.

That's what the Bill in our story thought. This book is based on the belief that today there is an imperative to change and that this imperative comes from powerful forces that are producing fundamental changes in corporations and public institutions throughout the industrialized world. Those forces are pushing their way into the water field and they cannot be stopped. In fact, they should not be stopped. *If harnessed thoughtfully they can be forces for great good.*

Smart water utility managers are striving to understand these forces, to put them to work, restructuring their organizations and operations into new models of efficiency, increasing cost effectiveness, and providing responsive customer service. And they are doing a good job of it. What is taking shape in our industry is nothing less than a new concept of what a water utility is. The old way of thinking about a utility is no longer valid. If managers cling to the old way they risk losing the confidence of consumers and owners and regulators. That is already happening, as many variations of the story of Bill attest.

This book was developed to provide managers with the tools to modernize and improve the performance of their utilities. It describes new ways to streamline operations, use new technologies and, *where appropriate,* private sector services to meet contemporary standards of public service. The authors offer experience-based guidance on how to lead your organization and all its functions into a new, more competitive, and productive way of operating, while maintaining the effectiveness and reliability your customers demand. Most importantly, in this chapter and the next we'll identify and explain the forces of change that are reshaping all organizations, public and private.

A BRIEF WORD OF BACKGROUND

In the transition from the old way of thinking about water utilities to the new way, it is absolutely essential that the core of the utility's mission be preserved intact. Our mission is to serve and to protect the public health and welfare by providing a safe, reliable, and sufficient supply of water to our communities. That mission is a public trust and must remain central in our thinking. What is changing are the ways and means through which we provide that service: the organizational and technical processes and systems, the fiscal strategies and mechanisms, and the managerial skills and creativity we employ.

Our mission is better served today by the tools of the day, just as it was well served almost a century ago by the tools new to that earlier era. It is important to remember that most of our municipal water systems began in the nineteenth or early twentieth century as private companies. These private utilities were motivated by the desire for return on investment through the sale of a commodity. While water is a commodity, it was increasingly recognized as having essential public values as well. The great fires that roared through the wooden structures of dense urban centers prompted an overwhelming public outcry for greater quantities of water to fight fires. A similar outcry arose in response to the typhoid and cholera outbreaks of the nineteenth century, demanding safer water. By the end of the nineteenth century more than 200 municipalities had changed from private to public ownership. Today more than 70 percent of the 59,000 community water systems serving about 90 percent of the population of the United States are publicly owned and operated.

How did this happen, considering the fact that most of the other utility services—electricity, natural gas, and telephone—are supplied by private, regulated utilities? Clearly, the dramatic issues of public health and safety made water supply a matter of compelling public interest in the view of many citizens. When fires drew people's attention to their water supplies, some cities, such as Boston and Seattle, realized that their basic supplies were inadequate for all needs. These cities wanted local government to take the responsibility and guarantee public water service for the future. In addition, the need for major capital investment led to the practice of long-term public debt to finance the development and expansion of water supply systems. This practice was also considered preferable to the higher water rates that would result from demands for early returns on investment in private water company operations.

The growing use of government as the preferred provider of water service was not adopted in all cases, however. Even today, about one-third of our community water systems are private, investor-owned companies. These companies survived because they demonstrated the same commitment to the public health and welfare, and state agencies were established to provide public oversight of the rates and operation of these investor-owned utilities. In the United States we can point to a number of great private water companies that have for many years maintained public confidence and the respect of their municipal counterparts.

The best of the investor-owned and publicly-owned utilities in the United States have manifested an admirable capability to reflect long-term public needs and interests. In particular the great water systems serving New York, Boston, Philadelphia, Chicago, Denver, San Francisco, and other urban centers were the engineering marvels of their time. For many decades they have set an example of dynamic vision and leadership in public water supply, developing the resources and infrastructure necessary to satisfy growing populations and industrial needs.

THE INTERACTION OF PUBLIC AND PRIVATE ROLES

In the past we usually thought of public and private water utilities as being entirely separate. In fact, private companies have often relied upon public agencies, regional and local, to assist them in providing water service. Examples include: Miami Conservancy District, which controls floods in the Miami River watershed of Ohio, the City of New York that provides water storage and watershed protection in the Catskill and Delaware watersheds, and the Santa Clara Valley Water District that provides recharge water in the San Francisco Bay area. They all contribute to the availability of privately distributed water.

Many public utilities, large and small, convey supplies that are used by private water companies. The Hetch-Hetchie system owned by San Francisco provides two-thirds of its supply to independent local purveyors along its delivery route from the mountains to the city. In southern California, the Metropolitan Water District provides an imported regional supply for 16 million people. This supply supplements groundwater sources used by many private utilities.

These and other public agencies that are faced with the need for special expertise in science and engineering have frequently turned to the private sector for assistance. Although some major metropolitan utilities are leaders in the science of water quality and technological applications, the water supply projects for many major metropolitan areas have been designed by private environmental engineering firms. With some exceptions, the designs of water treatment facilities have been completed by the private sector since the original technology developments for filtration and disinfection were introduced more than 100 years ago. In major crises such as floods and earthquakes, all utilities rely on private contractors to make repairs and supplement their workforce.

Smaller public utilities have found it difficult to cooperate by sharing services to reduce costs. However, there are many examples of functional regionalization achieved by private utilities that provide central services for widely scattered small service areas throughout common ownership. These services include: purchasing, technical consultancy for control and water quality systems, and engineering design for utilities that could not perform these services on their own or do them efficiently with local contracting. This ability to provide central support has been an incentive for communities to shift from public to private ownership.

Public-private partnerships are not really new, neither are public-public partnerships nor private-private partnerships. These relationships have arisen in response to local needs and circumstances. They are examples of our ingenuity in solving problems and ensuring reliable water service and reasonable rates.

THE CONVERGENCE OF PROBLEMS OLD AND NEW

The early and middle decades of this century witnessed prodigious progress for the water supply field. Major water-borne diseases were conquered; the fire threat was greatly reduced; and sufficient supplies were developed to meet the needs of population and industrial growth. Public and private investments to achieve these goals were by and large successful. Unfortunately, success of this nature can foster complacency, and so it did, among many public officials, water managers and professionals, and the public at large.

In many communities, the utility deals daily with one or more of the following problems caused by complacency.

Management Conservatism

"The way we've been doing things works; let's not risk failure or money by trying new approaches." With but few exceptions, this philosophy limited funding for research and development of treatment and other technology.

Labor Work Rules and Representation

Organized labor and civil service systems created to address social ills also exhibit an inflexibility that does not match today's competitive realities. Some public utilities have both union representation and civil service provisions that limit productivity and flexibility by restricting management's ability to realign task assignments and job classifications. For example, job classifications have sometimes remained fixed despite improvement opportunities provided by information technology and equipment improvements.

Deferral of Improvement and Rate Suppression

Replacement and upgrading of plant equipment and distribution systems have often been deferred for years, leading to system failures, a large backlog of work to be done, and inflated costs. Political fear of negative public reaction to rate increases has often delayed and reduced necessary revenue flow to support growth, solve operational problems, and to minimize long-term costs.

Diversion of Revenues

Some municipalities continue to draw on water revenues for other purposes, thus destroying the link between revenues and costs, including the capital resources for infrastructure development and replacement.

New challenges have now emerged that add to these problems. Chief among them are:

Biological Contamination Incidents

The cryptosporidosis outbreak in Milwaukee won the most attention, but other incidents including those in Washington, D.C. and Las Vegas have provided a new impetus for watershed protection and treatment improvement. The potential for similar events is a matter of concern.

Public Health and Detection Levels

Since the 1970s, health researchers have identified new threats in drinking water. The gas chromatograph and mass spectrometer enabled the identification of numerous chemical compounds at the level of parts per billion; new microbial test methods have identified pathogens of increasing concern; epidemiological and animal studies have led to more demanding regulatory goals for these pathogens, disinfection byproducts, and synthetic organic chemicals resulting from industrial and land use practices. The 1986 and 1997 amendments to the Safe Drinking Water Act mandated stricter requirements to address these concerns.

Water Quality Perceptions

Activist and media-driven demands for lower and lower levels of health risk make it difficult to produce a product that is generally believed to be safe even if it meets regulatory standards. The growth of bottled water and home treatment device industries has arisen, in the view of many, at least in part from a lack of consumer confidence in tap water.

Technological Innovations

New equipment, methods, and strategies are increasingly available to utility operators, including granular activated carbon treatment, ozone and other disinfectants, membranes, instrumentation control, and computer applications to meet these changing goals.

Excerpt from the summary of the 1995 AWWARF survey entitled "Consumer Attitude Survey on Water Quality Issues"

"...for the majority of water utility managers there is a clear gap between their expectations or understanding of public opinion and the actual perceptions, attitudes, and preferences of most customers and citizens. To summarize the most common misperceptions:

• *Most managers underestimated public demand for greater involvement in the decision-making process of water utilities.*

• *Most managers also underestimated the amount of information the public perceives it is getting about the quality of its drinking water.*

• *Most managers overestimated the importance that customers say they attach to the price of water when asked to identify the most important thing they want from their local water utility.*

The cumulative effect of three decades of these developments has stunned many utility managers. They have been caught between federal and state demands on one side, public health and environmental pressures on the other, and the conservative, resistant attitudes of their boards on still another front. Some reactions have been quite positive: support for the American Water Works Association Research Foundation (AWWARF), the National Water Research Institute and federal research programs began to grow; the government affairs program of the American Water Works Association (AWWA) expanded into an increasingly effective role; the Association of Metropolitan Water Agencies and the National Association of Water Companies joined hands with AWWA in a fundraising and action program to represent the industry more effectively. But as the final decade of the century appeared on the calendar, new forces from outside the industry have begun to make their way into the water field with unexpected, dramatic impacts.

A convincing insight into the problem was uncovered by an AWWARF public opinion survey released in 1995. One component of the study addressed utility managers, who were asked what their customers thought about a number of key issues concerning water service. The same questions were asked of the consumers surveyed. To their dismay, many managers learned that customers do not harbor the attitudes and values the managers assumed them to have and the customers expressed significant willingness to pay more for better water service. This misalignment of perceptions reveals that consumers, influenced by other forces in their lives, have moved on to a new set of expectations unrecognized by their water providers. While that misalignment points out an obvious problem, its greatest significance may be that utility managers have also missed the other driving forces at work in society today.

Professional commentators on the fields of socioeconomics and organization management have been reporting since the late 1980s on fundamental shifts in values and practice in the general public and in the business world. The driving forces behind these changes have been well documented and they are considered to be profoundly influential. Indeed they are expected to have a greater impact on society than did the industrial revolution. The water industry will not be exempt from these forces of change. It has been, however, slow to recognize and respond to them.

THE IMPERATIVE FOR CHANGE BEGINS

It was Prime Minister Margaret Thatcher who blew a trumpet blast of institutional change in the United Kingdom when she ordered the privatization of that country's 10 regional water authorities in 1989. Organized just 15 years earlier, these authorities were recognized worldwide as models of the value of regionalization of water service and its integration with wastewater services, flood control, and related functions. Once privatized, these new and well-funded private water service companies embarked on aggressive improvement programs, raising rates and the public's ire in the process. Although they are subject to rate regulation, new water quality standards, and requirements for capital investment, they have nevertheless been the subject of considerable public controversy. Criticism has focused on excessive salaries, rate increases greater than the rate of inflation, and water shortages.

Large, well-established and diverse French companies have been pursuing worldwide opportunities for privatization contracts for many years, and have shown recent success in Mexico City, Buenos Aires, Eastern Europe, Manila, and Sydney. The UK water service companies that have successfully generated capital to support investment in major improvements and rehabilitation now invest in other systems and businesses outside the United Kingdom.

During the 1993–1996 period, some larger cities such as Indianapolis, Oklahoma City, San Juan, Puerto Rico, and New Orleans, adopted short-term O&M contracts, primarily for wastewater facilities. But there were also contract operations for water systems in Jersey City, Puerto Rico, and Houston. In Canada, Edmonton, Alberta converted its water utility into a private corporation with an aggressive business mission. At the time of this writing, the City of Scottsdale, Ariz., was the only major American city that had fully privatized its water utility, but the city later repurchased it. Atlanta and other cities had announced interest in 20-year, long-term, operations contracts. Many dozens of smaller

• More people say they were willing to pay for water quality improvements than most managers expected.

• Most managers believe the public drinks bottled water for health and safety reasons, but they are just as likely, if not more likely, to drink it for other reasons or as a substitute for other beverages. Nevertheless, there also appears to be some tendency to underestimate the importance the public attaches to 'health and safety' when it defines quality. Many managers said the public cares more about how the water 'looks and tastes.' Finally, managers appear to underestimate the positive impact that a utility contribution for drinking water research would have on its image among customers."

11

communities have entered contract arrangements for their water operations.

At the same time an alternative approach was emerging in Charlotte-Mecklenburg, N.C., and Jefferson Parish, La., among other cities, which conducted a managed competition for an operations contract and awarded it to its own employees. Seattle and Puerto Rico devised innovative, competitive "Design-Build-Operate" contracts for large water treatment plants. (Case studies reporting these projects may be found in appendix A.) News of these developments flashed through the North American water community, initially raising only eyebrows, later—genuine interest.

The presence of worldwide service providers has highlighted a number of the primary driving forces for change, including the following.

Globalization of Business

Major business enterprises here and abroad are becoming more international. They range from cars to corn, from fast food chains to computer chips, from electronics to the entertainment media. In the water field, we see the two giant French conglomerates, Compagnie Génerale des Eaux and Suez-Lyonnaise des Eaux, seeking business opportunities worldwide, as are the large British water companies. Their activities in the United States and Canada are prodding municipal utilities into new ways of thinking and acting. Globalization provides utilities with greater technical opportunities that have been created by the collaborative research and development activities being conducted primarily in the United States, Western Europe and Japan.

Technological Developments

Continuing advances in computer applications, communications technology, monitoring and information systems, automation, and instrumentation are transforming the way business practices and plant operations are carried out. Treatment and analytical methods are also far ahead of where they were in the 1980s.

Competitive Challenges

Public water utilities are no longer untouchable monopolies. The private sector can compete for the right to operate water service. This fact is already motivating municipal utilities to seek greater efficiencies and cost effectiveness. We have, however, only seen the beginning of the impact of competition in the water field.

Customer Service and Responsiveness

The competitive challenges just mentioned are, in reality, competition for customers. Organizations that are perceived as responsive to the needs and interests of their customers will survive; others won't. This is true for all industries—supermarkets, airlines, manufacturers, restaurants, and water suppliers. These challenges are new for many water managers but they must begin to look at the customer as the primary focus of utility operations.

Scarce Public Resources

There is a growing reluctance to commit public resources to support government services when capital is available for private investment. Public agencies must demonstrate that they are as cost effective as the most hard-nosed competitive business. Public support must be gained through the persuasiveness of superior service and a superior product.

Communications Impact

There is another factor of considerable influence. The rather amazing speed with which news races through the industry seems to lend additional impact and urgency to it. Word of new developments, such as Seattle's "Design-Build-Operate" project, flashes across our E-mail screens; it appears next Monday in *Waterweek* and other industry publications; we hear about it at every meeting we go to and read about it at each stage in every journal we pick up. We are so inundated with news of privatization, operations contracts, and competitive improvements that the industry's atmosphere seems to be charged with the excitement of extraordinary change. It is difficult to sort out the actual influence of this phenomenon but there seems to be little doubt that it catches people's attention and arouses their interest. We are all subject to this stimulation, including members of boards and councils, public officials, and citizens at large. These driving forces, which are affecting all organizations, public and private, have become an irresistible imperative for change. The manager who ignores them, who doesn't like them, or wants them to disappear, does so at the risk of his or her own peril. The choice involved is not whether or not to accept these changes, but how to adapt to them productively and how to achieve a higher level of service for our customers.

SUMMARY

The public's expectations for water service are changing. These expectations focus on public health protection and cost efficiency. Fortunately, new tools, both physical and procedural, are available to do the job. They include: organizational change, stakeholder participation, new models for public and private sector contribution to performance, and new technologies, particularly for treatment, automation, and information systems. When combined with the traditional conservative strength of water utilities, a new practical program for change can be created by each utility. If the program is based on a realistic assessment of the utility's performance and its customer's needs, the imperative for change can be met.

The utility of the next decade will be substantially different from the utility of 1980 or even 1990. The model of a centralized, isolated bastion of water expertise will give way to a more open and sophisticated, lean and flexible, customer-oriented service agency. In the next chapter we'll offer a more detailed look at the performance imperatives for water utilities in the twenty-first century.

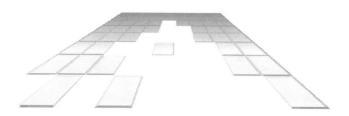

CHAPTER TWO

The Performance Demands of The Twenty-first Century

The first chapter in this book described the forces that are driving our industry to change. This chapter provides important background on how to begin thinking about what the utility of the future should look like and how to get there.

We emphasize that the following discussion is important because we want you to wrestle with the concepts, principles, and reasoning that underlie our message about the imperative for change and what to do about it. Admittedly what we are talking about is complex and, being somewhat new in much of North America, it may seem difficult to implement. Be assured that there is nothing here that a thoughtful person cannot readily understand and carry out. These ideas merit the attention of readers who are concerned about good management and quality water service.

There are no magic formulas, no quick solutions, no easy path to re-creating water utilities for the new era. The path we are urging you to consider requires hard work by you and everyone around you. *The beauty of this rugged road, however, is that it will lead you to the kind of utility that you, your customers, and your owners want to have in the next century. The outcome will be a utility that best suits you, your community, and your circumstances.* We offer a process, a strategy for change, and some tools that you and those around you can use to create the ideal utility for your customers.

If this comprehensive approach to organizational change stirs up a bit of nervous anxiety in you or your staff, we hope you convert that concern into the energy and determination needed to work your way through the process of change and all the work to come. The utilities that are going before you—and many organizations and corporations in other fields—would all assure you that the payoff is well worth the effort. They would confirm that there are no shortcuts. They would also urge you to have an open mind and to master the thinking and the tools of change to build your own success.

AN OVERVIEW OF THE NEW DEMANDS

The performance demands being made on utilities today are based on greatly increased public involvement and have produced four primary imperatives:

1. The focus of utility must be on the customer, the customer's interests, and the customer's needs.

2. The utility must demonstrate solid commitment to its essential purpose: reliably protecting the public's health and safety.

3. The utility must perform in all respects with competitive efficiency and cost effectiveness.

4. Establish an open attitude promoting public participation and knowledge of the utility's work and performance.

The ways and means by which these demands and imperatives are carried out will differ widely from utility to utility. Local circumstances vary; so will local solutions. Nevertheless, these four basic demands must be met on terms acceptable to the customer.

THE CUSTOMER'S CENTRAL PLACE

In the past, utilities defined success in their own terms, which meant by the terms of engineers and managers. Their challenge was to build a system that worked well and that would last for generations to come. They were focused on predominantly technical concerns such as water pressure and storage volumes, main breaks, contracting and overruns, expenses, and revenues. In building and operating systems utility managers made choices that they believed were consistent with what the public wanted. Remember the AWWARF Customer Attitudes Survey mentioned in the last chapter? Customer opinions were very different from what utility managers thought they were.

For example, customers valued water quality highly and were willing to pay more for it, whereas managers believed customers thought the opposite. That misperception will not produce what the customer wants most. Managers were convinced that their customers thought they received enough information about their water, when in fact the customers polled want much more information. Similarly, customers want to be involved in decisions that affect them, although managers are most often of the opinion that customers are satisfied with their level of involvement. This survey demonstrates the need for managers to make a greater effort to determine what the attitudes of their customers actually are.

When you ask yourself how well your utility is meeting customer needs and the four performance imperatives, guard against being satisfied with your own answers. What matters most is what your customers think. Your own initial appraisal, and that of your board, council, or commission is also important. But when the insiders' opinions clash with your external customers, you know you have more work to do.

What we are driving at here goes far beyond customer service in its narrow sense to include every interaction of your employees or activities with customers. It covers what you do to minimize the impact of main repair and other construction projects; the promptness and consideration with which you handle leaks and other problems; the behavior of meter readers, payment clerks, and others who meet the public; the neatness and cleanliness of your grounds and facilities; the impressions made by work crews in the field; the user-friendliness of your policies and procedures; your accessibility beyond normal work hours; your communication with customers on issues that affect them; and your responsiveness to customer concerns and problems. All these and more shape your customer's impressions of your attitude and lead to the need for a quality-driven employee involvement process.

Lessons from the world of business demonstrate why it is simply good business to place the customer first. Remember the shock to American car manufacturers when the Japanese responded to public needs with smaller, less expensive cars that were cheaper to operate and maintain? Have you noted the widespread use of UPS, Federal Express, and DHL as competition with the U.S. Postal Service, despite their higher prices? Wal-mart has become the largest employer in the country because of an aggressive marketing strategy that has won millions of customers away from Main Street, U.S.A. Their success is due to attention to customer interests (low prices), customer convenience (ample free parking), and customer services (staff, store layout, and fast checkout).

The IBM Executive Management Institute has analyzed the phenomenon of customer-focused success. They characterize the traditional approach to business as "make and sell." We make a product and we try to sell it. The thought process began with building a better product then moved to convincing potential users to buy it. The contemporary approach is "sense and respond." Companies begin by sensing (by identifying through research) what the customer wants and then responding by providing the product or service the customer wants.

In the water field we have long operated on a make and sell basis. We produce the product and the customer buys it. In fact, we have a monopoly and the customer must buy our product. Where else is he or she to go? There used to be no reasonable answer to that question. Now there is.

Will You Guarantee Your Service Delivery?

In the United Kingdom, the government's Office of Water has published eight measures of utility performance on which records must be kept to ensure service that meets customer expectations. If the

Bottled water has become a billion dollar business, competing with the one or two percent of our production that is actually consumed. We may think that competition for one or two percent of the market doesn't matter, but those bottlers, who charge up to a thousand times as much as we do, are winning more than big bucks. They are winning public support on the basis of a safer, better tasting product. How did we as utility managers allow this to happen? We are positioned in a distant second place, producing "ordinary tap water" with its alleged contaminants, disinfectant taste, and other suspicious shortcomings. Second place does not win much respect or support.

A similar conclusion follows the story of home water treatment devices. We know that they are usually not necessary, or in many cases desirable, but the people who buy them think they are. And that "necessity" of having to purchase special equipment does not increase customer respect for the public water supplier.

And now, as true competitors arrive on the scene and approach our owners claiming they can produce better water at lower cost, how much support can we expect from customers who have already turned elsewhere for help? *If we are not positioned in the minds of all or most of our rate payers as a high quality organization, producing a quality product at a fair price, we are not going to get much support from our customers or the public at large.*

More than that, many consumers are keenly aware that they are at the mercy of a monopoly. They must use what we send them and pay the price we charge. That involuntary exposure to a product they consume makes many customers resentful, suspicious, and distrustful. If they do not have good reasons to trust us, why should they?

Therefore we must provide our customers with ample reasons to trust us. We must demonstrate in every way that we are genuinely committed to serving them well—with a quality product, with cost-effective operations and efficient management, and with responsive policies and practices. Chapter 8 delves into this topic in greater detail. The accompanying discussion, "Will You Guarantee Your Service Delivery?" describes a unique new program used by Thames Water PLC in the United Kingdom.

It is worth noting here that when we speak of customers with one word, we include different classes of users with wide ranging quality and quantity needs. Major categories are: domestic, industrial, business, public services (for example, fire fighting and street cleaning), recreation (including golf courses), wholesale customers, and others. Each of these customer groups requires attention and responsiveness to their issues and terms. In addition, we face the fact that most of our customers constitute a silent majority, which means we don't hear from them very often. It does not mean that they don't want to hear from us, or that they don't have opinions that should matter to us. The vocal majority we do hear from are often advocates of various causes and interests. We

utility fails to meet certain of these "Levels of Service," the affected customers must be given a payment of £25 (about $40 US dollars) each.

The Thames Water PLC has developed a "Customer Guarantee Scheme" that goes beyond government requirements with a set of performance standards of its own. When they fail to meet their adopted service criteria, they automatically pay the customer £10 (about $16 US dollars). The Thames Water guaranteed standards are accepted and summarized below:

· Keeping Appointments: *When the utility makes an appointment to visit a customer at a certain time or within a time period, it will pay the customer £10 if its representative arrives outside that period.*

· Questions about the Bill and Other Payment Arrangements: *Inquiries about the accuracy of a bill must be answered within 20 working days. Requests to change payment arrangements must be acted upon quickly or denied within 10 working days.*

- Responding To Complaints: *Complaints about water or wastewater service will usually be responded to, in full, within 10 working days. Provision is made for situations requiring more time. An automatic payment of £10 is made when the promised time is exceeded. An additional £10 can be claimed if the automatic payment does not arrive.*
- Interrupting and Restoring the Water Supply: *Planned interruptions are guaranteed to be completed within the announced time or else affected customers can claim £10. The utility pledges 24 hours notice in writing when a planned interruption is going to last more than four hours, and will pay customers £10 each if they fail to do so. Guarantees are also made concerning main breaks.*

need to put these voices into balance so that the community as a whole is well and truly served by its water suppliers.

How utilities sense and respond to customers, consult with them, and involve them in decisions that affect them will be discussed in the next chapter. Here we have been trying only to establish the validity, the good business wisdom, of putting the customer in the center of our attention as the ultimate reason for all we do.

For the utility manager to adopt this philosophy is, of course, not sufficient to convince or transform the entire utility staff. One of the central aspects of realigning your organization is, as you will see later, a program to enlist your staff's understanding and commitment. The customer must be the focus for all utility personnel.

THE CORE RESPONSIBILITY OF UTILITIES

Water utilities are different. They have unique responsibilities that distinguish them from other public services and other enterprises. The public health aspects of water supply are matters of life and death. Accordingly, water is a public good that whether publicly or privately supplied is the ultimate responsibility of the community. Ensuring the safety of drinking water and the reliability of the community's supply are serious matters of public trust that must be vigilantly and conscientiously implemented. This critical obligation of the utility must be supported by its owners, scrupulously honored by its staff, and well known and trusted by its customers.

Those who look upon such statements as pious rhetoric do not understand the critical role drinking water plays in public health protection. That also means they are not likely to be persuaded by a lengthy discussion. Let us point to the calamity that struck Milwaukee in 1993 when approximately 400,000 people got sick, thousands were hospitalized, and about 100 died because of biologically contaminated drinking water. No one wants that to happen. However, without regard for the specific causes in Milwaukee, we must admit that that sort of major incident indicates a failure of the system and its safeguards.

The 1991 Oakland Hills fire consumed more than 3,000 homes and reminded us that the fire protection aspects of water supply are not just in the history books. Vigilant, optimal operations is a concept that goes far beyond the installation of treatment processes; it implies comprehensive attention to all aspects of operating the system. This includes staff competence and training, source protection, meticulous monitoring, emergency preparedness, well-maintained quality equipment, and a treatment train and distribution system that are operated at high levels of efficiency. From the source to the customer's tap, all involved must share in the effort to ensure the system's reliability and the water's quality and safety.

The issue of risk management arises when we ask (as a matter of public responsibility) just how much risk protection we should provide. Of course we want more than is necessary for normal conditions; we want to ensure that the system can cope with all serious, anticipated challenges. But how far should we go in that direction? How much extra protection should be provided? This is an excellent example of a question that must be answered in each community through consultation with customers, for there is no universally correct answer. It is a judgment call that must be made by those affected by the decision and those paying for it. Experience in Cincinnati, Seattle, Des Moines, and East Bay shows a consistent pattern of customer willingness to invest more to achieve "higher than standard" water quality.

Finally, on this point, we want to emphasize that while the core responsibility of utilities cannot be relinquished, it can be delegated. This is the principle that allows a public water utility to contract with a private or public sector firm for all or part of its operations. That municipality or water district must ensure, through the contract and through oversight, that the contract operator will perform in ways that fulfill the utility's health protection obligations. There are a number of alternatives available to a publicly-owned utility in considering how best to structure, manage, and operate its water system. However, under any arrangement it is the utility that bears the ultimate responsibility to its customers and the community. The matter of contracting for services with the private sector or another public agency is covered in detail in part 3.

• Flooding from Sewers: *If a sewer overflows and effluent gets into customer buildings, customers can claim a refund of their sewage charge for the year, up to £1000 (about $1600 US dollars). A few exceptions apply.*

• Low Water Pressure: *If water pressure drops below 70% of normal minimum levels twice in a 20 day period for more than an hour, the utility will pay claims of £25.*

All UK water utilities must keep records on the time it takes to answer telephone calls and written complaints from customers and on water use restrictions imposed on customers.

Sources: "Our Guarantee To You," a draft pamphlet to be published by Thames Water, and Appendix I, Levels of Service Indicators, published by the UK Office of Water (OFWAT).

THE OBLIGATION TO PERFORM COMPETITIVELY

The third performance imperative for utilities in the twenty-first century is that they must operate at a high level of efficiency and cost effectiveness. The standard of performance here is "Best of Class." Put another way, that means a competitive comparison with whoever is doing the best job in specific utility functions or in overall operations. Customers do not believe they should tolerate anything less than the best. And indeed, why should they?

In this matter, public utilities suffer the burden of public stereotyping about all government agencies: they are thought to be unresponsive, bureaucratic, arrogant, wasteful and inefficient, overstaffed, and ineffective. These harsh accusations certainly do not apply equally to all public institutions, but we must admit that to one degree or another they may at times apply to water utilities and companies as well. To some degree, these criticisms may be true of many of us. If our customers perceive us in this nasty light, that perception must be addressed before we can expect public support.

If your customers were to challenge you by asking whether or not you can demonstrate that your utility is as efficient and cost effective as it can be, what could you reply? This is precisely the position utilities are in today. They must be able to withstand competitive scrutiny, to look their governance and customers in the eye and say, "I can show you that no one could do this work better and more cost effectively than we do."

In truth and candor, that is a high expectation of any organization in any field, whether public or private. Nevertheless, it is the standard of performance we must strive for. Our customers will not tolerate vague or secondary goals such as "We're meeting all federal and state requirements, we're better than we used to be, and our rates are lower than they are in other cities."

PUBLIC PARTICIPATION IN UTILITY AFFAIRS

The fourth characteristic of the utility of the future is the participation of the public in utility affairs. Customers no longer accept the tacit notion that they should take what they are given and pay what they are charged, no questions asked. People believe they have a right to a voice in matters that affect them. This includes the quality and safety of their drinking water; the acquisition and protection of water sources; the level of protection provided against contamination, fires, floods, droughts, earthquakes, and other natural disasters; the cost effectiveness of the organization and service they are paying for; the policies and procedures they must abide by, and so on.

Managers attuned to their customers have used standard, effective techniques that are within the reach of most organizations. These include advisory committees (ongoing or established for specific projects), focus groups, frequent surveys, a home page on the Internet with feedback capability, asking customer callers questions, open houses with feedback opportunities, talks and discussions at meetings of various organizations, ongoing media relations, customer information programs, youth education activities, and other such techniques.

A key method of public involvement is asking a few well-placed, well-informed customers to serve as representative stakeholders on the planning or steering teams you may establish as part of an organizational improvement program.

The processes for achieving the levels of performance described in the preceding pages are detailed in subsequent chapters in this book. For our purposes here we need to make only three further points: (1) there are many ways to improve service performance, but it takes a lot of work to figure out the best way for your utility; (2) the debate over privatization can be reduced if it is better defined, and (3) the solution for your utility will be one of compromise, not the adoption of an "ideal" model.

THE PATH TO COMPETITIVENESS

To become competitive, ready for comparison with the very best, requires a long-term commitment and much detailed work by many people. Your customers, governance, and staff must be involved. Luckily, there are many choices and alternatives available to you and many ways of doing things. The challenge is to find the right choices for your organization, given its circumstances. Many self-improvement efforts are carried out internally, around the existing workforce and scope of services and activities.

Some organizations conclude after much study and analysis that more dramatic changes would be best, such as outsourcing certain functions or even the entire operation. What is wise and valid in one community may not suit another at all. You are faced with the difficult but satisfying task of working your way through the many questions involved to get to the best answers for you.

Whatever approach you take, whatever system of planning, consultation, and decision making you use, we urge you to insist that all involved address the task with open minds, searching for innovative solutions that best serve your customers. Experience has proven that you can trust this process to produce better results than you can now foresee.

THE MEANINGS OF PRIVATIZATION

The word "privatization" has been used to describe many developments and it is often given a negative connotation. It is a word commonly misunderstood and misused. For the purposes of this book, let's give it and its more frequently encountered cousins a working definition. Since the British claim ownership of the English language and used "privatization" more than 10 years ago, we can safely accept their usage, which we understand to be as follows.

Privatization. The transfer of the assets, responsibilities, and functions of a government agency from public ownership to private ownership and management.

This definition is consistent with *The New Shorter Oxford English Dictionary. So in this book, privatization means transfer of ownership.*

There are several approaches to performance improvement. The following definitions start with internal performance improvement and describe increasing degrees of external involvement in the utility's operations.

Internal Performance Improvement. Activities the utility undertakes with its own staff, perhaps using consultants or peer advisers, using appropriate benchmarks to set and achieve competitive performance goals.

Operations Contracts. Used to delegate to private or public contractors short- or long-term operation and maintenance activities necessary to run a particular facility, function, or utility service.

Managed Competition. When a utility staff competes with the private sector for a contract to operate and maintain the utility.

Outsourcing. Often used to describe the letting of a contract by any organization, public or private, to another organization for specific services or functions; for example, meter reading, meter installation and replacement, billing services, and laboratory services.

Figure 2-1 uses these definitions to illustrate improvement options with increasing degrees of delegation. It is important to note that public agencies often provide services to other public agencies; this includes water, billing, laboratory service, and even operations and maintenance services. There are also similar arrangements between private water companies. In these cases, the use of the term "privatization" is not appropriate.

Regionalization. The merger of adjoining utilities into a single, larger organization for common operations and management. This is often encouraged among smaller water systems to achieve economies of scale. Perhaps someday it will develop in metropolitan areas that now come close to this concept in certain respects. Objectively it is an attractive and cost-efficient model, but local pride of ownership, politics, and job protection are obstacles that are difficult to overcome. Some of the benefits of size can be obtained by sharing contract services provided by a central contractor or utility.

→ → → → → → → Increasing Degrees of Delegation → → → → → →				
1	2	3	4	5
Internal Performance Enhancement	Managed Competition for Identified Function	Selective Outsourcing of Services	Operations Contracts for a Function or Facility	Privatization through Ownership

Figure 2-1   Improvement options

Such elements of service may be large or small. It could be a contract for janitorial work, meter reading, or a regional contract for lab services to groups of facilities. The California Water Service Company provides technical assistance for control systems, laboratory services, and engineering for more than 40 small utilities in various parts of California. Similar services are available both regionally and nationally. In fact, a municipality with an existing capability that is centrally located could provide services to reduce its unit costs by contracting with smaller community systems to achieve better performance in any number of utility functions. However, aggressive marketing of such services by public agencies is commonly not part of our utility experience. Perhaps it should be. For instance, the Metropolitan Water District in Southern California is considering providing treatment plant operator services to its member retail agencies.

This book tries to avoid bias for or against any of these organizational arrangements. *Our focus throughout is on whatever management approach is best for each utility in the opinion of its customers, owners, and managers.*

SUMMARY

In this chapter we have described the imperatives or demands that will be expected of water utilities at the turn of the century. These demands constitute high expectations by consumers that will have to be met by managers and owners. We believe that the water utility that is shaped by these performance imperatives will be quite different in many respects from the typical utility of today. In the next chapter we offer some guidance in assessing your performance as a baseline for launching a program of improvement.

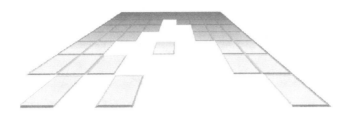

CHAPTER THREE

Initiating The Improvement Process

In chapters 1 and 2, we discussed the imperative for change and the elements of a new focus for utility management. For some industry leaders in both large and small utilities, these ideas have already been translated into programs that are underway. For others, a general plan for attacking the change process may be beginning to take shape, but these individuals recognize that the program for change and improvement needs to be developed in concert with their organizations and other stakeholders. Chapters 3, 4, and 5 are primarily for managers and organizations in the second circumstance—those of you who know where you want to go, and are looking for approaches on how to get there. Your task is one of building an overall strategic improvement process capable of transforming your organization into a new model.

THE STRATEGIC IMPROVEMENT PROCESS

Improvement can be achieved through many different means. In the past, when a "command and control" approach to utility management was more prevalent, embarking on a process of change toward an objective of better customer service and higher performance levels would have been viewed more in terms of logic (if the organization were a machine, how would I repair it?)

than in terms of the people of the organization (how can I convince people that we need to improve performance and get their insights about how to do it?).

This traditional approach to management is no longer effective in solving day-to-day problems in most organizations, let alone in tackling major organizational change and improvement. The leadership may decide that there is a need for change, but employees want and deserve a say in where the organization is going and what path it should take to there. In most cases their ideas on both topics will greatly improve the end product. After all, many employees know their areas of expertise better than the leadership does.

It follows then that the degree of success in meeting the customer and performance challenges of the next decades will be directly attributable to the commitment by management to allow staff members to be more involved in devising solutions and planning activities. The recognition of the imperative for change, the identification of opportunities for improvement, and the creation of goals aimed at producing effective and meaningful change must arise from within the organization. Involvement of all stakeholders, from operations and maintenance staff to other agency administrative support services, to governance and customers provides a powerful resource that, if successfully tapped, functions as the springboard for effective and long-lasting improvements in efficiency and competitiveness.

The process for strategic improvement of the organization therefore begins by recognizing that effective and enduring improvements are best achieved through the dedicated commitment and participation of employees and other stakeholders *in all phases of the improvement process*. The five major components which make up the improvement process build on this foundation. Each are the subject of more extensive discussion in later chapters.

1. *Demonstrating the imperative for change.* Chapter 3 describes the need to establish this imperative early in the process. Chapter 4 outlines how the assessment process and strategic plan help the organization as a whole to understand this imperative. Chapter 14 provides a far

more detailed picture of how the change process can be introduced and managed.

2. *Developing an effective process for involving the organization and other stakeholders in the improvement process.* Chapter 3 introduces the need for employee and other stakeholder participation in all aspects of the improvement process, and the use of a team-based approach to achieve this involvement.

3. *Assessing the gap between your current and desired performance.* Chapter 4 introduces the concepts and a broad framework needed to conduct an organization-wide assessment of performance. Chapter 12 provides a more complete discussion of methods, tools, and examples that can be used in conducting an assessment in your organization.

4. *Developing a process for more detailed functional assessments and a plan of action for the overall improvement process.* Again, chapter 4 introduces the concepts and framework for using a strategic planning process to identify work process improvements and to develop a plan for the change process. Chapter 11 helps you to take this framework and to develop a strategic plan that meets the needs of your organization. Chapter 13 provides insights, tools, and the "how to" of work process improvements.

5. *Implementing, monitoring, and modifying the plan.* Chapter 5 outlines the need for effective implementation, and provides the details of how to achieve it.

These outlined steps begin with the recognition that effective organizational transformation requires that the problem, the new vision, and the plan of action all be commonly understood and "owned" by the organization at large. The first step in the process is achieving a general acceptance of the imperative for change.

DEMONSTRATING THE IMPERATIVE FOR CHANGE

Every significant change will have its supporters and detractors. Because the status quo is difficult to overturn, building a coalition that supports change is an important task early in the process. This process usually begins with a recognition that a new direction is required by the utility's leader—the general manager. Decision makers, elected or appointed, and the organization's leadership team should be exposed to the general manager's vision with the goal of achieving their support. Paradoxically, elected leaders may actually initiate major change in some cases. Change is difficult under any circumstances, but in situations where the leadership is not united behind the goals, and the organization remains unconvinced about an imperative to depart from the status quo, the change process will be dogged with criticism and problems.

With this in mind, in all organizations, public or private, it is essential that an imperative for change be demonstrated *in advance* of embarking on a multifaceted change and improvement program. The majority of stakeholders must recognize that:

1. A meaningful gap exists between current performance in the organization and the performance of competitors or industry leaders in one or more functional areas.

2. The potential for substantial and cost-effective improvement exists in each functional area where a gap exists.

3. Change is essential in order to ensure the continuation of the organization and the confidence of the public.

Addressing these elements requires the effective communication of the ideas described in chapters 1 and 2 regarding the emergence of customer and performance imperatives in an environment of increased competition—both to your political leaders and to your organization. We will now discuss how to do this.

BUILDING SUPPORT WITHIN YOUR POLITICAL LEADERSHIP

In the absence of an immediate competitive threat, or an initiative by elected officials or regulators for change, building the case that change is essential for the survival of the organization requires the utility manager to be bold. For example, it may be necessary to engage key members of the utility's governing board in a substantive, retreat-like discussion of the imperatives to change and the examples of other utilities, using a facilitator and/or knowledgeable, thought-provoking speakers.

Approaching elected officials or other leaders to serve as allies in forging change will certainly be unusual to many utility managers, will undoubtedly be threatening to others, and could be viewed by some as inappropriate. By seeking their agreement, it might be said that the utility manager is making the case for the organization's mediocre performance. Yet how else can one create an imperative for change in a public monopoly? The challenge is to communicate a positive perspective in which the organization is simply responding in a responsible way to the demands of the times and of the public. If elected officials are acting as surrogates for the discipline that competition produces, and if improvements leading to better service and lower costs are available, it is the responsibility of elected officials to act and the responsibility of the utility manager to inform them.

Once the imperative for change is understood within your political leadership, it is essential that the impending change be understood and "owned" by the organization at large. The most effective way to accomplish understanding and ownership is through extensive stakeholder involvement in defining and executing the metamorphosis.

BUILDING STAKEHOLDER PARTICIPATION

An effective way of involving stakeholders is though a *strategic business planning process.* A well-executed strategic planning process will allow the organization to more fully embrace the imperative for change and improvement, to build a new vision around the theme of improvement, to set goals that drive improvement and to develop a work plan to help achieve key improvement initiatives.

Within the planning context, employees and other stakeholders can participate through teams, workshops, focus groups, and individual interviews, in the identification of areas in need of improvement. These points of involvement give stakeholders a

"Every category of employee holds and wields some portion of the organization's power."

Terry L. Cooper—
The Responsible Administrator

way of defining proposed changes in terms that benefit from their knowledge and experience. They also emerge from this process well informed and willing to help guide their peers through the process of change.

Without a formal mechanism like the strategic planning process, developing stakeholder support may simply be a matter of laying out a plan and ideas and getting support from a selected group of leaders; or, it may be a more complicated process of developing consensus around a set of objectives. *The end result should be that you emerge from an initial consensus-building stage with a clear statement of why you are initiating the changes and what is expected to be gained from the process.* The statement or list should be realistic because it is likely that decision makers will hold organizational leadership responsible for making certain that the objectives are achieved.

USING TEAMS AS THE BUILDING BLOCK OF THE INVOLVEMENT PROCESS

A variety of methods can be used to encourage staff members to participate in the improvement process. One approach relies on workshops and individual interviews at all personnel levels to get and keep the staff involved. Through the workshop method, staff at all levels can express their ideas and suggestions on how to improve a given area based on their familiarity with the details of how that area or function operates. These workshops are a powerful means to get staff included and informed on competitive strategies and issues. The workshop approach is well suited to smaller organizations or to more limited improvement efforts.

Another approach to involvement uses the *team approach* to plan and then later carry out the basic tasks of improvement. We recommend this simple and easy-to-understand structure, and will build on the team concepts in later chapters. The formal improvement process can begin by creating a central steering committee that we will call the *steering team* in this and other chapters. Ideally, it includes the top managers of the organization; a few representatives of the rank and file in various departments (selected because of their experience, judgment, and positive attitude); representation from your governance and your customers;

and possibly a union representative. A significant challenge is to construct a representative team and yet a workable one in terms of numbers.

This steering team (or whatever name you give it) can then decide what other teams are needed, for what purposes, and when. In this way you can manage every planning activity this book proposes. Your steering team can set up focus groups to gather input from a larger sampling of your stakeholders. It can establish work *process teams* in various functional areas of the organization to help assess your performance, develop new performance goals, streamline your work processes and procedures, and devise action plans. In short, this team is the organization's brain trust in planning and coordinating a comprehensive improvement program.

Process teams need to be made up primarily of those employees who will be affected by the improvement process in their work area, and by those who have a vested interest in the utility's success. Thinking about team members in these terms will help you decide who needs to be involved, to what degree, in which activities and when. Obviously the list will include appropriate representatives of your customers, senior staff, employees, and your governing body. Perhaps there are others with a special stake in your utility, such as wholesale water customers, major businesses and industries, a labor union, the mayor's office, or other city officials.

It is critical that the right individuals within the categories described be recruited for each team. These individuals must be representative of the organization, respected by those they represent, and be innovative. If it can be said that the strength of the overall improvement process lies in the use of teams, it follows that the strength of the teams lies in the people that comprise them. It is also critical that employees trust the recruitment process to be objective and believe the team approach to be genuine.

Once the process teams are formed there may be some employee reluctance to engage in identifying new ways of doing business that could lead to layoffs or loss of job security. (This fear of change is discussed in chapter 14.) Unions usually become involved at this point, and if not handled thoughtfully may go to a governing board or leadership with a request for job security. In some cases, depending on the level of cooperation with labor, it is wise to agree to some type of job security for employees who engage in seeking new ways of doing business that may lead to greater efficiency and cost savings. Employees cannot be expected to come up with creative ideas for efficiency that will result in unemployment.

SUMMARY

Recognizing the imperative for change, identifying opportunities for improvement and creating goals aimed at producing effective and meaningful change must happen within the organization, as its own initiative. Involvement of significant stakeholders, from employees to customers, provides a powerful resource that if successfully tapped, functions as the springboard for effective and long-lasting improvements in efficiency and competitiveness.

A team-based approach to the improvement process is an effective mechanism for involving stakeholders in the improvement process and for using the knowledge and creativity needed to produce meaningful change. In the next chapters we will use team concepts in describing a process for assessing and planning the overall improvement process.

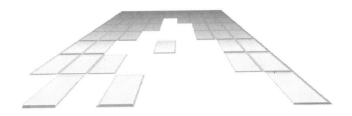

CHAPTER FOUR

Assessing Your Performance and Setting Your Course of Action

In chapter 3 we described the need to establish the imperative for change and some of the involvement approaches that are critical to achieving long-lasting change and improvement in your organization. In this chapter we describe the next steps in the improvement process—performance assessment and creating a plan for change and improvement. We encourage the use of a team-based approach for accomplishing both of these steps. More specifically, in this chapter we provide an overview of:

1. *Conducting a self-assessment, determining the gap between your organization's current performance and potential performance, and identifying key functional areas in need of additional focus.* (This step, with team-based involvement will assist in demonstrating the imperative for change to both management and staff.)

2. *Identifying specific improvements in those functional areas shown to be promising in the gap assessment.*

3. *Planning and prioritizing these improvements in a manner that reflects your organization's capacity for change and its other commitments.*

This chapter also shows how a strategic plan can serve as the framework for all these steps and the larger challenge of reaching consensus on a vision for the organization that is built around the concepts of change and improvement. Later chapters discuss the process in greater detail. What follows is an opportunity to master the principles and basic structure of this important stage of your move toward improvement.

We begin the discussion by building on the premise that all of the organization's stakeholders ultimately need to be engaged in: (1) the process of understanding the need for change; and, (2) fashioning the organization's approach to improvement and change. Understanding the need for change is satisfied through the combination of a message from the leadership, as described in chapter 3, and through knowledge and involvement of employees in the assessment or "gap analysis" process which is described in this chapter. Fashioning the organization's approach to improvement and change is best accomplished through the strategic planning process that includes more detailed assessments within specific functions, also described in this chapter.

Both steps will benefit greatly from the type of team-based approach introduced in chapter 3. Accordingly, we will first describe a team structure that meets the objectives of full involvement of staff and stakeholders in both the assessment and subsequent planning steps, and then proceed to the assessment and planning discussions.

INITIATING THE STEERING AND PROCESS TEAMS

Ideally your improvement process will begin by initiating a *steering team*, which will help guide the overall improvement process. As described in chapter 3, their job is to work with the utility leadership in steering the overall effort including: assessing the performance gap, building a detailed plan of action, and then implementing that plan. Initially the steering team can take the lead in the self-assessment process, calling for external peer review, and deciding on the need for consultant services and training. These teams need to be thoughtfully selected, as described in chapters 3 and 14. Once formed, the steering team can organize and direct the functional or *process teams*, which are created to identify opportunities for improvement in specific functional areas, such as in the treatment works or in maintenance and operations. Again, the process by which individuals are selected and the ultimate membership of the team is very important.

ASSESSING THE GAP BETWEEN CURRENT AND TARGET PERFORMANCE

With a team structure or an alternative method for involvement in place, your organization is ready to confront the fundamental question underlying the improvement process: *Are the operations of my utility as effective and efficient as they could be?*

People in the organization development field don't like the word "operations" in that question. They talk in terms of "core processes," the groups of interdependent, related activities that may well cross over several organizational units and together constitute the basic functions of the organization. Thinking in terms of core processes makes it easier to track how your utility is actually performing and it helps to identify where changes would improve efficiency. Thinking in terms of your organizational chart is not nearly so helpful.

To assess how you are doing in core processes requires measures or standards of performance. You will need measures of your performance and you will need to be able to compare them to the other utilities. If you are not the best (and perhaps you are in some processes), who is? How well do they do? What do they do differently that produces better results?

The dilemma facing the water industry is that while there is an urgency for utilities to assess their performance against an industry standard, many of the concepts, methods, and tools required to accomplish the task have not yet matured for the industry. Earlier efforts to establish standards have been burdened with arguments about comparability and approaches to measurement. Yet comparisons are possible if undertaken thoughtfully and in the right context as described.

LESSONS FROM THE WATER INDUSTRY
BENCHMARKING EFFORTS

Numerous benchmarking efforts have been undertaken during the past 10 years in the water industry. (See chapter 12 for more information about these efforts and their status). An important lesson emerges from these efforts and from benchmarking efforts in the manufacturing sector—you must use a mix of approaches to performance measurement to adequately assess your performance. Your assessment should include information from three different and complementary perspectives: *internal, customer, and competitors/peers*. A balanced, three-pronged assessment will help identify the performance gaps in your organization.

The employees of your organization, those closest to the work, are a wealth of performance information. With their internal, firsthand knowledge and experience, they can tell you what is working well and what isn't. Your customers provide a second perspective. Part of the new performance demand is recognizing their importance. Feedback from them, therefore, should be a significant part of your performance assessment. Finally, assessing your performance requires you to compare your organization with industry leaders, *whether they are competitors or peers*. You can set reasonable performance targets for your operations by using your best judgment to interpret their information.

We strongly urge you and those in your management and planning group to not get distracted by the difficulties and debates surrounding performance measurement or "metric benchmarking." While there are problems of measurement and comparability associated with just about any performance measure, comparing yourself to others will start to provide insight into your performance—insight that can factor in some of the more subtle factors that will make the comparison more substantial.

The hard fact is that each and every utility must grapple with its own performance measures and performance level that is reasonable in each case. Information from other utilities, when available, is helpful, but the final performance goals are a judgment call for each utility. The lack of perfection in this analytical step is no reason to stop trying. Taking a few steps in the right direction is better than taking none at all.

FRAMEWORK FOR ASSESSMENT

The following steps have at their core the customer, employee and comparative perspectives of self-assessment.

1. *Get the organization involved with the gap analysis and its ramifications.* The assessment process will be a period of uncertainty for all those involved in the processes under review. Acknowledge this fact. Use the gap analysis as a tool to build confidence in the imperative for change and improvement. Sometimes developing an assessment team that is responsible for overseeing this effort and that reports to the steering team is effective.

2. *Ask your employees to identify the functions and activities where improvements are possible and where performance gaps are greatest.* Use focus groups, process teams, workshops, or other means to create a discussion forum for specific improvements and process changes. This approach is most successful when the process being looked at is at a fairly detailed level. For example, a process team would probably have a difficult time assessing maintenance and operations as a whole, but could easily look at the hydrant replacement process and develop recommendations for work process improvements. Chapters 6–10 address specific process improvements in key functional areas including treatment works, maintenance and rehabilitation, customer services, instrumentation and control systems, and utility-wide energy efficiency.

3. *Survey your customers to determine what is important to them, how they view your performance, and what they believe are the performance gaps relative to expectations.* Customers will often surprise you by giving you an entirely different perspective on areas where you are not performing well. There are two components to the customer perspective: first, whether you have provided a service mix and quality that *they* define as important, and second, whether they are satisfied with the products and services. Both components are important.

4. *Compare yourself with your competitors and peers.* This can be done in two ways. First, through metric comparisons with competitors and peers, where quantitative measures of bottom line performance are compared in a manner that permits equal comparisons. Second, through process benchmarking, where the practices of your utility in conducting a particular process are compared against those regarded as "best in class" for that process. The metric comparisons require the following additional steps.

5. *Adopt a common functionally based framework and accounting system.* Organize your processes into functional or activity-based groupings that provide sufficient detail to make comparisons meaningful and for which data can be assembled. In the process, isolate costs assignable to support for nonutility municipal functions that frequently confuse comparisons between utilities.

6. *Choose the relevant set of performance measures for your utility.* This needs to be done considering the data available on your own utility and on your referents. The measure you choose should help you measure progress toward your goals and outcomes that are important to your customers. Your measures should reflect a balance of quantitative, objective measures and qualitative, soft measures.

7. *Make comparisons to utilities in comparable circumstances.* Even when comparisons are made at the functional level between utilities, it is not always possible to make an "apples to apples" comparison. "Unitizing" measures at the functional level (e.g., treatment cost per gallon of finished water) helps but is sometimes overwhelmed by other mitigating factors (e.g., the need to pump as part of the treatment process in some circumstances and not in others).

8. *Choose appropriate referents* (i.e., comparable organizations or functions that you wish to compare yourself with). For example, you might choose for comparison

competitors, recognized leaders, top quartile of the industry, and nearby utilities of similar size and circumstances. Don't forget that the leaders for a particular function (such as customer service) may be from other industries.

CHOOSING YOUR KEY PERFORMANCE AREAS: WHERE TO START

The three-pronged approach to assessment previously described uses an employee-based gap analysis, a customer-based gap analysis, and comparisons with competitors and peers, and will go a long way in helping you identify those functional areas where there are gaps in your performance. Not all of these gaps will be of equal importance to your organization, and you will need a basis for limiting action to a few key areas—at least to start with. We suggest three criteria to help you prioritize your focus, although your own needs and circumstances will ultimately provide you with a set of criteria appropriate to your organization.

- Is the function a large part of the overall operation?

- Is the function prone to performance drift that requires periodic correction?

- Is the function one that has significant potential to be supplied or performed differently (e.g., automated or outsourced)?

While the answer to these questions will be different for each utility, five functional areas are likely to be key areas in many utilities. They include:

- Treatment works

- Maintenance and rehabilitation

- Customer services

- Instrumentation and control systems

- Utility-wide energy efficiency

For some utilities, this list may not include one or more of the key functional areas in which improvement should be initiated. For example, in smaller groundwater utilities that do not have treatment or transmission facilities to maintain and operate, finance and administration may emerge as a key functional area for improvement. A detailed discussion of improvement programs in each of the five areas listed is provided in part 2.

THE STRATEGIC PLANNING PROCESS:
A FRAMEWORK FOR ACTION

Getting a large organization moving in one direction, especially if it is a new direction, is a complex process. The performance assessment described in chapter 3 informs you about your organization's strengths and weaknesses and suggests specific functional areas that would benefit from a focused improvement process. We suggest that these areas would likely include the treatment works, maintenance and rehabilitation, customer services, and information and instrumentation systems. Developing a plan of action for change and improvement in these areas and for your organization as a whole is the next task. This plan of action should close the gap between your current performance and the goal of where you want to be through a sequence of prioritized actions. If the plan is to succeed, it must be understood and endorsed by the organization as a whole, not just by those who are the proponents of change.

The development of a plan of action for change that is organizationally owned and understood, is most easily accomplished in the context of a strategic planning process. Through the strategic planning process the organization can:

- Work together to create a common vision and mission for the utility,

- Develop goals and objectives to achieve this vision,

- Identify strategies to close the gap between your goals and your current performance in the key areas you have identified for improvement,

- Develop an action plan for translating these strategies into concrete steps to improve your operations

To do all this you need a planning process that fully involves employees and stakeholders at all levels. This involvement ensures that:

- Employees, customers, and directors know where the organization is headed,

- Employees feel secure in promoting change and improvement,

- Employees know how their work contributes to the organization's goals and objectives,

- The action plan will be understood and translated into action,

- The organization has adequate resources to get there

The extent of employee involvement in planning will determine the time required to complete the plan. Some organizations think they do not have the time required for a fully participative process, but we urge you to use this approach as fully as possible. Not only will your plan be better, but its implementation will be more assured.

THE STRUCTURE OF A STRATEGIC PLAN

The old adage remains true: If you don't know where you're going, it doesn't matter which road you take, because you're never going to get there anyway. If you want to begin a program of utility performance improvement, you need to know where you are heading and what changes are necessary to get there. More than that, your employees, governance, and customers need to share the same ideas about the utility and its future, or at the very least, to understand what is planned for the future and why the organization needs to make changes toward that direction.

If you need any further convincing, just ask 10 or 20 of your employees at random, from all ranks and departments, what they think the organization's goals and values are. What you will get is proof that there are a number of different ideas among them as to what the utility is trying to accomplish.

Correcting this confusion and giving everyone a common idea of what they are working toward is the purpose of developing a strategic plan. Typically the planning process calls for the organization to thrash out a series of elements.

- *Vision:* A brief statement that encapsulates a picture of what the stakeholders want the organization to be.

- *Mission:* An expression of the organization's purpose for being.

- *Values:* A listing of the values or principles that motivate the organization and the way it operates.

- *Goals:* Long-range statement of desired states which, when taken together and achieved, constitute success in fulfilling the organization's mission and vision.

- *Strategic plan:* A road map for accomplishing the vision.

- *Strategic objectives:* End states that must be reached to accomplish each goal.

- *Action plan:* A set of concrete steps intended to carry out each strategic objective.

- *Evaluation process:* Process through which the plans are monitored and evaluated against a variety of measures including quality, performance, timeliness, and cost.

The value of having clear, specific, and brief statements from the vision on down through every level of the plan is self-evident. Everyone on the staff can see where and how his or her work contributes to the utility's mission and to the fulfillment of its vision—a critical need when the organization is in the midst of substantial change. Everyone can identify with the shared, unifying vision of a greatly improved, high performance organization. Some organizations, like the Pennsylvania American Water Company, print their vision, mission, and values on a wallet size card for all employees to carry.

There is also great value in the plan because it defines the sequencing and priorities for improvements in each of the functional areas. Sequencing can and should be paced in a manner that corresponds to the organization's needs and capacity for change. Most organizations have evolved to their present condition over many decades. Even under the most ambitious programs for change that have broad stakeholder acceptance, the process of widespread change will span a number of years. Chapter 11 provides a more complete discussion of the strategic planning process and techniques for success.

IDENTIFYING STRATEGIES TO IMPROVE PERFORMANCE

Building on the overall performance assessment and gap analysis described earlier in this chapter, the next step is to develop a more detailed stakeholder understanding of potential performance in the processes underlying those key functional areas identified for improvement.

BUILDING THE LIST OF STRATEGIES

Working within each of the functional areas such as treatment or customer service, begin the process by developing a master list of key processes as the subject of improvement. This is most readily accomplished by creating teams from among those who are responsible for a given functional area. They should be readily able to identify the key performance gaps and the key processes needing improvement. Improvements need to be rated in terms of their potential, costs, and feasibility for later use in setting priorities.

Working in a team structure where the process teams report back to the steering team ensures that agreement of the major stakeholders is achieved, including the leadership team and the employees who will be affected by the proposed improvements. Allow employees time to participate in this process, either as a team member or in a review function, so that they understand that changing process will lead to a better way of doing work. They must perceive that there is "a win" for them in the process and that it is in their professional interest to contribute to creative

Example: Cincinnati Water Works

Vision: *Cincinnati Water Works will be the Standard of Excellence in the Water Utility Industry.*

Mission: *To provide our customers with a plentiful supply of the highest quality water and outstanding services in a financially responsible manner.*

Values Statement: *Above all the Cincinnati Water Works values our customers; they are the sole reason we exist. Anticipating and exceeding their expectations guides our strategic planning, drives our decision making process, and prioritizes our actions. To that end, we recognize that successful customer relationships directly depend on our employees. The people who work here are the Cincinnati Water Works, and we value their loyalty, contributions, accomplishments, and their dedication to our customers. Cincinnati Water works employees, in turn, commit themselves to the following values that will*

45

enable us to realize our vision—to be the standard of excellence in the water utility industry.

- *Quality Drinking Water*
- *Involvement in the Community*
- *Innovating and Creativity*
- *Integrity and Professionalism*
- *The Environment*
- *Efficiency and Cost Effectiveness*

Sample goal: *provide outstanding customer service*

Sample related objective and actions:

Improve the timeliness of customer service as measured by internal benchmarks

- *Develop and implement response time standards for all customer service activities*
- *Monitor performance against customer service standards and continuously implement improvements*
- *Improve customer complaint inquiry tracking system with appropriate solution*
- *Develop ways to provide customers*

solutions. At the same time, managers need to loosen their traditional reins of authority and trust the process. This risk taking can be very rewarding.

If there is a possibility to provide incentive bonuses or pay bonuses for improvements through process redesign, employees will be more engaged in a search for innovative solutions. Other means of acknowledging or rewarding individual and team contributions should also be considered to foster a positive attitude among the entire staff.

SETTING PRIORITIES

Once the list of strategies is developed, it will be evident that not all of the strategies are of equal value to the organization, and probably only a fraction can be undertaken in the short term. Therefore priorities must be established to determine which functional areas, and which projects within them, are first in line and which will be addressed later in the process. The steering team is a good place to develop a prioritization process. In this process look for big wins and recognize that early successes are critical. Accordingly, try to recruit the best and most creative employees to work on the first projects. If early efforts are judged as successful, other enthusiastic employees will be more willing to sign on for the next round of efforts.

For example, your earlier gap analysis may have suggested that the organization was in need of significant improvement in five functional areas: customer service, watershed and reservoir management, water treatment, internal services, and energy usage. Through the previously described approach, process teams would have developed a set of strategies for improvement in each functional area. The leadership of the organization, working with and through the steering team and the process teams, would then make choices about which functional areas would be targeted first, second, and third, and which strategies within each functional area were top, second, and third priority. Using this process, the organization can develop a set of actions that are prioritized and sequenced to meet a pace of change that is appropriate and feasible.

DEVELOP AN ACTION PLAN FOR EACH PROGRAM AREA

Once the steering team and the process teams have done the creative thinking that is required to determine the directions the utility will go, they are ready to complete the strategic planning process through the development of an *action plan*. The action plan is designed to translate broad strategies into a set of concrete and measurable actions. For each strategic action the action plan includes a series of tasks, required resources (labor and nonlabor), timelines, and ideally, performance measures. Performance measures help document the success of each strategy and when viewed together provide an overall measure of success for the improvement effort.

In addition, the plan should identify who is responsible for carrying out the actions of change and who is accountable for seeing to it that the tasks are moving ahead on schedule. These elements of the plan should be seen as positive—that is, as a way that everyone can measure and enjoy success and as a way of clarifying who has the responsibility and authority for getting the job done. The process for developing an action plan is described in more detail in chapter 11.

with more accurate times for service delivery
· *Improve the participation of construction and operating services with external parties*

SECURING STAKEHOLDER APPROVAL OF THE PLAN

If external stakeholders are not involved in the steering team, it is essential that the action plans be reviewed with key stakeholder groups, and be presented as proposed plans rather than as completed work. This is usually done in the form of a briefing, with an opportunity for review and comments. The objective is to demonstrate that the plan for bridging the gap between the present situation and the desired change is sound, practical, and timely. If the plan is not acceptable, the steering team, working with the process teams, needs to engage in a serious, substantive discussion of shortcomings, alternatives, and concern to elicit practical guidance. Such exchanges often lead to fuller understanding and agreement.

SUMMARY

The key to success in strategic planning is to communicate that the plan is not the goal and purpose of the exercise. Strategic planning is merely the first step in an active process of change and improvement. Participants should understand that all the preliminary steps of self-assessment, priority setting, work process analysis, and so on are critically important actions that constitute intelligent planning and decision making to put the utility on the road to solid improvement.

This chapter sketches the elements of assessment, a strategic plan, and the action plans necessary to address the most potentially productive improvements in key utility processes. The next chapter outlines an approach for ensuring that your action plans are carried out and that your intentions are achieved.

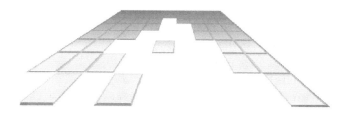

CHAPTER FIVE

Ensuring Effective Implementation

When General George C. Marshall retired as chairman of the Joint Chiefs of Staff, a position he held during the momentous days of World War II, he was asked what the most difficult aspect of his job was. "To get anything done," was his astounding reply. Despite military discipline and the urgency of the war effort, the nation's most powerful general officer found it tough to get anything done. What was true for General Marshall more than 50 years ago is true for us today.

This chapter is about how to get something done. Implementation requires having systems of planning and communication as well as having oversight and monitoring to assure that a plan or project stays on track and on task. In the context of the last two chapters, which have dealt with preparing for change, framing and assessing the organization's performance, and determining a plan of action to implement these objectives, this chapter offers guidance on *implementing* the product of these efforts—that is, the activities and commitments developed through the strategic planning process. Whether your organization is large or small, young or old, the principles and processes of implementation are basically the same.

Successful implementation also requires demonstrated and ongoing commitment to the plan by the leader of the organization.

Employees are quick to sense when an organization engages in an "empty" planning process—where the end product is a document that sits on the shelf and has no practical significance to the day-to-day realities of the organization. *To create an organization that is committed to make continuous improvements, to monitor its progress toward a goal, and to seek new ways of making itself more efficient, the entire group (including management) must sense that a captain is at the helm who is continually assessing the speed and direction that the ship is moving.* While this level of attention to progress and accountability may make some individuals nervous, most members of the organization will view it positively—it serves to reinforce a system in which they can succeed by remaining aligned with the rest of the organization, its leadership, and its customers.

PURPOSES OF EFFECTIVE IMPLEMENTATION

Getting the organization moving in one direction through the strategic planning process is a time consuming and sometimes difficult process. As described in chapter 4, it requires that:

- Employees and customers know where the organization is headed and why

- Employees are secure in promoting change and improvement

- Employees know how their work contributes to the organization's goals and objectives

Chapter 4 describes several methods for accomplishing these objectives. The list in chapter 4 also includes two additional elements:

- The organization has adequate resources to get there

- The action plan will be understood and translated into action

Accomplishing these two points is the art of effective implementation and the focus of this chapter. Success requires that:

- The principles of the plan or project are embedded in the management systems of the organization

- There are monitoring, oversight, and accountability mechanisms in place to measure progress and facilitate course correction

- Incentives have been established to motivate staff to follow through

The purpose of having effective implementation strategies is to assure that you will reach your goals and make enduring changes. The first rule of thumb requires building in attention to your goals in all aspects of organizational decision making. *The organization's leadership team must commit itself to unity in support of carrying out every step of the plan according to schedule.* They must create and maintain the expectation that the plan will be fulfilled because the organization is serious about accomplishing the intended changes and achieving the desired benefits. Without such leadership, implementation will break down and the staff will see that as proof that management was not serious in the first place.

At the same time, excellent leadership or management of an organization means that it is not dependent on particular individuals or personalities to move in the desired direction. Organizations should be structured so that if a leader leaves a job other people in the organization will carry on the goals and aspirations of the plan. Too often one hears of an excellent organization that has an outstanding leader, but when the leader leaves, the vision and progress of the organization collapses. One of the values of having a strategic plan and a system of implementation mechanisms is to ensure that the plan or project will be carried forward because its value is embedded in all employees.

Effective implementation also requires that accountability measures as well as organizational systems are linked to the plan. What follows are suggested tools for how to tie those systems together. Although these tools may not cover every conceivable situation, they will give you an idea of how to link whatever systems you have into an overarching accountability structure to carry out the organization's adopted plan.

IMPLEMENTATION TOOLS

We suggest that you use the following collection of tools to keep your implementation activities moving forward. Five key tools are outlined here.

1. *The strategic plan*, with its associated action plans and oversight process including:

   - commitments made in the strategic planning process to timelines, deliverables, performance measures, and budget for specific improvement actions

2. *A communication plan*, both internal and external

3. *Accountability mechanisms* including:

   - an ongoing organizational commitment of sufficient resources to carry out the improvement plan in the context of other requirements

   - regular sessions to report on progress and problem solve

   - visible graphical representations of progress posted in prominent locations

   - individual performance evaluations and accountability contracts

   - reward/recognition systems

4. *Ongoing mechanisms for employee involvement and training* including:

   - leadership, technical, and cross-functional training

   - training in the use of team and collaborative approaches to work

5. *Systems integration and management*

The theme song that must be heard throughout all these activities is your organization's vision, your reason for doing all of this. Employees should see each element described here as a step toward being more competitive, a step toward top quality performance and public service.

THE STRATEGIC PLAN

To focus your organization on achievement of performance improvement requires respect for your strategic plan. As stated in the last chapter, the plan details your vision, mission, goals, strategic goals, and actions to accomplish the objectives. The two major methods of ensuring implementation are to develop action plans and to have a regular reporting/monitoring process on the progress and problems along the way.

Action Plans

Action plans establish the specific steps to be taken to accomplish a goal or strategic objective. The plan should be developed by those most responsible for accomplishing the objective. It should include information on what is to be accomplished, who is responsible for carrying it out, who the partners are in the project, what the products will be, when they are due, measures of good performance, and the budgetary/resource commitments.

For example, if one of your strategic objectives to improve customer service is to establish a central service center to handle all customer calls, then your action plan would include the following elements (among others): a study of the telephone, computer, and other information system requirements; an analysis of the personnel needed and their training requirements; the development of the cooperation needed from other departments and arrangements for it; a plan to publicize the service; the methods for getting customer input on what services they want; and testing the system before full start-up. In each activity you must know who is going to do it, when, with what resources, and with what checkpoints or decision points.

Note that much of the work involved is gathering information on facts, alternatives, and the experience of others in and out of the water industry. This information is used by the planning team to choose the alternatives best suited to your utility's needs and resources. Their final plan goes forward to the steering team or to management as a recommendation for approval.

Oversight Meetings

With a set of action plans in hand, the steering team can have regular meetings with the process teams and have them report on current progress or on any problems. The regular meetings keep pressure on the employees to get the work done while giving them a place to discuss successes or problems being encountered. These meetings send out a message to all the staff that the plan is being implemented, that the "bosses" are paying attention. They also give the organization an opportunity to measure its progress toward the established goals. They provide opportunities to make course corrections, change time lines, or change resource allocations if the original estimates were not correct. The frequency of such meetings is a judgment call and can usually be left to those involved. Generally, in the early stages of a comprehensive effort meetings are more frequent than in later stages when staff members are more comfortable with the process.

COMMUNICATIONS PLAN

In order to implement a strategic plan effectively, employees and customers must know the purpose and goals of the plan. This requires that there be both an internal and an external communication strategy. A communications plan is vital to remind employees and other stakeholders of goals and progress and to help them focus on their priorities, responsibilities, and deadlines.

Internal Communications Plan

Any strong, effective organization must spend time thinking about how to communicate regularly and effectively with employees. This process has become more complex with new technology because the avenues through which people prefer to receive information varies. Individuals differ on whether they want printed material or information provided via a web or computer network. In addition, there are different levels of communications technology available in different parts of each organization. Field employees may or may not have access to computers and some organizations may not have information systems that talk to each other. Today's communication plans usually include print material such as newsletters, a manager's weekly report, an "Intranet" site, regular meetings (both large and small) with leadership, and posted notices. A successful communication strategy includes a variety of routes or methods to send a consistent message to employees. There is little risk of too much communication in our organizations, although care must be exercised to ensure that there are few themes and that they are simply articulated. The cause of many problems is lack of communication, which makes employees feel left out, threatened, and not valued.

External Communications Plan

Modern organizations must send word about their goals and plans to their external customers. Customers want to know where the utility is headed and what they can expect, thereby establishing accountability. Customers might be consulted in determining the most appropriate strategies for an external communications plan. Try using surveys and focus groups to determine what methods will most successfully get information to customers. These can include print materials such as mailings, bill stuffers, and brochures; and electronic media such as television and radio public service announcements, and an external web site. The external customer messages must be in accordance with the strategic plan goals, and should be consistent with the messages being sent internally to employees.

External stakeholders involved in developing the strategic plan should also be kept informed of its progress along the way. Those stakeholders represent important constituencies of the utility, and in fact, there may be more of them who should receive information about your activities than were originally involved in shaping your plan. In this instance the information you disseminate should be restricted according to the stakeholder's level of interest. Overwhelming public officials and business leaders with detailed reports, for example, is not a wise strategy.

In all of your communications, internal and external, it is best to be accurate. Be positive, of course, but be honest and clear above all. Never cover up problems. Lavish praise on those who deserve it. Use these communications opportunities to reinforce the utility's vision and commitment to service.

ACCOUNTABILITY MECHANISMS

Serious attention should be paid to organizing internal resources and accountability mechanisms to accomplish the goals set forth in the plan. The following are examples of ways that an organization can emphasize ideas and concepts in a plan in a number of its routine management systems.

Budget/Resource Allocation

In the annual operational planning/budget development processes that follow in the years after the strategic plan and associated improvement program commitments, the leadership must continue to emphasize to the organization these commitments and priorities—even if it means cutting other important or traditional activities out of their budget request. The reference for priorities comes from the strategic plan and its sequencing of improvement activities in comparison with other activities. As managers develop their requests and plans/budgets are reviewed by organizational leadership, managers should be required to report on the resources they have requested to fulfill their goals. Are they sufficient? It is frustrating to come to the end of a budget development process and discover that managers have not included the resources they need to accomplish departmental goals.

Performance Measures

Cities and high performance organizations are moving more and more toward using performance measures to track their level of accomplishment. The strategic plan should include performance measures that can be used to gauge progress toward goals and objectives, both those that are improvement-related initiatives and those that are not. Having measures of performance is particularly important in an environment where managers are being challenged to both change things and to continue producing quality services.

It can be difficult to identify the appropriate measurements, and the tendency is to select activities that are relatively easy to measure. An organization should avoid using easy measures and concentrate on what it is really trying to accomplish and how to measure success. These measures can be both quantitative and qualitative. Some measures are clearly a necessity, such as customer satisfaction surveys, which an organization needs in order to determine if customers are happy with the range and quality of services being delivered. The entire leadership team of an organization should be involved in identifying and selecting the key array of measures, and then should be held accountable for achieving them. Finally, the selected performance measures should be checked against the budget so funding is available.

Performance Evaluations/Accountability Contracts

Most organizations have some type of performance evaluation system. Another tool for implementing a strategic plan is to tie the evaluation system to the objectives in the plan. If an employee's performance evaluation is clearly related to the goals of the organization, it is more likely that the goals will be achieved. In addition, both large and small organizations are moving toward the development of accountability contracts with high-level management. These accountability contracts specify the chief executive's expectations of senior management. A portion of the accountability contract is related to accomplishing the strategic plan and meeting the performance measures set out to demonstrate achievement of results. A strong accountability system that includes performance evaluations and accountability contracts enables the general manager to put in place a reward system to recognize managers who achieve the agreed upon goals.

Monitoring Meetings

The value of monitoring meetings held by the steering team was mentioned earlier in the chapter as a mechanism for ensuring accomplishment of the strategic plan. These meetings can be expanded to reports by management to the general manager on budget progress and achievement of performance measures. If managers know they will be held accountable and have to report in person to the general manager on progress toward achieving budget and performance goals, they are likely to pay regular attention to what work is being done to accomplish these goals.

Graphic Reports/Posters

A technique for focusing organizational attention on accomplishing objectives is to have large charts or graphs posted around the facilities, reminding staff of their goals and progress. These can list the specific objectives a group is to achieve within a period of time or include charts that measure progress on accomplishment of tasks. They can also be regular monitoring reports that managers give to the staff and post to show unit progress. For example, in customer service units regular statistics can be posted showing progress on numbers of calls answered and customer wait time. Certain information, such as individual performance measures, should not be posted without first consulting employees and affected unions.

Reward/Recognition Systems

Tying all the performance and monitoring systems to some type of reward and recognition system is most likely to encourage good results. It will also produce some resistance among employees who may fear being held accountable for performance. Any reward or recognition system needs to be reviewed and understood by unions before it is implemented.

Systems can be as simple as award certificates, listing of kudos in the employee newsletter, giving employee of the month or year awards, and so on. They can be more complex by tying vacation or pay to accomplishments. For organizations with simple recognition systems it is important to recognize teams as well as individual employees. Team recognition tends to foster teamwork and spirit and help employees move away from individual competitive performance. Private organizations often have compensation systems that are geared to performance. This practice is beginning to appear in public organizations. Having a strategic plan with performance measures attached and accountability contracts tied to performance is a way to link together the goals and the productivity of an organization.

SYSTEMS INTEGRATION AND MANAGEMENT

Thus far we have tied successful implementation to integrated processes for planning, budgeting, resource allocation, performance assessment, and personal accountability. *The integration and automation of these processes is as important if not more important to successful implementation as the individual processes.* Achieving a fully integrated system in a modern organization is usually an elusive goal because organizations are so complex and change so frequently. However, as senior management integrates and implements the systems to support the goals of the organization, employees will most likely begin to understand and follow their direction.

This activity will be greatly aided if you are lucky enough to have a few strategically-minded systems personnel on staff, or if you have access to good contractors who can play the same role on a contractual basis. These employees will find a stimulating challenge in advising the management team on how to more fully use the systems, data, and technologies available in the utility.

SUMMARY

Implementation is not easy. It requires leaders to be creative about the tools available to encourage and inspire employees. It is also a process of ensuring that employees understand the vision and mission of the organization and have their objectives communicated to them until it seems second nature to focus on them. Accountability should be thought of as a friendly system: one in which employees are given clear direction and where implementation strategies are simply the guide or compass to achieve the common goal. Most importantly, employees must not think of this entire process as an additional duty on top of their real work; they must see the reorganization of the utility and the realignment of its performance as an assignment critical to the organization's future success. Indeed, they are building their utility's future.

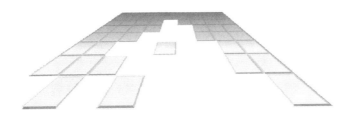

Enhancing Performance in Key Functional Areas

Foreword to Part 2:
Enhancing Performance in Key Functional Areas

Water utilities employees carry out their work through many hundreds of activities. Ultimately every activity offers the potential for improvement, for who can say that there is no possibility of doing a function better than it is being done now? Common sense tells us that it would be foolish to try to improve every utility activity at once. We have to make choices, set priorities, and decide where best to begin our program of comprehensive improvement.

The five chapters in this part of the book address the utility functions that offer the best opportunities for improving performance and reducing costs. There are other targets, of course, and one utility's priorities may not be the same as another's. That is why each utility must go through the assessment and planning process for itself. Nevertheless, many will discover that some of their priorities will be found in the functional areas discussed here.

Treatment works and maintenance rehabilitation (chapters 6 and 7) are obvious areas for many utilities to pursue improvements, so no special persuasion is needed here. Similarly, energy efficiency (chapter 10) has been widely recognized as a long neglected opportunity for cost reducing changes.

Customer services (chapter 8) has become something of a buzz-word in North America. Many hear it and stop listening on the spot because they think they know what it means. In this book we have tried to reflect the thinking of those organizations and water utilities that have taken customer responsiveness to a much higher level. It is not a matter of answering the phone politely and sending out bill-stuffers of information. Rather we propose, as chapter 2 argues, that utilities must reorient their focus, their system of values, to place the customer at the center of all their thinking and planning. This is not public relations; it is a new way of understanding the nature of the utility. It has extensive implications that require a lot of thought.

Information and control systems (chapter 9) are also rich veins of opportunity that have not been well mined by many water utilities. We may think we are advanced in information technology, but the fact is we are in the early, swift-moving stages of a technological revolution. There are resources now available that greatly enlarge what we are able to do with and learn from technology. If there is a great leap forward in water utiltiy operations, it will be found in information and control systems. While this chapter will be of greatest value to information technology (IT) managers, others will also find informative insights here.

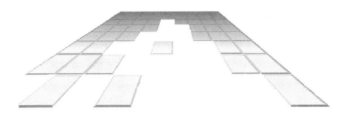

CHAPTER SIX

Treatment Works

Today the major functional area being targeted for enhancing utility performance and controlling costs is the water treatment system itself. While it is essential that other functions within a utility be considered for opportunities to improve efficiency, the treatment works can be as much as 35 to 40 percent of many utility operating budgets and deserves special attention.

There are many opportunities to enhance performance within the treatment function. However, when seeking ways to improve efficiency and reduce costs, the manager's first priority is his treatment plant mission to produce water of a quality that protects public health. Balancing the target of efficiency, as measured by cost containment or reduction, against the degree of risk a utility is willing to assume becomes a paramount consideration in establishing any realignment effort at the treatment works. In fact, when assessing the gap between current and target performance the manager must address these questions:

Are my treatment plant O&M activities as effective and as efficient as they can be?

and

If not, what should I do about it?

A foundation for addressing these questions can be found in Chapters 3 and 12 on performance assessment and tools to assess performance. In essence, the manager should attach a performance objective to his mission of public health protection.

As a manager there are two basic options you can follow to improve the efficiency of your treatment works: internal improvement of existing practices and processes using existing public workforces, or contracting out the function to an external (and typically private) service provider.

This chapter is oriented toward improving internal performance and builds upon the concepts and strategies presented in part 1, chapters 1–5. Here we explore specific tools, strategies, and processes that the manager may apply to improve the efficiency and effectiveness of the treatment works. Although we focus on internal improvements, the reader should keep in mind that selective outsourcing of certain activities may be advantageous.

Appendix A contains several case studies under category 2, Internal Improvement, which will enable you to benefit from the experience of other utilities that are in the process of internal improvement.

SETTING UP YOUR PERFORMANCE GOALS

Once you have determined that your treatment works operation and maintenance activities, or work processes, can and should be more effective and efficient, a framework for your own performance goals needs to be developed. You can use information contained in other chapters on performance assessment, establishing a course of action, work process improvements, and people and change to help you establish goals.

The information presented in this chapter is comprehensive. It may even be intimidating for smaller utilities, which should consider which approaches best fit their needs and how to adapt them.

PERFORMANCE IMPROVEMENT METHODOLOGY

Self-improvement can be achieved through many different means. The degree of improvement, however, is often directly attributable to the motivation and support of staff in the change process and the commitment by management to allow staff to be more involved in decision making and planning activities. This is the basis for the concept of a forward-looking utility model, which is predicated on the reality that achieving effective and long lasting improvement is best accomplished through the dedicated commitment and participation of the employees of your organization in all phases of the improvement process.

It is also important to consider that improvements in your treatment operations are best undertaken as part of a utility-wide effort to address all operations and functions in a coordinated manner. Uncoordinated change efforts going on in the treatment works, in field operations, and in administrative units will quickly become counterproductive, confusing to employees, chaotic to managers, and frustrating to all stakeholders.

This is a reminder that a treatment works improvement process relates organizationally to other change processes in the utility. Ideally a central team (whatever name is given to it) establishes other teams in the major functional areas under review. The water treatment or treatment works team (whatever it is called) will coordinate and set up all treatment-related teams, e.g., technology, plant operations and maintenance, or staff organization. In turn, these teams may need sub-teams or work groups. The idea is to involve representatives of those performing an operation in your utility's effort to help find ways to perform more efficiently and cost effectively. This system marshals your utility's expertise and builds confidence in the outcome of the process.

OPPORTUNITIES FOR IMPROVED EFFICIENCY

We have organized the remaining contents of this chapter on opportunities for improved efficiency as follows:

1. Overview of opportunities summarized into three areas:

    • Operations and maintenance

    • Technology

    • Organizational factors and constraints

2. Opportunities for enhancements in each of the three areas

3. Best practices evaluation in each of the three opportunity areas

AREAS WITHIN A TREATMENT FACILITY THAT MAY BE OPTIMIZED

Opportunities for efficiency improvements and/or cost containment or reductions exist within numerous areas at a treatment facility, including:

- Labor allocation and skill level
- Work force consolidation and functional cross training
- O&M procedures and practices
- Energy and chemical consumption
- Solids handling and disposal
- Equipment application and performance
- Applied control and information technology
- Treatment processes
- Management of assets and inventories
- Support systems and services
- Laboratory services
- Purchasing requirements
- Integration of SCADA and other information technologies
- Selective outsourcing of Noncore functions

The three most significant opportunities for more cost efficient practices are found in labor, power, and chemical costs. You will arrive at the best balance for your facility by carefully evaluating each activity involving them and then implementing a plan for their optimization.

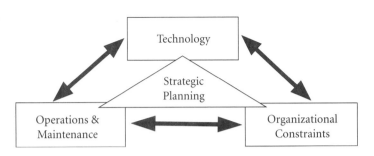

Figure 6-1 Three opportunity areas where strategic improvements reside

When we consider how each of these factors may influence the treatment facility, it is clear that each has some affect on the other. For example, reductions in labor to achieve a partially attended operation requires much greater investment and reliance on applied technology, more efficient support systems, greater responsibility for those who are on shift, and ultimately some level of increased risk with respect to quality of service and water quality. Therefore, we cannot change one factor without examining its dynamic relationship with other elements. This cascade effect relationship is significant and should not be overlooked.

Figure 6-1 illustrates three opportunity areas within a treatment works that define significant enhancement opportunities. In a three-pronged analysis that functions as part of a strategic planning process, the interaction between these opportunity areas becomes evident.

The following discussion presents some of the major control factors within each opportunity area that can affect performance.

Operations and Maintenance

This topic focuses on how the facility actually functions and is maintained. Figure 6-2 presents some of the major target elements that a manager may examine as opportunities for enhancement.

Clearly the human element is one of the principal controlling factors governing the efficient operations of the treatment works. Factors such as shift scheduling and coverage, workforce

Staff Reductions Yield Significant Savings

The City of Tampa, Fla. (Case Study 2.4) conducted an assessment of its operation of two water treatment plants and has established goals to increase efficiency and lower the cost of operations while improving service.

The anticipated savings are:

- *Reduced chemical costs (10%) $400,000/year*
- *Reduced corrective maintenance costs $100,000/year*
- *Staff reduction (30 positions) $1,200,000/year*

The implementation of the reengineering program is expected to take place over two years, and be completed at the end of 1999. This completion date coincides with the scheduled completion of the automation and equipment replacement construction projects and the two-year technical training program to develop a cross-functional staff.

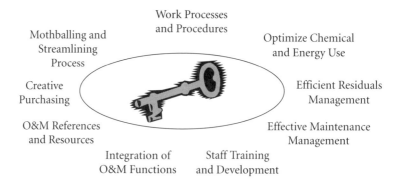

Work Processes
and Procedures

Mothballing and
Streamlining
Process

Optimize Chemical
and Energy Use

Creative
Purchasing

Efficient Residuals
Management

O&M References
and Resources

Effective Maintenance
Management

Integration of
O&M Functions

Staff Training
and Development

Figure 6-2 Opportunity elements of the operations and maintenance area

flexibility, skill levels, management/union relationships, and personal productivity and motivation (including performance rewards) are powerful background factors influencing how efficiently the job gets done. Personnel and record keeping and monitoring practices are principal determinants of how effectively a facility is controlled and risk is minimized. Operation and maintenance practices evolve in response to treatment process requirements, the availability of tools and resources, the reliability of equipment, and the skill level and degree of individual accountability. All of these factors are subject to improvement.

Technology

The effectiveness and reliability of technology, or lack thereof, greatly influence the evolution of O&M practices and staff allocation and skill requirements. Figure 6-3 identifies the elements of opportunity for improved effectiveness and efficiency in the technology area.

Technology refers to the "hardware" in use within a treatment works such as: motors and equipment; instrumentation and control systems; pipes, valves, and piping appurtenances; HVAC systems; and analytical laboratory equipment. In addition, technology refers to design and operating elements that result in the physical plant processes and size, including process configuration,

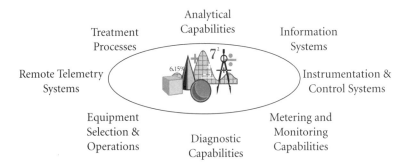

Figure 6-3 Opportunity elements in the technology area

and capabilities. Finally, technology includes information, vehicles, portable tools, and other equipment that may be used by staff members to operate and maintain the facility. Technology also affects the degree of exposure to risk.

Technology, the processes, equipment, and especially the instrumentation and control are closely related to the staffing and level of expertise required to operate the facility.

Organizational Factors and Constraints

Organizational factors determine the degree to which resources may be effectively allocated and efficiently used. They include management structure, organizational systems and support functions, labor and union issues, general policies, interdepartmental relations, budget allocation, and resource sharing. For example, union and civil service constraints can make it difficult to implement significant changes in labor class redefinition and to improve workforce flexibility. Figure 6-4 presents some of the more common organizational factors.

Although organizational factors are sometimes difficult to change, they must be considered part of a utility improvement program because they greatly affect the function of the treatment facility.

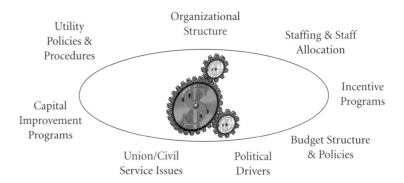

Figure 6-4   Organizational factors and constraints

OPPORTUNITIES FOR ENHANCEMENT

In the following sections we identify activities in which many utilities have found opportunities to enhance their efficiency. This overview will reveal their potential applicability to your utility. These have been grouped into the same three areas previously described: operations and maintenance, technology, and organization.

ENHANCEMENT OPPORTUNITIES—OPERATION AND MAINTENANCE

Routine Work Processes and Monitoring Procedures:

Work processes and procedures generally evolve from habit or in response to a previous problem. Once initiated, they often become the normal routine and may require additional steps, extra forms, and extraneous data that are unnecessary to meet the performance objectives of the facility. Examine opportunities to reduce complexity and frequency of procedures. Some ideas include evaluating monitoring rounds sheets, the appropriateness of data collected, and the frequency of collection.

In reassessing work processes and procedures your planning teams or work groups should ask themselves: *"Why do we do each procedure? What would happen if we didn't do it? Is there another way to do it? Can it be combined with something else? If we were inventing this system from scratch, would we do this? If so, why and how?"*

Optimize Chemical Consumption

Chemicals are often overdosed by the operator as a margin for safety. Experience shows that chemical consumption is typically an area in which significant savings in cost may be found. Slightly overdosing chemicals, while an easy approach to reducing the monitoring requirements of a process, can result in impacts to other processes as well. For example, overdosing coagulant will increase sludge quantities and may also reduce filter run times. Both can increase costs of operation. Relating and controlling chemical dose to flow and source water quality characteristics may be a much better practice.

Examine whether or not present chemical dosing procedures and application points are doing the job efficiently and effectively. If automatic pacing equipment is not now in use, it should be considered. Use of monitoring tools such as jar testing and/or streaming current detectors can help optimize chemical dosing. Another option is the potential use of alternative chemicals that reduce dosage and costs. Continually ask yourself, "Am I practicing the best procedures and using the best chemical that I can?" One useful approach is to have an independent chemical use audit performed by an expert other than the plant staff to assure unbiased thinking in a search for an optional approach that may be preferable.

Note that optimization of chemical consumption may also require significantly higher operator and management attention to properly track and monitor the processes.

Optimize Energy Consumption

Energy consumption can be optimized. Even small reductions result in savings over the year. Monitoring energy consumption by unit treatment process is an excellent tool for assessing areas that may be candidates for optimization.

Examine motor types and other energy users for efficiency and usage. Would replacing an older, less efficient motor offer cost recovery in energy savings within a short (1 to 3 year) period? Can energy use be modified so that peak demand charges can be reduced? Could an energy conservation program reduce energy costs? Consider installating meters to monitor energy consumption by unit treatment process.

A complete discussion for creating an energy efficient utility is included in chapter 10.

Efficient Residuals Management

All water treatment plants produce one or more waste streams. The two major sources of these streams are the continuous or periodic removal of residuals from the sedimentation basins and the backwashing of filters. Where membranes are used another significant waste stream is the rejected water. The waste streams contain suspended materials that are removed during the treatment processes and treatment chemical residuals.

The conventional strategy for efficient handling and disposal of waste streams and residuals is to reduce the quantity of the solids generated and to utilize a low cost (capital and operating) method of treatment and disposal.

The opportunity to reduce solids is largely associated with the quantities and types of coagulating chemicals used. Hence there may be opportunity to reduce both chemical and residual disposal costs with periodic reevaluation of the treatment processes and careful control of the quantities of chemicals used. For most plants landfilling the residuals on or off site will continue to be the most cost effective disposal method. In many plants operators initially resist these changes because it is easier to operate the old way. However, when they are brought into the process of streamlining plant operations they usually become enthusiastic about the challenge and the savings that result.

Residuals treatment and disposal also present an opportunity to consider contracted services with either private contractors or other public utilities. Case Study 5.3 is an example of a utility outsourcing of wastewater sludge disposal. Similar considerations could be applied to water treatment residuals.

Effective Maintenance Management

Maintenance management is a critical factor in treatment plant operations in improving overall work process improvements. Converting from a reactive maintenance mode to a preventive mode— such as predictive, programmed, or usage based—can greatly affect equipment life, reliability, and labor/parts/lubricant costs.

Maintenance should be managed as aggressively as process operations. The use of maintenance management tools such as computer-based maintenance management systems, inventory control, labor tracking and predictive diagnostic tools are essential to achieve "best in class" results. Supportive training and resources are required to make these systems an active part of the treatment works in facilities that have operated in a traditional reactive mode.

Staff Training and Development

Staff must be continually trained and their professional skills developed and expanded. Otherwise the utility will not achieve significant improvements in efficiency and effectiveness. The staff governs the cost of operation. They also govern manager's successfulness in achieving the organization's goals. They must be given the tools, resources, and skills to conduct their work.

As new or improved technology is introduced to improved treatment performance and/or reduce labor requirements, training becomes even more important in the success of the operations and employees.

Training budgets are often the first area cut in the utility budget. This is the one area that should never be cut and, similar to police and fire training needs, operators and maintenance personnel must receive continual training in order to properly operate the complex chemistry and other equipment within the treatment facility.

Integration of Operations and Maintenance Functions

Many utilities, especially larger ones, have traditionally segregated their operations functions from maintenance functions. Union and civil service rules have often prevailed in keeping the

O&M functions separate, with labor classifications and titles commensurate with clear separation. This can result in significant inefficiencies.

Seek for ways to integrate the two functions. One solution is to utilize operations personnel in light maintenance functions such as lubrication, painting, and cleaning. Maintenance staff can also be used to offset operations tasks such as rotating and exercising equipment. The ideal situation is to eliminate the traditional operations and maintenance classifications and move toward a single classification that embodies both elements. This provides maximum opportunity for efficient use of personnel, as well as new opportunities for employees to learn new skills and to gain additional qualifications.

O&M References and Resources

A staff requires resources to make decisions and find information. Resources mean anything printed or on a computer that can be used to assist in operations, decision making, learning, or planning. At a minimum, the staff must have the following available to them:

- Updated operations and maintenance manual

- Equipment maintenance manuals

- Training documents, such as AWWA's Water Supply Operations series

- Facility plans and specifications

- Trade publications include *Journal AWWA,* maintenance management publications, and equipment journals such as *Waterworld.*

Computerized information resources are powerful tools for storing and finding information. Combining the O&M and other facility-specific design and general reference information into an online information access system is extremely easy to do at a relatively low cost.

Creative Purchasing

Creative purchasing opportunities are found in the area of bulk items such as chemicals and consumables. One large, potential cost savings is in volume discounts and effective negotiation of contracts.

Volume discounts are difficult to obtain unless you have means to store chemicals and supplies without exceeding shelf life or unless you can enter into agreements with other communities who use the same products. Intercommunity purchasing agreements are effective tools to reduce costs of higher volume consumables.

A review of purchasing procedures and specifications that may be overly restrictive or burdensome on vendors, which results in higher costs, will be beneficial.

Mothballing and Streamlining Treatment Processes

If a facility has excess capacity and does not need to operate all systems, why do so? There is a cost associated with operating systems, whether it be electrical, chemical, mechanical, or in the cost of labor.

Shifting production to a sister facility with lower operating costs (when demand allows such an approach) may be a way to lower a utility's total O&M costs. It is not unusual to shut down a portion of an entire facility if seasonal water quality or demand allows this to be done.

A useful tool for assessing potential performance improvements in your utility is an evaluation grid that lists the opportunities for enhancement vertically against a horizontal ranking of performance indicators from best to poorest. A sample grid for the operations and maintenance opportunity area is shown in Table 6-1. A few revisions may make this table more appropriate for your utility.

The table can be distributed to your staff or to a broader number of stakeholders requesting individual evaluation of present practices. Accumulation of the individual evaluations form the basis for group discussion. Initial emphasis should be placed on those that fell shortest of the best in class designation.

Table 6-1   Best Practices Evaluation: Operations and
Maintenance Area

Factors	Best In Class	Effective	Conventional	Basic	Ineffective
Routine Operations Monitoring Procedures	Streamlined procedures for essential data only. Routine review and updating of data needs and monitoring procedures.	Streamlined procedures for essential data only. Occasional review and updating of data needs and monitoring procedures.	Formal procedures for gathering data. Some data are nonessential.	Formal procedures for gathering data. Some data are nonessential. Data gathered for busy work purposes.	Few data are gathered. Data not used.
Chemical Consumption	Proper dosing with continuous monitoring.	Proper dosing with frequent monitoring.	Dosing based upon routine monitoring.	Overdosing to compensate for infrequent monitoring.	Dosing with no monitoring.
Energy Consumption	Energy management strategy with highly efficient equipment and aggressive energy monitoring.	Energy management strategy with mostly efficient equipment and energy monitoring.	Mostly efficient equipment with some energy monitoring.	Mostly inefficient equipment with little energy monitoring	No energy management strategy with inefficient equipment and no energy monitoring.
Training and Staff Development	Formal, well funded, nearly continuous staff development and training program matched to individual needs.	Formal, well funded staff development and training program matched to individual needs.	Moderately funded, but not program managed; staff development and training matched to group need.	Insufficiently funded, occasional formal training provided.	No formal training funded; only on the job training provided.
Integration of Operations and Maintenance Staff Functions	Both functions integrated seamlessly with extensive staff cross-utilization.	Both functions integrate smoothly with good staff cross-utilization.	Separate functions with cooperative formal, protocol-based interaction. Some staff cross-utilization.	Strong segregation of functions often with strained relations. No staff cross-utilization.	Strong segregation of functions with highly strained relations and little interaction.
References and Printed Resources	Full spectrum of immediately available information in a multimedia computerized database. Nearly continuous updating.	Full spectrum of immediately available information. Updated frequently.	Facility reference materials and trade magazines available. Not updated.	Few reference materials available.	Lack of reference resources. Information kept locked up.

table continued on next page

Table 6-1 Best Practices Evaluation: Operations and
Maintenance Area (continued)

Factors	Best In Class	Effective	Conventional	Basic	Ineffective
Creative Purchasing of Chemicals and Consumables	Aggressive use of volume discounts and effective price negotiation.	Use of volume discounts and intercommunity bulk purchase agreements.	Purchasing based upon best price offered by competitive bid.	Purchasing based upon extensive solicitation of alternative vendors and price quotation.	Little or no efforts to obtain best pricing.
Mothballing and Streamlining Treatment Equipment/Processes	Operating only necessary and effective systems.	Operating systems based upon need with frequent review.	Operate systems based upon need with periodic review.	Operate systems and occasionally review need.	Operating all systems regardless of effectiveness.

AWWA has published a self-assessment system under its
QualServe program. You might look at this approach as well to
decide which assessment method is best for your purposes.

ENHANCEMENT OPPORTUNITIES—TECHNOLOGY

Treatment Processes

Treatment processes are often initially selected based on treatability studies and/or engineering criteria to meet the expected range
of flows and changes in raw water quality. This initial process
selection during design must anticipate operational complexity,
and consider direct and indirect cost impacts. For example, a process design that doesn't effectively handle seasonal load/demand
variations will require more operator attention and higher operating costs. The potential range of operating modes must also be
considered. The design must provide the appropriate amount of
flexibility without building in unnecessary complexity.

Once the treatment processes are placed in operation, the
treatment staff must experiment to determine optimum operating modes, which may require daily, seasonal, or cyclical adjustments. The full range of impacts must be considered including
power and chemical consumption, equipment wear and tear,
residuals generation, and labor costs. It is prudent to periodically

step back and consider whether process adjustments or changes may offer opportunities for improved efficiency. Here, too, an audit by an independent expert or a peer review by staff from another utility may provide insights not being considered by your operation staff. They can work together to pool their wisdom and experiences.

Analytical Capabilities

Most facilities focus on the required and routine laboratory work. While they are necessary, there are many other techniques available to help an operator optimize treatment processes. Bench-scale work such as jar testing, treatability testing, process simulation, and pilot studies also indicate required process adjustments or impending problems and identify opportunities to improve efficiency.

Operations, administrative, and laboratory staff must work together to analyze and understand the full range of possible process conditions in a facility. Taking advantage of available analytical techniques gives the staff additional tools to optimize process conditions while reducing operating costs.

Information Systems

The broad category of information systems includes data storage, transfer, and management systems for various functions such as maintenance management, laboratory operation, process monitoring, regulatory reporting, reference information, billing and collections, customer service, and staff communications.

Facility management must be forward thinking since the need for some applications may not be intuitively obvious. However, to become and remain efficient, management must invest in information system hardware, software, and personnel training. This will ultimately pay off as staff members become more productive at their jobs, find the information they need readily available, and reduce the duplicated efforts in data entry and information reporting. The most productive and cost effective approach is to evaluate the facility needs and establish a plan that will guide the organization through a process that may take several years. Chapter 9 discusses this important area in more detail.

Instrumentation and Control Systems

Treatment facilities have traditionally included a certain level of instrumentation and control, with the extent increasing in recent years. Much of it is designed to protect individual equipment or to provide discreet system monitoring and control. Most new facilities, and many older ones, now have comprehensive distributed monitoring and control systems designed to provide extensive equipment and process automation and to centralize data collection, monitoring, and alarm indication.

Evaluate the existing and potential level of monitoring and control while considering present and future desired staffing levels. The use of technology handles routine, repetitive tasks and frees personnel to perform more important functions. Don't confuse automation with staff elimination. In most cases staff can be reduced, but importantly, they can be allocated more effectively to other assignments. These areas are also discussed in chapter 14.

Metering and Monitoring Capabilities

Effective and accurate metering of all flows into, within, and leaving a facility is essential. Flows are used for balancing, tracking, recording, calculating, and determining cost projection. They are useful for troubleshooting and cross-checking other instruments and measurements. We should have accurate meters on all key flow lines, whether they are process or chemical.

Upkeep and correct application and installation of flowmeters are essential. Generally most meter problems are traceable to ineffective calibration and maintenance, improper installation, or misapplication.

Diagnostic Capabilities

The evolution of portable or bench top diagnostic technologies to monitor and troubleshoot mechanical and process performance has resulted in the availability of powerful tools to assist plant staff. Investment in and proper use and calibration of diagnostic tools greatly aid operations and maintenance functions.

Some maintenance diagnostic technologies include: thermographic analysis, vibration sensing, flue and exhaust gas analysis, infrared thermography, wireless load metering, ultrasonic leak detection, proximity detection, and oil analysis.

Operations technologies include: particle counting, streaming current detection, microscopic analysis, jar test, Hach kit, and other analytical bench top tests.

An efficient facility effectively uses available diagnostic technologies to monitor and manage its performance objectives and makes sure that staff are adequately trained in the use of diagnostic equipment.

Equipment Selection and Operation

Having the right equipment for the job will make both operator's and mechanic's jobs easier, thus saving labor costs, and it will ultimately cost less to operate and maintain the facility. For example, eliminating over-sized equipment will save money because smaller equipment requires less power.

Evaluate the size and capacity of existing equipment. Once a facility is designed and constructed, equipment selection must be periodically reevaluated if process conditions change due to changes in demand/load or regulatory requirements. It may be more cost effective in the long run to replace a unit even if it still has some useful life left. Make sure design and specifying engineers fully understand the intended operation, as well as the staff's needs and past experience.

Remote Telemetry Systems

Remote monitoring of the distribution system, storage tanks, wells, and system pump and booster stations will reduce labor costs by eliminating the need to drive to various sites to simply check on a system. The staff can focus on the critical equipment and provide better response time when emergencies arise.

Evaluate critical systems and prioritize if necessary. Consider remotely monitoring day-to-day operation of flow stations, pump stations, or storage tanks. Address potential problems such as loss of service, flooding, and overflows.

Building Operations and Maintenance

There are numerous innovative techniques to improve the efficiency of building operation and maintenance including: material selection, energy recovery systems, and use of grant and loan programs for innovative system applications.

Consider building and HVAC systems not only in new and upgrade designs, but also in routine operation and maintenance. Boiler and chiller operation can be optimized to maximize efficiency and to reduce costs in the same way that process operation can. Technology applications similar to those previously discussed can monitor heating and cooling loads and thereby assist in this optimization.

Table 6-2 is a grid for use in evaluating your enhancement opportunities in technology areas and can be used in a manner similar to Table 6-1.

Table 6-2   Best Practices Evaluation: Technology Area

Factors	Best In Class	Effective	Conventional	Basic	Ineffective
Treatment Processes	Application of proven process technologies that best fit reliability/performance, economics, and operability.	Application of proven processes that adequately balance reliability/performance and operability criteria.	Application of proven process technologies that generally balance reliability/performance and operability criteria.	Application of proven process technologies that poorly fit reliability/performance and operability criteria.	Application of unproven, overly complex, labor intensive and expensive treatment technologies.
Analytical	Highly advanced analytical capabilities and QA/QC program efficiently used for better understanding of process.	Advanced analytical capabilities and QA/QC program efficiently used for better understanding of process.	Competent analytical capabilities and formal QA/QC program used for monitoring only.	Basic analytical capabilities and informal QA/QC program used for monitoring only.	Limited analytical capabilities with no QA/QC or data management program.
Information Systems	Extensively used network-based information management systems and highly skilled network/software administration.	Moderately used network-based information management systems and skilled network/software administration.	Extensive use of PCs with limited information management system capabilities. Most staff members skilled in PC use.	Moderate PC use with no information management system. Few staff members skilled in PC use.	No information management system or limited PC computer use. Staff unskilled in PC use.

table continued on next page

81

Table 6-2   Best Practices Evaluation: Technology Area (continued)

Factors	Best In Class	Effective	Conventional	Basic	Ineffective
Instrumentation and Control	Extensive use and highly effective maintenance of I&C technologies to offset labor and increase system reliability.	Many automatic controls and effective maintenance of I&C technologies.	Moderate amount of automatic controls and monitoring devices adequately maintained.	Few automatic controls and monitoring devices with adequately maintained instrumentation.	Manual control only with limited or poorly maintained instrumentation.
Metering	Extensive use of calibrated and reliable metering to track all flows with a proactive maintenance.	Use many technically reliable flow metering devices given regular maintenance.	Use few technically reliable flow metering devices with adequate maintenance.	Use few unreliable flow metering devices with limited maintenance.	Limited use of flow metering with poor maintenance.
Diagnostic Technologies	Many, plant-appropriate diagnostic technologies available with a highly trained staff.	Some diagnostic technologies available with an adequately trained staff; outsource as needed.	Some diagnostic technologies available with an adequately trained staff.	Few diagnostic technologies available with an untrained staff.	No diagnostic technologies available.
Equipment Selection and Operation	Properly sized and effectively operated, energy efficient equipment serving its intended operation.	Properly sized, energy efficient equipment serving its intended operation.	Improperly sized, somewhat energy intensive equipment serving its intended operation.	Improperly sized, energy intensive equipment serving its intended operation with limited useful life remaining.	Over or under-sized, energy intensive equipment not serving its intended operation, limited useful life.
Remote Telemetry	Extensive use of remote monitoring systems to offset labor and increase system reliability.	Remote monitoring systems to offset labor and increase system reliability.	Remote monitoring systems in critical locations only, with manual monitoring at other locations.	Little use of remote monitoring systems with mainly manual monitoring.	No use of remote monitoring systems with manual monitoring.
Building/HVAC	Use energy efficient HVAC and lighting systems and aggressive management program.	Use energy efficient HVAC and lighting systems with a proactive management program.	Use energy efficient HVAC and lighting systems with an adequate management program.	Use energy intensive HVAC and lighting systems with an adequate management program.	Use energy intensive HVAC and lighting systems with no management program.

ENHANCEMENT OPPORTUNITIES— ORGANIZATIONAL CONSTRAINTS

Organizational Structure

Organizational structure evolves in response to the number of activities and services an organization performs. Accordingly, span of control issues, coupled with workload volumes and the tasks involved, drive the traditional structure. The process of transforming a vertical hierarchical structure into a more equally structured, or flatter, organization has effects throughout the organization. Also, as an organization is flattened, the effectiveness of the process is related to the number of departments or units that also must decrease.

Opportunities for enhancement include:

- Merging similar departments/functions

- Elimination of midlevel positions

- Outsourcing services (e.g., payroll, accounts receivable)

- Reducing bureaucratic rules and empowering individuals

It is important that as the organization is flattened, the amount of coordination and communication between horizontal levels increases. There is a continual need to motivate key staff to stay in the organization since replacement is usually difficult and expensive.

Staffing and Staff Allocation

The public model is often one of low risk and conservatism. It has evolved to respond to high quality service delivery needs in an unpredictable environment. Accordingly, the traditional approach to staffing and allocation is characteristic of formalized, centralized bureaucratic structures that have rules, clear division of labor, and clear definition of hierarchy and decision-making responsibility. The consequence of these factors is often staffing levels that are higher than that of best in class operations.

To enhance staff allocation and size we must first focus on organizational characteristics. Can the organization become less centralized and more flexible? If so, there may be opportunity to reduce staff by granting individuals more decision-making responsibility and span of control. Perhaps there are opportunities to reduce staff by examining tasks and the necessity of individual positions. Finally, some positions may be legacy based; they are there simply because they have always been there. Identifying these positions provides an opportunity to adjust staff numbers.

Incentive Programs

Alternative personnel reward systems are needed to motivate and reward employees when they assume more responsibility or achieve higher levels of performance. Employee recognition and reward is an essential component of change. Examples of employee reward systems being used by Charlotte (Case Study 4.4) and Colorado Springs (Case Study 2.2) are included in appendix A. Some rewards include: individual and group incentives; gainsharing; performance-based and skill-based pay increases; other than cash awards; recognition; and improvement/suggestion awards. See chapter 14 for further discussion of this topic.

Budget Structure and Policies

The structure of the budget to enable cost tracking is important as an organization seeks to identify costly areas of operation and actual versus apparent costs. Budget and cost accounting information must track actual costs to enable informed decision making. This information must be conveyed to staff so that they understand actual costs and cost/effect relations in what they do. Without cost data, personnel become unaware and indifferent about the actual costs of their actions.

Political Drivers

Political factors, especially as they might affect the size and capabilities of the operations staff, are a reality for a number of public utilities. Sometimes staffing, procedures, and organizational structure may have been shaped by political reasoning rather than by actual functional needs.

Overcoming politically driven inefficiencies is difficult to do. However, politically driven issues may be reduced by communicating actual cost information, using benchmark comparison data, and explaining how others accomplish similar work more efficiently for less.

Union and Civil Service Issues

Union and civil service rules govern how much employees may be used. Clear distinctions between position responsibilities and labor classifications are barriers to reorganization. Therefore union and civil service leaders must be involved in any change process if it is to be effective. Getting and maintaining union/civil service support in the change process is essential. The characterization of management versus union must be overcome in favor of a more cooperative model.

Capital Improvement Programs

Capital improvement programs and other long-term needs should be considered when assessing organizational and staffing needs. It is expensive and inappropriate to undergo an extensive restructuring process or employee cut-backs only to go on a hiring binge to properly staff a facility at next year's plant upgrade completion.

Utility Policies and Procedures

Policies and procedures from other departments and agencies have a significant impact on the effectiveness of a function. One major area this is manifest in is the public purchasing system. Purchasing rules, policies, and inefficiencies can have a tremendous effect on performance effectiveness within a plant. Some communities retain central control over computer systems. Many cities dictate budgeting and accounting systems, exercise control over travel authorizations, and impose other administrative restrictions that limit the flexibility of water departments and others. It may be time to pursue reconsideration of these practices when they affect performance.

Table 6-3 is a grid for use in evaluating your opportunities for performance enhancement in organizational matters and can be used in a manner similar to Tables 6-1 and 6-2.

Table 6-3   Best Practices Evaluation: Organizational Area

Factors	Best In Class	Effective	Conventional	Basic	Ineffective
Staffing and Staff Allocation	Highly motivated and skilled, results-oriented staff who are fully able to be cross-utilized.	Motivated and skilled staff with a moderate degree of cross-utilization.	Somewhat motivated and skilled staff with basic degree of cross-utilization.	Generally unmotivated and somewhat skilled staff with low degree of cross-utilization.	Largely unmotivated and somewhat skilled staff members with very low degree of cross-utilization.
Organizational Structure	Flexible, flat organizational structure with few bureaucratic controls.	Flexible, flat organizational structure with some bureaucratic controls.	Slightly flexible, vertical organizational structure with many bureaucratic controls.	Inflexible, highly vertical organizational structure with many bureaucratic controls.	Inflexible, hierarchical organizational structure with high degree of bureaucracy.
Incentive Programs	Many formal incentive programs available to all employees.	Some formal incentive programs available to all employees.	Some formal incentive programs available to some employees.	Few formal incentive programs available to some employees.	No formal incentive programs.
Budget Structure and Cost Accounting Systems	Systems track costs for all plant activities against budgets and are continually reviewed by all required staff.	Systems track costs based upon activities rather than on cost centers with some staff participation.	Continual tracking of cost based upon cost centers with some staff participation.	Occasional review of cost center budgets with limited cost staff participation.	Annual cost center budget preparation and review only with no staff participation.
Political Drivers	Proactive communication with political influences at all times.	High degree of communication with political influences in only times of need.	Some communication with political influences only in times of need.	Little communication with political influences.	No communication with political influences.
Union and Civil Service Issues	Very effective union/civil service relations with management. Effective and open communication.	Effective union/civil service relations with management. Good and open communication.	Moderate union/civil service relations with management. Formal communication only.	Marginal union/civil service relations with management. Marginal communication.	Weak union/civil service relations with management. Poor communication.

table continued on next page

Table 6-3 Best Practices Evaluation: Organizational Area (continued)

Factors	Best In Class	Effective	Conventional	Basic	Ineffective
Capital Improvement Programs	Continual long-term needs planning done by facility-knowledgeable staff.	Frequent long-term needs planning done by facility-knowledgeable staff.	Occasional long-term needs planning done by facility-knowledgeable staff.	Occasional long-term needs planning done by executive level staff.	Infrequent or no needs planning done.
Administrative Policies and Procedures	Only balanced, necessary policies and procedures used that are frequently reviewed and streamlined.	Only necessary policies and procedures are used that are occasionally reviewed and streamlined.	Few unnecessary policies and procedures that are sometimes reviewed and streamlined.	Some unnecessary policies and procedures that are seldom reviewed or streamlined.	Many unnecessary policies and procedures that are seldom reviewed or streamlined.

A WORD ABOUT STAFF OPPOSITION

When employees first learn about a manager's plan to improve the utility's performance and cost effectiveness, they usually feel a surge of anxiety that is the emotional equivalent of an acute case of tinnitus, a ringing in the ears that blocks out much of what is being said. Their overriding reaction is: *What is this going to mean to me?* And they don't expect the answer to make them happy.

Some common statements that indicate a fear of change include the "We cannots" or "We tried it before, but . . ." Even better (and more ominous) indicators are the "it will never work" and "it's too much work" forecasts. The manager must pay special attention to correcting these sentiments before they become self-fulfilling. The good news is that some, if not most, of these concerns can be controlled and eliminated. Many utilities have found ways to change negative concerns into positive experiences.

The fear of change is discussed in detail in chapter 14, so we need not dwell on it here except to make two points of particular relevance to those who work in water treatment plants.

1. Resistance to change is a natural instinct and a traditional instinct for those with the responsibility for drinking water quality and safety.

2. It is that responsibility to the public that should open our minds to new and better ways of fulfilling our public trust. Striving to be the best is a wiser, safer policy than merely preserving the ways we know well.

The improvement process advocated in this book uses the expertise and judgment of the utility staff, as well as the lessons to be learned from the best of class. They are *our* changes, introduced because we want to be the best.

SUMMARY

With water treatment at the heart of a utility's work, searching for performance improvements here is truly essential. The treatment works are not the only utility function in which efficiency improvements are usually achievable, but it is the function that is receiving considerable attention by many utilities. Utilities can demonstrate its achievement of the highest level of effectiveness and professionalism in its treatment operations. There are numerous opportunities to do this, as this chapter has outlined.

AWWA has joined with the US Environmental Protection Agency (USEPA) in the Partnership for Safe Water, which is a four phase program providing for membership, accumulation of baseline data, self-assessment, and third party assessment for the purpose of continuous improvement in water treatment performance. This program can be supported by the concepts discussed in this chapter.

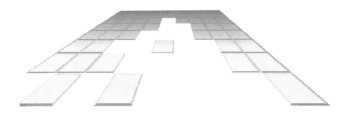

Maintenance
and Rehabilitation

Most water utilities are challenged by aging infrastructure. This is not a phenomenon unique to older East Coast utilities. It also occurs in the middle and western states. Portions of most systems were constructed more than 100 years ago, resulting in rising maintenance costs and large infrastructure replacement programs, particularly where replacements have been postponed. While the physical assets of most water systems—pipelines, water mains, reservoirs, pump stations, meters, hydrants, and treatment facilities—may have an average expected useful life of 25 to 100 years or more, replacements are not keeping up. A recent USEPA survey estimates national water infrastructure requirements at $140 billion over the next 20 years.[1]

Likewise, AWWA reported in its 1992 water industry data base that among 1,045 large water systems an average of 2,310 miles or only 0.5 percent of pipe systems are replaced annually. AWWARF's 1994 report, *An Assessment of Water Distribution Systems and Associated Research Needs*, estimates there are 880,000 miles of water mains in service. If these miles of water mains were to be replaced at the rate of 1 percent per year at a unit cost of $100 a foot, national

---

[1] USEPA Office of Drinking Water. *Infrastructure Needs Survey*. First report to congress, January 1997.

costs would total $4.6 billion per year. This equates to a replacement cycle of some 100 years. Again, this is not a phenomenon unique to older East Coast utilities, but is also observed in large and small utilities across the United States.

Utility annual budgets and capital improvement programs reflect funding decisions for repair, replacement, and maintenance of water systems. System expansions, facility upgrades, and new regulatory environmental and safety requirements have introduced new additional maintenance and rehabilitation demands. The best managed utilities are proactive about water system maintenance and rehabilitation while assessing the condition of the system, planning necessary replacements and improvements, becoming more efficient in maintenance work processes, achieving a higher ratio of preventive versus reactive maintenance, applying new technology, increasing levels of hands-on or direct labor, and focusing on employee performance, safety, skill training, and efficiency. In addition, new contracting approaches are used where efficiencies can be achieved.

This chapter expands and applies many of the concepts and approaches outlined in chapter 6 to the general infrastructure maintenance needs of typical retail and wholesale water utilities. It provides an overview of the maintenance and rehabilitation challenges faced by utility managers and describes the features and barriers of various approaches to meeting some of these challenges. The approaches described include:

- assessing the condition of assets and infrastructure

- developing a basis for repair versus replacement decisions

- evaluating maintenance practices

- developing strategies for improving performance

- creating financing plans and strategies for maintenance

MAINTENANCE AND REHABILITATION CHALLENGES

Utility managers are operating in a world that is more challenging than ever before. There are increasing requirements, constraints, and limitations that make it difficult to maintain and rehabilitate water systems even though there have been major advances in equipment, including the use of trenchless technology. These challenges apply to large and small systems alike.

Aging Infrastructure

Water systems, even with excellent maintenance programs, deteriorate with age. This leads to increased levels of repair and maintenance and eventually capital replacement programs for all water system components.

Rising Maintenance Costs

As systems age, facilities become more costly to maintain. Additional revenues are needed to cover ongoing maintenance expenses. This often leads to demands for higher water rates and increased staff. Material, equipment, emergency response, information collection, and overhead expenses typically increase correspondingly.

Increased Performance Targets

In the face of rising costs, water rates, and user fees, water boards, commissioners, and elected officials are increasingly scrutinizing short- and long-term financial and management requirements. Utilities are expected to keep rates low, estimate costs of multiyear maintenance and rehabilitation plans, cash finance major portions of capital improvements, achieve solid bond ratings, assure positive net income and cash flow, and deliver low debt-to-equity ratios while:

- preparing comprehensive systems and financial plans
- developing activity-based costing and budgets
- being efficient in all operations
- flattening and streamlining the organization
- achieving favorable customer satisfaction ratings
- assuring noninterruption of water service to customers
- minimizing water quality complaints
- reducing staff

Increased Data Requirements

In an environment of rapidly changing and increasingly complex technology and greater customer expectations, more sophisticated data technology systems are needed to manage maintenance activities. Other industries can teach us. Maintenance management systems are commonly used, along with Geographic Information Systems (GIS), to manage assets, resources, and performance information.

Increased or Changing Standards

Work processes are affected by evolving safety, health, and environmental regulations and standards and by consumer expectations for proactive environmental stewardship.

New Workforce

Today employees are increasingly autonomous and subject to greater accountability. In addition, a changing and diverse workforce makes it advantageous for the utility manager to involve employees more collaboratively in organizing and conducting their work. While this represents an opportunity to many, for others the change in organization culture is discomforting and challenging. It also occurs in a setting that is often constrained by established labor management practices and rules, overly rigid job classifications and salary structures, and vertically dominating hierarchies. Younger employees seek an organizational culture quite different from the traditional municipal utility.

Competition

There is a growing competition from the private sector to perform services typically provided by public employees, to consolidate regionally, and to demonstrate competitiveness relative to other water service providers. Maintenance and infrastructure development are key areas that are usually targeted.

The challenges identified above will vary from utility to utility but are representative of the environment in which modern organizations operate today. Many of these challenges particularly impact maintenance and rehabilitation requirements and necessitate the development of a systematic approach to address the challenges in this area. The next sections of this chapter outline key features of such an effort as well as barriers to be overcome.

ASSESSING THE CONDITION OF YOUR INFRASTRUCTURE

Understanding the relative condition of all aspects of the water system is necessary for a quality maintenance program. Typically this involves a complete inventory of all assets, cataloguing by type, and providing a level of detail about the component parts to be maintained. For each component historical performance data, maintenance history, failure incidents, materials, installation, cost, age, material type manufacturer, and other information should be gathered (see Figure 7-1). Careful attention should first be given to what information is needed and what it will be used for. Large and small utilities are currently using a number of vendors and common approaches as part of maintenance studies and systems.

As part of this inventory, an assessment should be made of the asset's criticality. The risk of service interruption from equipment failure should influence maintenance intervals. Manufacturer and AWWA guidelines are available for common system components such as hydrants, pumps, motors, valves, and others. Data bases, GIS systems, and other software programs offer valuable tools to store and analyze this information and to produce planned and predictive schedules for maintenance staff.

There are several common needs to achieve improvement:

- utility asset data properly formatted and available for users, schedulers, and decision makers

- ongoing maintenance

- updating asset information

- using the data management system instead of relying on key staff

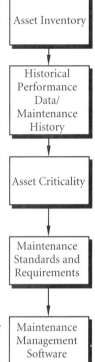

Figure 7-1 Maintenance inventory (adapted from an EMA study)

Infrastructure renewal represents the replacement of structurally deteriorating pipelines, i.e., water mains having a high likelihood in the future to leak and cause damage. A water main is a candidate for renewal when the costs of replacement are less than the net present value of projected future costs associated with its maintenance (including leak repair and collateral damage costs). The repair (leak) history of a main triggers the evaluation to see if replacement is "economically justified." This determination is part of the Pipeline Replacement Program (PRP).

The PRP evaluates water main leak histories on a regular basis to compile a list of replacement candidates. Each year a subset of candidate water mains are grouped into projects. Consideration is given to special factors when selecting candidates for the project packages including: spatially clustering the replacements to make for efficient

DEVELOPING A BASIS FOR REPAIR VERSUS REPLACEMENT DECISIONS

One of the most critical decisions to be made is whether to replace or repair major system components or assets. What are appropriate replacement cycles for water mains, meters, hydrants, valves, motors, and so on? On what basis should this type of decision be made, i.e., what criteria should be used? Should a replacement program be reactive to system failures and breakdowns or should it attempt to predict or provide replacement prior to failure? Should field crews be making repair versus replace decisions in the field based on observed conditions? What role does maintenance history, component age, materials, or soil conditions have on replace decisions? To what extent do the timing of other utility or street improvements have an impact on replacement decisions? Is there good life cycle cost information that could help with this issue? Are manufacturer or AWWA standards useful guides?

The answers to these questions should be carefully considered by water utility managers. The West Coast Water Utilities Benchmarking Group (WCWUBG) has developed, and continues to refine, a benchmarking survey approach that collects data about both the practices involved with repair versus replacement decisions, and the performance of specific components within its 14 member utilities. It is evident that improved models, practices, and eventually component performance will be outcomes of this work. In addition to this effort, several AWWARF studies are in progress on various aspects of water main repair and replacement, including financing and accounting practices, materials analysis, and replacement methodologies: These include:

- AWWARF Project #463 "Pipe Material Selection Manual"

- AWWARF Project #462 "Financial/Economic Implications of a Water Main Replacement Program"

- AWWARF Project #459 "Develop Decision Criteria to Prioritize Replacement and Rehabilitation of Mains and Appurtenances"

Common needs to achieve improvement include:

- accessibility of data on age, repair history, construction materials, and criticality

- cost information upon which to consider repair versus replace alternatives

- financing mechanisms for major repair and replacements

- budget or rate levels in the short- and long-term to sustain maintenance and replacement programs

- staff, equipment, materials, and systems to mount major maintenance improvements

- trained staff

- policy direction or commitment by utility leadership or elected officials

construction mobilization, high seismic hazard areas, unstable soil zones, inadequate flow conditions and customer requests. (Excerpt from East Bay Municipal Utility District (EBMUD) Fiscal Year 1997 Pipeline Master Plan).

EVALUATING MAINTENANCE PRACTICES

Work process reengineering, business analysis, work redesign, and methods improvements are terms used frequently by organizations to describe efforts undertaken to improve efficiency, eliminate redundancies, and engage employees and work teams in evaluating the way work is performed. Much of this work is derived from quality management initiatives, and a number of models have emerged to assist with this important work.

In the water business, like in many other industries, work practices surround large maintenance functions and are a major determinate of cost, efficiency, customer satisfaction, and adequacy of maintenance. In 1996 the WCWUBG embarked on the first in a series of studies focusing on water utility maintenance practices.

Recommended areas for improvement were identified in the categories of people, processes, system/technology, and cost control. While these areas for improvement applied to the specific utilities involved, managers of other small and large utilities may also wish to concentrate on the list of work improvement areas in Table 7-1.

Table 7-1 Areas of work improvement

People	Processes	Systems/Technology	Costs
• Skills training • Safety • Span of control ratio • Participation/Involvement	• Performance to schedule • Performance to plan • Planned versus unplanned work • Definition of critical components • Contract maintenance	• Extent of automation in data management • Access by hands-on labor staff to systems • Currency of system data	• Performance to planned budget versus planning based on budget • Income generating maintenance work • Activity based cost management where appropriate

Source: Paralez, L. Rose Enterprises, Inc., summary of analysis of Western Regional Water Utilities Benchmarking Group Finished Water Delivery System Maintenance Benchmark Survey, 1996.

Table 7-2 Operational performance levels

Doing the Work	Work Management	Maintenance Strategy
e.g., repair methods, work standards or standard maintenance procedures	e.g., scheduling, work order systems, backlog tracking, data collection and use, use of automation in work management, scheduling and conducting skill training, apprenticeship programs, and equipment and materials availability	e.g., rehabilitation and replacement programs, funding methods and allocations, decisions to contract or do work inhouse, levels of equipment investment, inventory practices, site delivery, investment in skill training, process performance standards, predictive model development, and investment in information management systems

Source: Ibid.

The maintenance survey instruments developed by the WCWUBG in cooperation with Rose Enterprises, Inc. have been designed to help participants evaluate their performance at three operational levels, as illustrated in Table 7-2.

As participants review their performance and business practices in each of these levels of organizational activity, it becomes evident that significant opportunities for improvement exist for all participants. This varies from utility to utility and depends on specific approaches taken, but cost saving or efficiency improvements may be as high as 50 percent.

The survey findings showed that most utilities are fairly diligent about data collection on their facilities. Many have data bases that contain leak history, failure rates, rebuild history and so forth. As part of strategic planning processes, most develop some type of operational strategy for maintenance, usually driven by funding constraints and availability, past practices, policy, and in some cases predictive models. Also, many utilities have developed work standards and apprenticeship programs and involve employees in work process improvements.

However, while these business practices taken in isolation would be considered good or best practices, the work of the benchmarking group allowed them to look at these practices more systematically to link strategies, work management, and doing the work. One example relates to facility repair versus replacement decisions. While volumes of data often exist for a water distribution system, it is not always formatted, evaluated, or included in work management systems to facilitate repair versus replacement decisions. Another example concerns the operation of water system valves. While activity based work standards may exist and crews may be efficiently deployed, the utility may not be adequately addressing the issues of criticality or frequency of operation. The City of Portland Water Bureau, for example, is now beginning to focus on the criticality of different valves, and may significantly reduce costs by operating certain valves less frequently. As in this example, changes in operational strategies often represent the greatest opportunity for cost savings.

Common needs to achieve improvement include:

- management commitment to work process improvements

- benchmarking or efforts to identify best practices for key work processes

- providing field crews with appropriate information to deliver optimal maintenance service

- involving field crews in maintenance practice improvement efforts or in making field decisions

- adequate planning, scheduling, reporting, and performance measurement systems

- evaluation of utility maintenance strategies

- adequate training and skill development

DEVELOPING STRATEGIES FOR IMPROVING PERFORMANCE

After better understanding assets and the adequacy of current maintenance practices, water utility managers may desire to employ a number of strategies to improve performance.

Implement Planning Systems for Maintenance

Utilities in Los Angeles, Colorado Springs and Seattle are organizations that have emphasized performance improvements in maintenance through enhancements to work planning and scheduling systems. One element of these programs includes the development of standards for each major maintenance activity and major unit of output. Standards can serve as a starting point for the creation of activity-based costing and analysis for all maintenance work. They can assist in resource planning; budgeting; tracking performance; automating work plans, time, materials, and equipment scheduling and reporting; and prioritizing work. Table 7-3 is an example of a basic work planning matrix that includes work activities, projected annual activity levels for a water distribution work group, work standards expressed in labor hours per unit of activity, and total annual projected labor requirements. Work management systems can plan and schedule labor materials and equipment for these activities and then report actuals against projections.

Table 7-3 Work Planning Matrix

Work Category	Projected Annual #	Projected Standard Crew Hours	Projected Annual Labor Hours
Emergency Work			
Water main leaks/breaks	120	16	1920
Service leaks/breaks	1200	11	13200
Hydrant repair (minor)	800	5.2	4160
Valve repair (minor)	250	4.5	1125
New Installations			
Small services	800	16	12800
Large services	100	75	7500
Maintenance & Operations			
Valve operation—large	1000	1.6	1600
Valve operation—small	9000	.8	7200
Replace obsolete hydrants	50	24	1200
Valve replacement	25	36	900
Service renewals—small	1000	15	15000
Service renewals—large	30	47	1410
Meter replacement—small	250	5	1250
Meter replacement—large	80	37	2960
Main flushing	1500	3	4500
Valve repair	100	6.7	670

Common needs to achieve improvement are:

- efficient and timely purchasing systems
- flexible, well-trained labor staff
- vacation and shift schedules coordinated to assure optimum staffing for maintenance activities
- work standards established for maintenance activities
- rational basis for staff deployment (geographic versus functional versus skill based)

- prepackaging of materials and scheduling of labor and equipment to support daily work plans

- willingness to modify traditional, hierarchical organizational culture

- reporting of actual performance against planned maintenance

Develop Information Systems and Applications of New Technology

Information technology offers systems and tools for improved management of maintenance functions. The following are specific strategies used successfully in a number of water utilities, both large and small, throughout the country for infrastructure improvements and maintenance:

- apply cathodic protection technology to reduce corrosion of pipe and appurtenances and increase life of system assets

- utilize GIS to locate, record, and monitor maintenance history and other characteristics or attributes of the water system

- develop maintenance management systems that contain such attributes as work activities, work orders, resource planning and scheduling, work prioritization, and cost accounting

- evaluate new metering technology including remote or touch read meters

- link maintenance records, work orders, assets, billing, and financial systems

- provide data bases that are accessible to engineering and maintenance staff for such elements as field inspection and maintenance history

Common needs to achieve improvement are:

- adequately trained and knowledgeable information systems staff

- information technology plan

- cost benefit analysis upon which to base investment decisions

- sufficient funding, budget, or rate levels for major system improvements

- top level commitment for technology and system upgrades

Measure and Improve Performance

There are a variety of specific strategies relating to performance measurement and contracting:

- measuring performance by activity, using both qualitative and quantitative indicators;

- determining essential maintenance functions and considering contracting for nonessential services, especially seasonal or nonroutine maintenance (see chapter 12)

- partnering or establishing material and/or service provider relationships with vendors or other utilities to achieve efficiencies (i.e., inventory warehousing, equipment sharing or other specialized maintenance work)

The following are examples of performance measures that are commonly used in the area of water system maintenance:

- number of main breaks per mile

- transmission and distribution operations and maintenance cost per mile of main

- transmission and distribution operations and maintenance cost per thousand gallons sold

- elapsed time to complete new service installations from date of customer request

The City of Evansville's (Case Study 3.1) private contractor partner has guaranteed that annual water revenue will increase by $190,000 each year during the first five years by implementing a water meter replacement program in which touch read meters will be installed. The meters increase flow measurement accuracy and reduce labor requirements for meter reading. The partner will also repair leaks and provide other means to reduce chronic water loss.

- percentage of water loss in system

- speed of response to customers reporting leaks

- number of hydrants out of service for repair

- proportion of operable valves

- scheduled maintenance as percentage of total maintenance hours

Common needs to achieve improvement are:

- clear mission, goals, action plans, and performance measures (see chapter 11)

- sufficient process or commitment for evaluating performance, work processes, and best practices (benchmarking).

Implement Employee Involvement and Development Activities

Employees have a major stake in creating high performing maintenance organizations. Specific steps can be undertaken to involve
employees in generating ideas for performance improvement.
One example is to identify core competencies desired for your
organization and provide specific and specialized skill training
and recruitment to achieve desired results. Others include:

- encourage employee safety awareness, training, and recognition programs

- support skill-based classification and compensation systems; encourage multi-tasking for employees

- increase hands-on labor while reducing overhead and
  supervision costs; emphasize employee responsibility and
  accountability

- flatten the organization, reduce levels of supervision, and
  encourage lateral rather than vertical hierarchy

- involve employees in goal setting, performance measures,
  and work process improvements

Common needs to achieve improvements are:

- recognition and resolution of labor/management issues such as cross-jurisdictional union issues, employee involvement committees, and bargaining strategies

- modify hierarchical, layered, and inflexible organizational structures

- create an organizational culture receptive to future goals and objectives

- knowledgeable or trained staff in basic quality management principles

ESTABLISH FINANCING PLANS AND STRATEGIES

Another major challenge is to fund or finance maintenance and replacement programs. Overall this requires a deliberate, fact-based approach that identifies maintenance elements (such as water mains, hydrants, meters), generates short- and long-term plans and financial impacts, and identifies budget, CIP, rate, and economic impacts. Repair versus replacement approaches need to be addressed using life cycle cost principles, risk assessments, system condition assessments, and overall water system assessment and financial considerations. Finally, alternative financing approaches need to be considered for replacement depending on the type of asset involved, including: funding from current revenues, bond financing, system development charges, user fees, local improvement districts, and sinking or reserve funds for replacement.

Common needs to achieve improvement are:

- sufficient water system data and project and maintenance information

- applying or understanding financing tools or mechanisms in light of specific local circumstances

- gathering good cost projections and financial and rate impacts

- committing to addressing infrastructure requirements

SUMMARY

Utility infrastructure represents an enormous investment. How well it is maintained determines the return on the consumers' investment. Achieving the full service life of your infrastructure and doing so through an efficient maintenance and rehabilitation program constitute perhaps the most important responsibility of utility management. The key is to commit to a serious program of work performance improvement, to provide the resources and technology needed, and to let success inspire your employees to carry it forward in a continuing quality effort.

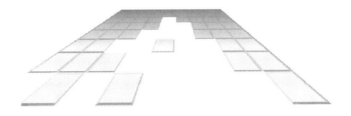

CHAPTER EIGHT

Customer Service

Chapter 2 described the performance imperative for the twenty-first century as having three primary bases, the first of which is:

> *"The focus of the utility must be on the customer, the customer's interests, and the customer's needs."*

Developing a customer service program that aggressively confronts this customer response imperative will be a big step for most utilities that have operated largely within a "build, produce, and sell" model of doing business rather than within the "sense and respond" model used by leaders in customer service. Yet the imperative is real and the rewards are great to those who step forward and meet this challenge. In meeting the challenge you will win the loyalty of your customers and in the process create real value for them and for your community through your partnership.

This chapter begins by reexamining what customer service is and the fact that customer service can and should be defined by the customer's interests and needs rather than by the organization that serves them. A preview of the likely components of service quality that would result from a customer-defined process is provided in this chapter to illustrate the major elements necessary for a comprehensive initiative to improve a utility's customer service. Although actions should be undertaken throughout a utility to support a customer response initiative, the majority of actions will be focused within the traditional customer service department or function within any utility.

Accordingly, the focus of the remainder of this chapter is the task of undertaking large-scale improvement in providing core customer services. The process of charting and executing the path to change and improvement is described in the context of assessing a customer service department in a large public organization. The ideas in this chapter provide a guide for creating the customer service component of your strategic plan.

The approaches suggested in this chapter are not intended to be a formula. For example, smaller organizations may face a less complex process of change, particularly with respect to labor issues. However, the approaches described provide a place to start, some important principles and guidelines to keep in mind, and some key steps to take. In the end, you will need to craft your own plan of action that reflects customer needs and your organization's capacity to deliver.

GETTING CUSTOMER FOCUSED— RETHINKING CUSTOMER NEEDS

Listening to Your Customers

America's service industry leads the world in providing excellent and innovative customer service. Companies that are the leaders in customer service, such as Federal Express, Nordstrom's, Land's End, and Disney, have shown that the process begins by recognizing that service excellence is defined by the customer, not by the service provider. Allowing the customer to define service excellence requires that you listen carefully to their needs, expectations, and interests, and recognize that these needs will change over time. With this as a starting point you can begin working on a number of ways to address their needs and to improve service in ways that they value. Through this process you will win their loyalty.

So how does one go about accomplishing this? A starting point is recognizing that there is more than one way to listen to your customer and that the whole utility, not just the customer service department, needs to be involved in the process. In

"Fabled Service: Ordinary Acts, Extraordinary Outcomes,"[1] Betsy Sanders suggests nine ways that an organization can listen and respond to customers to creating service excellence.

1. Listen to what customers tell you about your business.

2. Involve the whole company in the service process.

3. Respect that customers vote most clearly with their patronage.

4. Benchmark yourself against the winners (the ones the customers elect).

5. Hang out where your customers do.

6. Observe your customers using your products and services.

7. Relate your own experience to your customers'.

8. Organize opportunities to pay attention (focus groups, employee site visits, training employees to listen and communicate feedback).

9. Create customers' point of view for employees.

In most water utilities where the customer is largely captive, paying attention to each of these suggestions is critical and suprisingly easy. Critical, because they are captive and hence have no other means of having their needs met other than through your utility's attention to their desires. Easy (in relative terms), because your customers are unchanging. You know who they are and where they are located. The effort that you put into learning about their needs and in crafting a good relationship will not have to be repeated because of customer attrition.

Paying attention to customer needs and interests is particularly crucial with your large customers whose uses for water and service may be unique. In numerical terms these customers constitute a very small share of the customer base, yet account for a substantial share of water sales and revenues. The lesson from the

---

[1] Sanders, Betsy. *Fabled Service: Ordinary Acts, Extraordinary Outcomes.* San Francisco: Jossey-Bass, 1997.

electrical industry is that these customers, if ignored, can affect the changes necessary to have their needs met, including lobbying for the deregulation of an entire industry.

Showing these and other customers that you are serious about meeting their needs is also important. A symbol of this commitment can be helpful. Airborne Express publishes a "Customer's Bill of Rights" as part of their Quality Partner Plan. One of these rights is the customer's ability to talk to Airborne's top management about a service problem or complaint.

Elements of Service Quality and Customer Satisfaction

Research has shown that the process of working with customers to allow them to define service will result in a multifaceted concept of service quality. The SERVQUAL model of perceived service quality suggests that there are 10 underlying components to a customer's overall perception of service quality.[2] They are listed in Table 8.1 with two applications of the components: the first from the banking industry (supplied by the referenced authors) and the second from water utilities (suggested by the authors of this book).

The components provide a preview of the elements of a comprehensive customer service program that is likely to emerge from a customer-defined process. The portion of the strategic plan that addresses improvements in customer service would likely have strategic action plans associated with each one of these components. The majority of these actions would be focused on the core service delivery functions found in most utilities.

We will now narrow our focus in the remainder of the chapter to the task of undertaking large-scale improvements in providing core customer services. The remaining sections of this chapter will guide you through tackling the barriers, charting the approach, creating the plan, managing performance, and peering into the future.

---

[2] Rust, Roland T., and Richard L. Oliver. *Service Quality: New Directions in Theory and Practice.* Thousand Oaks, Calif.: Sage Publications Inc., 1994.

Table 8-1 SERVQUAL components of customer perception of service quality

SERVQUAL Components	Banks	Water Utilities
Tangibles	A. Facility and equipment B. Selection of financial offerings	A. Appraisal of product quality (appearance, taste, & odor) and pressure B. Appearance of facilities, including customer service centers C. Appearance of crews and equipment
Reliability	Accuracy and dependability	Consistency and dependability of service
Responsiveness	Speed of service	Response time to reported problems
Communication	Communication with customers	Clear and effective communication with customers and the community at large
Credibility	Reputation for honesty and integrity	Reputation for honesty and integrity, particularly in emergency response circumstances
Security	Financial strength and security	A. Safe and sanitary water quality B. Quality of public fire protection
Competence	Knowledge and competence of personnel	A. Knowledge and competence of central office and field personnel B. Perceived competence of public sector
Courtesy	Politeness and courtesy of personnel	Politeness and courtesy of central office and field personnel

table continued on next page

Table 8-1 SERVQUAL components of customer perception of service quality (continued)

SERVQUAL Components	Banks	Water Utilities
Understanding	Understanding of individual customer needs	Understanding of individual customer needs
Access	Convenience of location and operating hours	A. Convenience of telephone service function B. Convenience of service centers and operating hours

TACKLING THE BARRIERS—GETTING STARTED

Initiating a new customer service vision and organization requires confrontation of the same set of challenges that are described in other chapters:

- Fear of change

- Overcoming inertia from those who prefer the status quo

- Lack of resources

- More fear of change

- Lack of political will to make the change

- Lack of vision

- Lack of discipline needed to set tasks and schedules

- More fear of change

- Lack of perseverance and follow through

Overcoming these challenges requires ingenuity and a well thought-out process. Key elements of this process are:
Make sure that the commitment exists to make the change happen. This very first step can't be missed. The decision to proceed must include a commitment of resources, tolerance for

some pain during the change process itself, willingness to take risks that some things will suffer temporarily while the change process is underway. While customer service improvements are usually viewed as win-win, the changes require support from all concerned.

Work with your customers in designing the customer service vision and systems. As we discussed earlier, customers have an opinion of you and will tell you if asked. Find out how you are doing in satisfying their expectations. Let them tell you what they care about. Use this information to help shape your vision of what you want the utility to be.

Include other stakeholders in the planning process. Stakeholders left out of the planning process are much more likely to become antagonists—either openly or covertly. Think about who will be affected by this effort and bring them into it.

Define a service vision. There are two key pieces to a service vision. First, try to describe what customers will experience when the new service organization is in place. A vision describes your desired future in a way that inspires and expresses superior performance. This vision is needed so that employees and managers alike can focus on the same target. Second, sketch out a functional concept of a service organization that sets forth how the organization should work toward the vision you have defined. Be open to refinements and innovations that will come from your stakeholders/planners.

Enlist support. Your most valuable supporters or ardent detractors are your employees. Get them involved early. Give them substantial roles and give serious consideration to what they recommend. Establish employee teams for each major activity but make sure there is a core of people on each team who have the vision, are reasonably well organized, and are capable of completing tasks.

Get the labor unions on your side. Labor will be looking out for the interests of their members. Give assurances if you can that jobs will not be lost in the change process. Quickly identify and work out any personnel impacts. These concerns can deflect attention from the change process and possibly even derail it if left unattended.

CHARTING THE PATH—A TRADITIONAL APPROACH OR RADICAL MAKEOVER?

If you are tempted to create a traditional work plan and start setting schedules—stop! You'll do those things soon enough. First, figure out if it is reasonable to create the new service organization from what currently exists. Most traditional organizations—banking, utilities, financial management firms—have been organized by function for many years. Functional organizations work well for managing discrete tasks. The trouble is that most customers have needs that cross functional organization boundaries. Bad service happens when customers get caught in the quagmire of organizational boundaries, competing priorities, and conflicting policies. Look again at the vision being created; these are the things you and your customers care about. You must answer the question: *"Can we get there from here with the current organization?"* The service outcomes you strive for may suggest structural changes.

Here are service outcomes some businesses choose:

One stop service. Customers may call or visit a central service center where most business can be handled on the spot.

A single well-coordinated service center. Most transactions can be completed at the time of initial contact. Employees are trained to handle a broad range of basic business. There is a seamless way for customers to get help with specialized services. Customers are not sent to seek out specialists, but rather the initial contact agent makes a referral and facilitates completion of the service request.

Customers can access some services 24 hours a day. Telephone, geographic, and billing systems are functionally linked so a broad array of services is available electronically or by telephone.

The service organization is built around customer needs. Rather than organize by function, organize according to how you can best serve each customer group. Teams might be designed to reflect a cross section of skills needed to serve residential, commercial, wholesale, or key account customers. Look for examples of how this is done in other businesses.

Quick and effective response in emergencies. Service staff can quickly mobilize support for off-hours emergencies. This requires a call center capable of receiving customer calls, helping to identify geographic locations of service outages, getting special help to at-risk customers, sending crews to the right priorities, and making accurate, timely information available to the media and to customers.

Customer representatives have broad authority to solve problems. Service representatives are prepared to handle a broad range of utility services, perhaps for multiple utilities, and are trained to assess and solve problems quickly without layers of review and approval.

Special needs customers get the help they need. Low income, elderly, ill, and other special needs customers receive compassionate treatment and assistance. Staff is well trained to recognize and respond to symptoms of distress. Programs are available to help avoid service interruptions.

Key services achieve specific performance targets. The services key to the success of the utility have specific performance goals and are frequently reviewed, measured, and reported. Customers are asked what they think and if their expectations are being met. Water service itself must be seen as a customer service.

Special lines of business. So called "signature services" are those around which a utility expands its reputation. These could be electronic billing and remittance for large multi-meter customers, special services for absentee property owners, debit or credit card options for paying bills, special metering or submetering, conservation audits, or other services appropriate in your area as identified by your planning teams.

CREATING THE PLAN

The plan for your service revolution needs to address issues on four levels. Changes on all four levels will exist concurrently, with frequent interchange between levels.

Conceptual. The conceptual plan consists of a vision, service mission, customer feedback, and service model. These elements should be in writing, but only after they have been openly talked over, challenged, and argued about until the key players reach a place of relative comfort. Everyone doesn't need to agree, but they do need to be willing to try to make it work. One technique for resolving conflicts is that of "try and see". The idea is to reach a common starting point upon which people are willing to proceed. An important understanding is that if problems emerge that can't be solved, the model will be revised.

Perceptual.   Perceptions about the change are critical to success. The place to start is to use a process of employee and external stakeholder involvement in the change process. The desired result is to create positive perceptions and shared ownership. Establish a critical planning and coordinating team with a few work teams that have specific tasks to complete.

Set up a forum for keeping employees up to date on progress. Regular meetings to discuss issues and to report on status work well. Newsletters also work but are not a substitute for face-to-face dialog. Dissent, of course, should never be stifled. Bring it into the open and answer it. Seize opportunities to take suggestions. Use these occasions to compliment participation of employees.

Accountability.   Like any multi-task project, a major service overhaul needs a specific project plan that identifies tasks, due dates, and responsible persons. This document is explicit in detail about what will be done by when, and by whom. The project plan is where you schedule phasing of vital parts of the effort. Be sure to recognize and provide for interdependencies, such as for installation of technology improvements (telephone, billing, and geographic systems), changing schedules, and need for appropriate facilities, training, and process design.

Leadership.   The leadership element of your plan is essential for success. Look at the previous conceptual elements and identify centers of responsibility. The new service plan needs a cadre of leaders who have a vision, concept, desire, and the willingness and energy to move ahead. Team leaders, supervisors, and managers must be held accountable to pick up defined responsibility areas and start working the issues, problems, and details.

ELEMENTS OF THE PLAN—PULLING IT ALL TOGETHER

Be concerned about two major areas of your new plan. First, the tools of the service business are the foundation upon which employees are able to serve customers. Second, the utility must become a service organization. The specifics within each of these areas might well be assigned to separate work groups, with your control planning group providing the necessary coordination.

Tools of the Service Business

Telephone systems. Wise use of telephone technology will enormously extend the productivity of your agents and greatly facilitate the service they give. Your system should get calls to the right service center quickly, manage customer wait time, prioritize calls, handle some transactions electronically, offer customers a streamlined menu of information and choices including how to opt out of a telephone queue, provide location information about callers, and possibly perform automated call functions, meter reading, and more.

Review potential telephone system capabilities against your service vision. Secure the services of a qualified telecommunications consultant who has nothing to gain from selling services or equipment. Do an assessment of your current and projected call volumes, the specific services you would like to offer, and the interface you would like to see with your billing system. Your consultant should be able to develop at least two alternative packages (of services and technology) for your review. The packages available will be determined by the volume of your calls and the complexity of your routing requirements. Costs vary so careful review is worth the effort. Select someone from your staff to take charge of moving a telecommunications assessment forward and to oversee the multitude of details involved with installing new equipment and training users. Greetings and messages convey the image of the utility. Consider investing in a professional message voice and design your message to be friendly and concise. Always provide a means for callers to reach a real person.

Customer information systems/geographic systems/personal computers. Great service depends on accurate, timely information. Most of what you need probably resides in your customer information system (CIS). CISs are often structured on the basis of functions without serious consideration of how systems need to interrelate or what are appropriate internal efficiencies for users. To provide integrated, one stop service, it is more important than ever that customer information systems have functional links to geographic information systems. This is very important for managing emergencies and also for handling service interruptions efficiently, and locating service impacts.

Internal efficiencies are likewise important if only for the purpose of assuring that high volume transactions are reasonably streamlined and easy to use. If the number of key stroke levels needed for a transaction is high or cumbersome, customer representatives may look for shortcuts that prove unwise. If you can afford them, personal computers are more flexible and effective than terminals for accessing multiple electronic media and information sources.

Facilities. Work spaces define your image and affect the efficiency of your work teams. If you make investments in service improvements, don't neglect the space in which the work is done. For public service centers, choose accessible locations, preferably on public transportation routes. Consider the parking needs of your customers. Think about access for special needs customers and employees. Create comfortable waiting areas that are cheerful and well lit. Display information materials attractively. Plan space to serve customers who have quick transactions and maintain areas for those who need privacy for discussions of appeals and overdue accounts. Make sure signs are clear and visible. Do you need to provide essential information and directions in a second language? Security should be reviewed by crime protection specialists for the safety of customers and staff alike. Above all, involve staff in planning the space. Make it theirs. Be as generous as you can with ergonomic furniture and sufficient floor space.

Policies. Policies define your utility as customer friendly or bureaucratic. Review your policies with two critical questions: *"What does this policy do to improve service of this utility?" and "What does this policy do to protect the interests of rate payers in general?"* If the policy has little value or if the policy protects customers on rare occurrences, get rid of it or rework it. Many policies are needlessly rigid and impose unnecessary administrative work. They can also be responsible for bad service and create an unflattering image of the utility. Policies should be easily understood by all employees, make sense to the typical customer, and be available in a streamlined written form.

Business process design. A business process is a series of tasks necessary to complete a service for a customer. From the customer's perspective certain things matter: ease of access to the service, the manner in which the employee performs the service (courteous, efficient), and whether the service is done satisfactorily (timely, correctly). To improve a service, gather a representative group of customers who use the service and employees who perform the service. Each has part of the information needed to make the service effective and efficient.

The customers can tell you whether or not the service meets expectations. They can also offer suggestions about service improvements you are considering. The employees can identify tasks essential to the service. Try creating a work flow chart for the service—a series of boxes in a graphic representation illustrating how the service gets done. You'll find the bottlenecks. Redesign the work flow with employees. Identify tools needed to make a simplified version work. Add the tools, trim the policies, refine the plan. Try out the new plan and watch closely to see if the things that matter (access, courtesy, efficiency) improve. There's no mystery to improving business process; it's a continual cycle of listening, looking, trying, and listening again.

Training. No service plan is complete unless there is training that focuses on both the human and business sides of the customer business. A comprehensive training plan includes: service skill training, leadership training for supervisors, team leaders, and managers, practical problem solving, general overview training of each utility system, integrated technical training, business policies and practices, and individualized coaching in real service situations. Consider involving your employees in a substantive way in both designing and delivering a great training program. Or use a consultant-employee team that utilizes consultant materials, with employees as part of the training team. It will be your own employees who make or break the technical system and business training. Supervisory and leadership training and practical problem solving are good candidates for consultant-provided training.

The Service Organization

The service organization you design will emerge from the needs of your customers and your internal assessment of what is working and what isn't. General trends in many cities, in both the public and private sectors, are for consolidating services but not necessarily in a centralized place. That could mean, for example, that many services could be offered together in decentralized locations. Consider the following:

Call Center. Call centers provide a wide range of basic services (usually the high volume services of the organization), often for extended hours of operation. Call centers usually have links to other teams that provide specialty services on referral. The chief feature of a call center is that it makes its services available at the convenience of the customers, meaning one stop service including evenings, weekends, and holidays.

Of course, not everything is appropriate for one stop service. Hence it is important to have easy to use linkages to specialty service providers. Agents of the best call centers make sure that each customer's needs are fully addressed and that any follow-up services are arranged so that the customer need not try to navigate the organization on their own.

Specialty or field service teams. Specialty teams can support a call center by having technical experts available. They might accept a customer referral from the call center or provide advice to call center agents who need a question answered. Make sure there is an easy way for call center agents to make a referral or get assistance in a call they are handling. This is an area that will need the special attention of your telephone team. Consider placing your field services, including inspections, here.

Credit/collection. Some traditional collections functions, such as making credit arrangements with customers, can be part of a comprehensive call center. Consider training your customer representatives to handle this important service. Not having to handle credit arrangements frees up the collections staff to focus attention on seriously delinquent accounts. It's to everyone's advantage to handle small accounts in a way that customers get quick service. Management of long term or large dollar delinquencies should be handled by specialty teams or as a single function. Remember, credit arrangements and collections are integral to customer service.

Billing, accounting, and auditing. Sometimes thought of as "backroom functions," billing, accounting, and auditing (quality control) have a direct and important influence on customer service. Consider linking these functions to your customer service organization. Employees will see clear relationships between direct customer services they provide and the quality control functions. In the absence of this linkage agents may be less alert to potential inaccuracies and not as motivated to get things right when they notice errors. You want your agents to feel a personal sense of responsibility for account and meter accuracy. This is more likely to happen if these functions are part of a larger customer service organization.

Metering. Think of meters as you would think of cash registers that determine the accuracy of your charges to customers. Metering is often linked organizationally with the field organization. However, consider this: meter work is technically unrelated to other field operations and extremely interdependent with billing and auditing work. The success of any meter improvement program depends on the ability to identify meters that are becoming or have become inaccurate. This is work most likely initiated by billing and audit staffs based on usage histories. Consider a link in your organization to create a common set of priorities and an efficient work plan for increasing meter accuracy. Because meters that fail always measure low should motivate you to correct meter inaccuracy problems.

Remittance. Timeliness and accuracy of remittance processing is critical to good service. Efficient processing requires equipment that is appropriately matched to the type and volume of your payments. Unless you are an extremely high volume processor serving multiple organizations, consider hiring a competent contractor to provide this service. Remittance equipment must provide high speed, high volume processing, total accuracy, and reliable links to billing systems. Increasingly customers are demanding debit and credit card services. Commercial and wholesale customers will also want electronic remittance services. Make sure your technology keeps pace with customer demands.

Public Information and Education

Your customers depend on timely, accurate information that anticipates their questions and concerns. Use your customer survey information to design target campaigns. Even more important, make sure your plan provides clear, logical, understandable information about rates, how to access service and policies, and any special needs that customers are experiencing. Consider making public information staff and educators an important partner in your customer service organization.

MANAGING PERFORMANCE

Customer service is widely viewed as high pressure, low reward, burn out work. Because entry qualifications for new employees are less stringent than other occupations, it tends to attract employees on their way to doing something else. The most talented service providers tend not to stay. If they do, they often become burned out over time. Some challenges to giving continuously good customer service are large volumes of work, difficult people, few incentives, and lack of rewards for good work. Clearly, one of the most challenging problems facing the customer service manager is attracting and retaining competent employees. Make no mistake, it boils down to compensation, career advancement opportunities, working conditions, and performance incentives.

Performance measures that target the most important service functions. The key to great performance starts with identifying specific service attributes that customers care about most. Consider setting performance measures for elements coming from customer values (the things customers rate as most important). This approach will keep you focused on the most significant service areas.

If customers want fast service (most do), you must have a performance measure that targets wait time. If customers care about friendly, caring service (most do), you must devise a way to have customers rate their satisfaction on that element, and so on. It's not easy to identify the right measures. Measure the wrong thing and you may miss the information you need most. Or worse yet, you could be deluded into thinking everything is fine.

There is an old saying about performance measurement, *"What gets measured, gets done."* This means that whatever you choose to measure will dictate what the staff does well. For example if the number of customers handled per shift is measured and used as a positive measure of performance without a complementary measure of the quality of each transaction, unintended consequences are sure to follow. Be careful what you choose to measure.

Service performance goals that show clearly where you are now and where you desire to be.   After selecting the service elements that are most important to customers, look at each element from various perspectives. Each perspective will tell you something about the service element. Ask yourself whether the information it provides is worth the effort it takes to collect the data.

An example explains how this works. Let's assume that waiting time is important to your customers. Here are some different perspectives about waiting time:

- Percentage of customers being served within a goal time you select (example: percentage of calls answered within one minute)

- Percentage of customers being served within an acceptable range you select (example: percentage of calls answered in three minutes or less)

- Average wait time

- Longest wait time

Each of these four performance measures tell you something different about wait time. The first two give you a profile of how you are doing in relation to standards or goals you have set. The third measure gives you an average that is only useful as a statistical measure. It must be viewed only in the context of other measures. The last measure gives specific experience information about the worst service situation encountered for the time period. You may want to measure all of these aspects of wait time or be selective about what you intend to improve.

Philadelphia Water Department: Measuring Performance through Customer Surveys

The Philadelphia Water Department has integrated customer surveys with service delivery in order to gauge how well the department has provided a service. When a customer calls the department to lodge a complaint or request a service, such as to request the repair of a backed up sewer line, a survey will be sent to that customer approximately two weeks after a crew has responded to the request and visited the site. The survey is a page long and contains questions that focus on the initial contact the customer had with PWD customer services, the performance of the crew or representative that ultimately provided the service, and satisfaction of the interaction with the crew and or representative. In addition to tracking performance in sewer maintenance, the department also surveys customers that have interacted with other divisions such as: customer information/ services, inlet cleaning, meter, construction services, water main repair, and the drainage information unit.

INCENTIVES IN A BURN OUT BUSINESS

Incentives

Incentives can be compensation tied to performance such as time off, awards, recognition, or anything that motivates employees to perform their best. The most successful private sector companies find multiple ways to compensate and recognize employees for best performance. Public sector utilities have been slow to follow suit but the need is great if we want to attract and retain the best employees. The City of Scottsdale has been a leader in this area, providing a bonus program that awards up to 2.5 percent of each employee's salary if they meet or exceed performance standards. The bonus also can be awarded to employees who provide suggestions for work process improvements that lead to cost savings.

Job Classification and Compensation

Seek ways to create customer service job classifications that are defined broadly with both breadth and depth of assignment possibilities. Look for ways to create performance incentives in your classification/compensation system by setting specific performance ranges for each assignment level.

THE FUTURE

The only realistic way to plan for the future is to watch customer trends in your utility and to watch other businesses in your region for trends that will help you define customer expectations. Remember, the cycle is listen, look, try, and listen again.

To guess about the future one might see more electronic technology to streamline services coupled with more personalized service for those who have needs not matched to automation. There will be demands for more one stop shopping coupled with multi-utility service partnerships. More emphasis will be needed on programs to help customers who are having trouble paying bills. There will be increasing pressure to hold the line on rate increases, which will in turn require that services be operated more efficiently and cost effectively.

OUTSOURCING UTILITY FUNCTIONS

Some customer service functions can be provided by private sector vendors, but be aware of potential problems. Are there enough controls and incentives to ensure that the customer service you are purchasing reflects the type of image you want? Will you lose control of your data and find escalating uncontrollable costs? What happens when technology expertise no longer exists in your organization? Be wary of splitting functions where critical relationships exist. An unfortunate result could be that the customer suffers from the lack of coordination between the utility and the private service provider. These cautions do not mean that contracting for outside services is always wrong. For smaller utilities, for example, it might make good sense. In all cases, however, the customer's interests and the utility's reputation must be protected.

SUMMARY

Customers must be at the center of the organization. We must structure our organizations with customer service as our focus. All employees must recognize that they work for the customer and they must act accordingly. Building a service organization is a major commitment; it takes time, effort, resources, and solid, sensitized thinking. In the end we must have the correct answer to the question, *"Who are we working for anyway—the customer or ourselves?"*

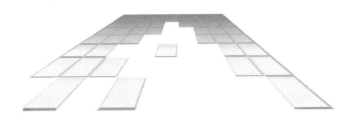

CHAPTER NINE

Information Technologies

Integrated information and control system technologies (IT) are valuable tools for improving efficiency in water utility operations. They play an important role in achieving and managing a higher level of performance and improved efficiency by increasing productivity at all sizes of water utilities. It is important to correctly apply and maintain technology that include: real time control systems for process and equipment monitoring and control and information management systems, (maintenance management systems for streamlined facility maintenance); laboratory information management systems for management of data; and information systems for maintaining knowledge and digitized documentation.

Many water utilities have used IT to improve reliability and reduce costs. IT systems provide the data needed by facility personnel for timely analysis and decision making, and support improvements through reorganization and staffing reductions.

There are many systems available to utilities to improve a variety of processes, functions, and activities. The systems installed at some water utilities are, however, not without problems. Unless the utility applies technology to enhance specific organizational needs, following a well-developed implementation strategy, full benefits may not be realized. Implementation and selection should be carefully planned as technology changes. System selection should be made to minimize the effects of obsolescence. The best results are achieved by the application of IT to achieve performance improvement and by its maintenance by providing users with technical and training support.

IT strategies used by high performance manufacturing and service organizations have recognized that IT should be:

- carefully planned from design and selection through implementation and post-implementation support

- accepted and supported by users in order to achieve intended improvements. Fear of increased accountability and/or distrust of the technology itself can result in ineffective system use.

- based on product life cycles, especially for computer hardware and software, which experience accelerating change. In this environment flexible and adaptable systems are important.

- designed to achieve the business purposes (utility goals), including reliability and efficiency

The following discussion on Information and Technology summarizes opportunities available to utilities to improve a wide variety of processes, functions, and activities. The best results are achieved by the application of IT to utility functions that have been strategically planned in anticipation of applying IT. The use of IT, including automation, is no substitute for improving the performance of a particular unit or function. However, improving your utility's performance is a long-term project, and there are benefits to be gained by applying IT to the existing way of doing things even though the result may not be the best achievable. For instance, remote controls and monitoring pumps and storage facilities can provide reliability dividends even though the pumping system may not be the most efficient or the local utility rate structure (see chapter 10) does not now reward time-of-day pumping.

OBSOLESCENCE

Until recently, many utilities have refrained from extensive application of IT because of high cost concerns, questionable reliability, and fear of change and obsolescence. Unlike most waterworks investments, IT users can expect that both hardware and software will be obsolete in three to five years. Obsolescence generally results from improvements in hardware and software. However, even if a system becomes "obsolete," it may provide many more years of effective service to the utility.

Early users of SCADA systems have achieved significant reliability improvements, particularly during crises, but they have paid a significant cost. For instance, East Bay Municipal Utility District invested more than $25 million in a SCADA system in the 1980s, but has had at least two stages of improvements since then. Today, major systems can be installed for a fraction of this cost and they are designed with features such as standardized databases and they operate in standard environments to enable application of improved technologies so that their obsolescence does not result in high replacement costs. In fact, updating and improvement costs can be included under the utility's operating rather than its capital budget. Experience has shown that delays in implementation due to concerns about obsolescence are rarely justified.

UNDERSTANDING INFORMATION TECHNOLOGY INVESTMENT

Technology Change and Life Cycles

The pace of information technology change is accelerating. If we examine the dominant phases of information technology over the past 50 years, we can see that:

- From the mid-1940s to mid-1960s information technology consisted of a very few centralized computers, which were used primarily for research and military calculations.

- From the mid-1960s to about 1980 mainframes and mini-computers were dominant and adopted widely for general business computing. Early control systems were designed and installed but were found to be unreliable and expensive to maintain.

- In the 1980s personal computers and local area networks rose to prominence, and personal productivity software such as word processing became widespread. Control system technology reliability continued to improve.

- In the past five years we have seen universal adoption of wide area networks (WANs) and Internet-based applications, which provide interconnection of computer systems within an organization or on a national and global scale. Applications such as electronic mail and the World Wide Web depend on these networks. Control system technologies have evolved to become highly reliable and easy to maintain systems.

We are now in another period of rapid change. The adoption of the worldwide Internet by businesses and individuals has made possible a whole new set of applications for collaboration, information sharing, and interaction. An area of rapid growth lies in information systems that use Internet technologies for serving a single organization. These private networks using Internet technologies are called Intranets and are now widely used by organizations as central information resources. Intranets have even spawned a new opportunity for multi-utility or organization networking, termed the Extranet. For example, organizations such as AWWA may someday link in an "AWWA-net" with its member utilities. Utilities may also desire to partner with each other and link in an inter-utility Extranet to capitalize on sharing information and resources.

Information technology life cycles are much shorter than typical capital investments made by water utilities. Life cycles for computers and peripheral hardware are roughly three to five years. A prediction known as Moore's Law suggests that the computing power of microprocessors doubles every 18 months (a prediction that has been uncannily accurate since the advent of the microprocessor). Similarly, software life cycles can range from six months to five years or less, and electronic software distribution over the Internet has made possible continuous software updating from manufacturers to users, thereby reducing the cost and impacts from obsolescence. Table 9-1 shows a comparison of typical life cycles for facilities and equipment.

Table 9-1 Typical facility life cycles

Facility or Equipment	Typical Useful Life
Reinforced concrete structures	50 to 100 or more years
Ductile iron pipe	20 to 100 years
Hydrants and valves	30 to 50 or more years
Centrifugal pumps	10 to 30 years
Control system computers	7 to 15 years
Maintenance management software	5 to 10 years
Personal computer workstations	3 to 5 years
PC operating system software	18 months to 3 years

NOTE: Obsolescence resulting from changes in process requirements may significantly shorten life of structures and mechanical equipment.

Planning Considerations

Planning and implementing systems in the face of rapid, accelerating change requires a unique combination of strategies and tactics:

- Develop and follow a strategic plan with objectives that will provide direction and sustain multiple years of technology investment.

- Design technology solutions to achieve objectives from the user's perspective. This means first understand user needs, skill level requirements, and how the technology will be accepted. Assessing all user requirements prior to design of IT systems greatly facilitates ease of implementation and improved efficiency.

- Select, scope, and organize projects to be implemented in four to six months, thereby avoiding long project time frames that have a significant potential for midstream obsolescence.

- Implement integrated technology that addresses specific needs, not perceived needs. Measurement of improvements from IT investment should be conducted.

129

- Provide technology support systems and personnel to sustain the investment and maintain user acceptance. Responsive and effective support to IT problems is critical to sustaining user confidence and continued reliance on the system as a reliable resource.

- Use realistic depreciation. Three year amortization is not unreasonable.

- Consider buy or lease strategies to fit utility budget structures.

A framework for planning and implementing technology projects is described in detail later in this chapter.

Automation to Enable Improvement

Traditional IT investments have concentrated on reducing or eliminating routine monitoring and laborious tasks, with the objective of controlling labor costs by freeing up individuals for other more important duties or by elimination of positions. For example, computer spreadsheets and databases replaced hand calculations and written data summaries, which, in turn reduce calculation errors and allow for collation of data for problem solving and trend analysis. Computer Aided Design and Drafting (CADD) systems replaced manual drafting. Automation manages routine water treatment process functions, thereby reducing the labor requirements for continual monitoring and adjustment to maximize process performance and treatment efficiency.

The versatile, multimedia information processing capabilities of today's personal computers, and the great variety of communications options now available, have created situations in which technology can enable a complete reorganization of the work people do, and of the people who do the work. This is use of technology to enable change. For example, maintenance management processes are greatly affected by implementation of a computerized maintenance management system (CMMS).

Automation approaches have fostered a tendency in some utilities to view information technology as the solution applied to an organization's existing problems. The lack of a particular system is offered as an excuse for inefficiency, and the objective at hand becomes searching for (and perhaps not finding) the perfect software or automation solution. Automation generally is not a substitute for a poorly designed or constructed process. Organizations that concentrate on complete automation ultimately may not get full value. The lesson learned is to automate only those processes that you know may be improved or will have less labor and resource requirements.

Justifying Technology Investments

Are there objective measures that demonstrate a positive return on investment in information technology? This question has been fiercely debated for the past 20 years, particularly with respect to service businesses such as water utilities.

Recent large research studies[1,2] document a positive return on technology investment for companies that concentrate on enhancing customer service, eliminating cost control as a primary objective, and adopting a tactical approach to new system development. These ideas can be adapted for water utilities that have a slightly different set of business objectives and that focus on high quality service at the lowest cost.

While cost of IT is often the issue, and should be carefully considered, noneconomic factors may frequently be the most important. A variety of questions other than economic impacts should be considered and answered, such as:

- Does the proposed project directly support the primary water utility objectives, mission, and vision?

- Is this technology project coupled with a work process change that can be accepted by the organization?

- What are the primary nonmonetary benefits? Will customers directly perceive the beneficial changes brought about by the system?

---

[1] E. Brynjolfsson and L. Hitt, "The Customer Counts," *Information Week*, September 9, 1996.

[2] E. Brynjolfsson and L. Hitt, "The Productive Keep Producing," *Information Week*, September 18, 1995.

Information Systems Vision

The Birmingham, Ala., Water Works and Sewer Board has developed an ambitious long-term vision for the implementation of information systems. The vision includes SCADA, a GIS, a comprehensive hydraulic model of the distribution system, and ancillary systems such as a Laboratory System (LIMS) and a Maintenance Management System (MMS).

Overall SCADA vision is complete integration

The vision is for complete integration by allowing data to be shared among the various information systems. The implementation of SCADA is seen by the board as the first step to achieving the overall information systems vision. The SCADA system provides a platform on which the other information systems can be built.

131

• Can this project be implemented, in full or by phases, in six months or less? Are there immediate benefits of early phases?

Economic impacts include:

• What is the full life cycle cost of implementation and its impacts on other activities?

• Is it necessary for or can it assist in regulatory compliance?

• How long will it take to realize the benefits of implementing the technology?

• What are the full life cycle projected labor and cost savings? Do these benefits depend on or overlap with other technology projects?

DEVELOPING AN INFORMATION TECHNOLOGY IMPLEMENTATION PROGRAM

This section presents an IT framework depicted in Figure 9-1.

Vision and Mission

Long-range IT planning is the basis for an effective information technology strategy. An IT team should include managers and some operating staff to create a definition of the utility's IT mission and vision.

Strategic IT Planning

A water utility master plan supports strategic IT planning. The IT plan should set the framework for your utility's IT projects over the next five years.

IT strategies are the bridges between specific operating goals and the expanding set of technology choices available to every water utility. They establish criteria for setting project priorities and selecting from technology alternatives. While it is not possible to define one set of strategies appropriate for every water utility, there are common current strategies that should be considered:

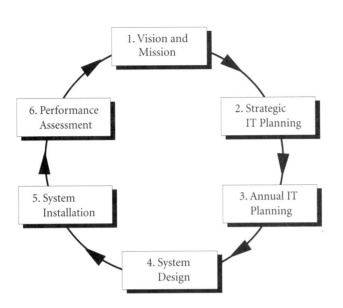

Figure 9-1 A framework for information and technology implementation

Fully networked infrastructure. Establish a computing and communications infrastructure based on Ethernet and Transmission Control Protocol/Internet Protocol (TCP/IP) networking throughout the organization. Ethernet is the most popular form of Local Area Network (LAN), the way in which computers are wired together in a building or group of buildings. TCP/IP is a set of communications protocols that make sure all computers on a network can exchange information efficiently, regardless of other differences such as different manufacturers or operating system software. In other words TCP/IP and Ethernet are the common language and communications pathways that all computers can use. A fully networked infrastructure will position the utility to take advantage of the vast majority of innovative hardware and software designed for the networked environment.

Wide Area Network (WAN) and remote access. Establish communications links so that all parts of the organization, including the smallest and most remote parts, are connected to the enterprise network, and the utility is fully participating in the

Human Resource and Finance Uses

El Paso Water Utilities (EPWU), Texas, is managed by a public service board, operates independently of the City of El Paso, and provides water and wastewater services to 700,000 people.

Upon completion of creating a strategic plan, EPWU developed an integrated Finance and Human Resources System (FHRMS). The system is based on client/server technology, provides a graphical user interface, and uses the Oracle relational database management system running on a UNIX platform.

The project included defining requirements for billing, accounts receivable, purchasing, accounts payable, job costing, inventory control, asset management, general ledger, financial reporting, human resources, payroll, and contract administration. In addition to defining software functionality, the project also included defining El Paso's hardware, support, conversion, and training requirements.

Internet. In general, any organization with multiple, separate locations will require a WAN, which consists of communications links between all locations. These links will make the multiple locations act as one large network. In instances where a permanent link is too costly, remote access via a dial-up link can be provided. This link will achieve the same networking functions while the link is active. Finally, *with proper security devices in place,* a water utility can confidentially connect its network to the Internet, the worldwide standard network. This strategy will enable the water utility to implement effective communications and collaborative computing applications.

Client/server computing. Select software and systems that enable client/server computing. Client/server computing refers to the way in which computing work is divided among multiple computers on a network. In the client/server model, a centralized computer, known as a database server, stores, organizes, and secures data, as well as enforces business rules about how data are added, modified, or deleted. On the client side, the computer work stations (typically personal computers) provide the menus, data entry displays, and reports associated with running a software application. With these types of client/server systems there is a great deal of software flexibility and the information assets of the utility are highly accessible, yet protected in centralized databases. Intranets are a form of client/server systems.

Standardized, consistent purchasing. Select and standardize on a minimal set of computer types, desktop operating systems, network operating systems, databases, and telecommunications methods. Enforce these standards through controlled, consistent purchasing. This strategy will help eliminate the high labor costs of supporting multiple, essentially equivalent computing environments.

Mainstream software innovation. When selecting software, concentrate on companies that are innovating in the emerging mainstream of computing. Do not select software on the basis of maximum features without a clear indication of imminent innovation from the software company. Look for strong and proven post-implementation product support and companies that have been in business for awhile. Recognize that all software products have a life cycle measured in years, not decades. With this strategy, utilities will steer away from market leading products that may be too mature and on the verge of technological obsolescence and decline.

Annual IT Planning

Regular, formalized planning of technology projects on an annual basis will provide a mechanism for satisfying the multiple, occasionally conflicting needs of groups within the water utility. This can be based on the planning and participation approaches discussed in other chapters.

IT System Design and Installation

Utility experience indicates that information technology projects can end up significantly over budget, delayed, or canceled. The following guidelines may be helpful to avoid many common pitfalls:

- Precede any IT project with a description of what the project will achieve, the anticipated system users, their roles, and the planned work flow. The IT project must be included in the IT master plan and annual planning activities.

- Seek out a project champion from among the intended users. Avoid making your IT manager or consultants the key proponents of change. Staff must believe that compelling needs, not technologies, are the driving force for a new or modified system. Consider having your staff visit other site installations with advanced applications.

- Create a project plan with frequent milestones and opportunities to assess the appropriateness of the technology. Pilot or prototype systems should be created and evaluated within the first six months of most projects.

- Avoid function and "feature creep." This condition occurs when a wide cross section of users are encouraged to build a laundry list of required features, and the list is updated well into the design stage. Someone needs to carefully define and finalize all of the user suggested features early in the design process and hold firm on adding additional features until after implementation.

· Minimize the risks and cost of customization by selecting from available software that can be adapted to your needs.

· Seriously consider abandoning most old system data, retaining only current information wherever possible. Typically, data conversion costs tend to outweigh benefits, especially when data are held for unspecified or unscheduled contingencies.

IT Performance Assessment

Utilities should not assume that IT projects result in early benefits. As with other complex initiatives, IT projects should be assessed using performance criteria and planned measurements. The results of performance assessment are then used to modify the project approach in a process of continuous improvement.

Traditionally computing professionals have used easily quantified measures tied to processing power, storage efficiency, and printing volumes. While these performance criteria may be used to assess the effectiveness of computer center management, they are not appropriate measures of business success or failure. Instead, IT performance measures must address the desired business outcome or result. These measures will vary with each organization and the specific technology projects being implemented.

Table 9-2 lists performance measures that may be appropriate for various technology projects.

REVIEW OF STRATEGIC TECHNOLOGIES

Utilities, like other businesses, have many opportunities to implement information and control technology. Considering all of the options, which technologies are likely to have the greatest beneficial impact on the effectiveness and efficiency of your utility? The following sections, adapted to accommodate your utility's size and potential return on investment, describe technologies that you may wish to consider:

Table 9-2 IT project performance measures

IT Project	Potential Performance Measures
Monitoring and control (SCADA) system	• Reduction in routine operations monitoring requirements • Reduction in operations staff numbers • Decrease in chemical and energy use • Improved treatment process performance • Higher finished water quality
Management information system	• Better cost tracking • Accurate inventory tracking and depreciation control • Reduced time and labor require-ments to produce special summary reports • Reduced purchasing approval time frames • Reduced invoice approval and pay-ment duration • Improved understanding by all per-sonnel of actual costs within their domain or function
Maintenance management system	• Increased labor "wrench time" • Decrease in outstanding work orders index • Improved accuracy between esti-mated and actual work performed in labor hours • Increase in PM/CM ratios • Improved inventory and tracking • Improved storeroom efficiency as measured by decrease in average time to obtain/order parts • Improved cost accounting of main-tenance function • Reduction in average equipment down time index for maintenance
Geographic Information Systems (GIS)	• Faster response time to distribution and service line faults • Better tracking of distribution sys-tem maintenance

table continued on next page

137

Table 9-2   IT project performance measures (continued)

IT Project	Potential Performance Measures
Customer information and billing system	• Reduce the average number of disputed bills per month • Increase surveyed customer satisfaction ratings • Reduce time to resolve customer inquiries • Reduce nonpayment to receivables ratios • Reduce billing labor requirements
Electronic Document Management System (EDMS)	• Fast information access to reference documents • Reduced file storage requirements • Greatly reduced incidences of lost files or references
Online operations and maintenance information	• Fast access to information • Increase in staff knowledge base and skill level • Improved facility performance • Faster troubleshooting response, diagnosis, and correction
Laboratory Information Management System (LIMS)	• Improved process control • Rapid access to data • Improved process performance • Improved data entry and database management
Computer-Based Training (CBT)	• Reduction in training costs • Improved employee effectiveness and performance • Improved cross-utilization of employees • Reduced duration to complete apprenticeship and operator-in-training (OIT) programs

Supervisory Control and Data Acquisition (SCADA)

Supervisory Control and Data Acquisition (SCADA) automation systems provide real time monitoring and control of equipment and processes. SCADA systems have been utilized by water utilities since the 1960s. Early systems used tone or pulse signals, radio, or telephone communication to monitor tank levels and to start and stop pumps. As microprocessors became increasingly powerful and affordable, remote telemetry units (RTUs) incorporated automatic control strategies. As personal computers came into common use, graphical user interfaces (GUIs) replaced lights and bar graph indicators, and programmable logic controllers (PLCs) replaced proprietary RTUs. Most current SCADA systems are based on the concept of distributed control, in which instruments for a given process or equipment cluster communicate with a close by programmable logic controller (PLC), or distributed control unit (DCU). This local monitoring and control device then communicates over a shared data highway (wiring) with other PLCs or DCUs and with a centralized monitoring and control work station. This hierarchy of local and centralized functions provides a great deal of flexibility and fault tolerance.

There are essentially three types of benefits provided by SCADA systems:

- Equipment startup, operation, and shut down sequences can be automated, which provides improved process control and efficiency, with resulting quality and cost benefits.

- Operators can be released from routine manual operations to attend to other critical tasks.

- Monitoring and control of energy and chemicals.

Many utilities that have implemented SCADA systems have been reluctant to implement significant automated control functions or to change operator work assignments. This may be due to a conviction that certain treatment processes cannot be automated, requiring instead a trained operator's eye to maintain acceptable treatment performance. Utilities may have also had poor experiences with early SCADA systems, particularly with costs of servicing proprietary software. Fortunately there are successful implementations from which to draw experience, including operating facilities that are highly automated and that reliably meet treatment objectives. These systems are frequently cited as references by the leading control system software vendors.

Automation Master Plan

The City of Houston, Texas, (Case Study 2.3) developed an Automation Master Plan that documents the direction of automated systems for SCADA and wastewater collection and treatment automation for a five year period. The plan developed a vision for wastewater operations, integrating several systems through the wastewater operations group, with the SCADA system as the enabling technology. Fourteen projects were identified in the plan over a five year period.

SCADA Benefits to Daytona Beach

The City of Daytona Beach, Fla., recently completed the contractual installation of an upgrade to the Ralph F. Brennan Water Treatment Plant Supervisory Control and Data Acquisition (SCADA) system. Installation and startup were

139

completed in January 1997. After the startup of this new system the city has four main areas that have benefitted:

Real time monitoring of treatment plant and satellite facilities operation:

• *Alarm system notification of unfavorable conditions or equipment failures*
• *Observing changes in treatment conditions*

Operation of plant processes without operator assistance due to computer logic control language:

• *Maintaining optimal system flows and pressures*
• *Efficiency in pump selection and operation*
• *Chemical application with greater reliability and accuracy*
• *Emergency conditions—opening elevated tanks and starting remote booster station due to system pressure drop*
• *Operating wells to meet treatment flow requirements and to meet operational requirements as to drawdown*

Opportunities also exist for going beyond a conventional SCADA system implementation and using additional technology to optimize treatment performance and the operator work definition. Many SCADA systems are now capable of being integrated with other network based applications such as:

• Advanced logging and notification systems that use alarm filtering, response rules, and paging or E-mail notification to reach mobile or remote workers

• Online operations guides and support material keyed to SCADA system events to provide context-sensitive reference material

• Simulation and decision support models, including those based on neural network technology (a form of artificial intelligence), that receive continuous real time data feeds, and that continuously adapt their algorithms to ongoing plant or system experiences

• Remote access methods that open up the full range of SCADA system monitoring and control functions to remote users with conventional computers and modems

In any application SCADA systems are only as good as the quality of data they obtain and how the control logic is programmed. Therefore SCADA technology requires technical support from the manufacturer and sound calibration and programming support.

Management Information System (MIS)

Effective management requires responsive and accurate information to enable decision making and monitoring of the status of the utility. Management information systems have evolved to meet the full information needs of management including cost tracking and control, purchasing and accounts payable, personnel tracking, asset inventory and tracking, accounting management, and use of database access to produce special reports in a timely fashion.

Maintenance Management

For many water utilities there is a significant opportunity to improve the effectiveness of operations and maintenance by implementing a computerized maintenance management system (CMMS). CMMS systems help streamline a variety of laborious functions including supplier identification, quote management, proactive spare parts and supplies ordering, and reliability-based preventative maintenance planning. There are many excellent CMMS software systems on the market. However, the cost of purchasing hardware and software for CMMS applications are only a portion of the total costs to implement a fully functioning, data-rich system that is utilized to the fullest extent of its capabilities. CMMS technologies are one of the greatest means to change a maintenance organization. Not only is the administrative aspect of maintenance management affected by implementation of change, but also procedural and staffing areas as well.

Information Access and Document Management

With the recent rapid adoption of Internet technologies, businesses have refocused on providing employees with online access to information previously locked up in paper manuals or in computer formats that require special proprietary software to access the information. Intranets are simply internal Internets—networks of computers using Internet technology but the information is secured and separate from the public Internet. The potential benefits of Intranets are significant because the technology can be applied in many areas including:

- Management information (as previously described under MIS)

- Online publishing of operations and maintenance procedures, as well as access to supporting materials from manufacturers and contractors. These links can extend across the Internet to include current information.

- Publishing and updating of policies, benefits information, and related human resources materials

levels and hours of operation

Reporting
- *The ability to select a wide range of operational data to maintain a high level of operation*
- *Generate regulatory reporting requirements. (Department of Health and Rehabilitative Services, Florida Department of Environmental Protection, United States Environmental Protection Agency)*
- *Increase maintenance productivity in areas of preventive and corrective maintenance*

Personnel Requirements
- *Reduce necessary manpower requirements by nearly 50 percent. Automate routine tasks. Examples: sampling, desludging of softeners, and chemical adjustments*

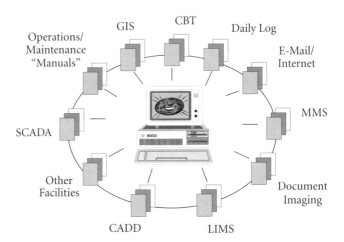

Figure 9-2 Intranet as a strategic center for information

- Access to accurate employee and location phone/fax directories

- Shared meeting calendars including room management and audio/visual resources

- Computer-based training with centralized administration and instructor support

- Gateway access to SCADA, Maintenance Management, and other similar systems. In this area Intranets are poised to dramatically alter the way in which software is designed, deployed, and executed in a networked environment. Quite simply, the web browser may become the universal interface for most specialized software functions, as shown in Figure 9-2. This is an especially significant trend for geographically dispersed organizations because Intranet based systems allow efficient remote access and management.

Customer Service and Billing

Customer service and billing systems are used to enhance superior customer service, support, and communications. These functions are typically tied together because they draw upon a common database of customer information, and because a substantial proportion of customer issues are related to billing. Many organizations have identified customer service and billing as a focal point, using customer service and satisfaction as one of the primary indicators of business success. The many opportunities include:

- Enhancing response to leaks in customer and utility systems

- Automatic payment of customer account bills through direct bank account debits

- Providing the customer with continuing audits of usage

Geographic Information Systems

Geographic Information Systems (GIS) use databases and specialized display technology to analyze and present spatially related data. Typically the data or analytical results are displayed on maps of a system or site. The GIS database consists of two major components:

1. The topological database, which represents the physical world (such as ground surface, roads, pipes, property lines, and hydrants) as points, lines, and shapes. A key feature of the topology is that every physical object has a relationship to its neighbors. Objects are connected to other objects. This is especially important when the GIS database is used to support simulation modeling.

2. The attribute database contains information on the objects in the topological database. For example, an attribute table on pipes contains information such as material, age, and estimated roughness. Attribute databases tend to overlap with databases for other applications, particularly for customer service and maintenance management. Accordingly, the most powerful GIS will be flexible enough to draw upon multiple data sources simultaneously.

The benefits of GIS are found in the applications defined for a given GIS. In other words, GIS is a tool with many potential applications, but as a base system benefits may not be sufficient to justify the substantial cost of data collection associated with implementation.

Potential utility applications for GIS include:

- Maintaining accurate system maps and related information on system features. This is the traditional starting point for many utilities.

- Analyzing watershed or water source management and protection alternatives.

- Providing a graphical interface to system simulation models.

- Providing a graphical interface for customer service and maintenance management functions.

- Optimizing routes for meter reading and main flushing.

- Geographically analyzing system data such as rates, lost water, main breaks, pipeline flow rates, and water quality.

INFORMATION TECHNOLOGY BEST PRACTICES

This section describes examples of IT use by water and wastewater utilities that may be considered best practices. Because information technology is continuously improving and the potential for innovation is virtually unlimited, current industry practices should be studied. A good starting point is the AWWA home page on the World Wide Web (http://www.awwa.org), which provides direct information on water utility practices and links to numerous water utility web sites.

The following examples illustrate innovative approaches. However, successful use of information technology depends on following a good planning and implementation process.

Examples of Maintenance and GIS Integration

Computerized maintenance management systems (CMMS) have long been recognized as a powerful tool to help streamline maintenance functions. More recently, following on the heels of large municipal installations, water utilities have implemented GIS to provide more accurate distribution system maps and to provide more convenient, intuitive access to data that are spatially distributed.

In some cases the CMMS and GIS systems have been integrated. The purpose of integrating these systems is primarily to enhance system maintenance and customer service functions. For example:

- Work orders for maintenance, meter reading, or emergency dispatch can be referenced to a GIS location so that maintenance workers can directly retrieve map views pertaining to the work site.

- Customer service staff can retrieve the status of work orders for a particular address or view the incidence of complaints or work orders, by type, on a color coded map display.

- Maintenance supervisors and engineers can use the spatial display capabilities of the GIS to schedule more effective maintenance or replacement programs based on system features and maintenance history.

Example of SCADA System Enhancement

SCADA systems provide real time information to operators on processes and equipment. In typical SCADA system implementations the operator interface is a series of graphical displays depicting the operating facilities on a summary or detail basis. Properly operating equipment and processes are depicted in one color; malfunctioning equipment or out-of-limit process parameters are depicted in another contrasting color. Alarm conditions are shown on the graphical displays and as a chronological list, usually color coded by alarm type.

While adequate, the typical SCADA interface does not significantly change the role of the operator as a passive observer of system events. By combining SCADA systems with general purpose database and communications capabilities, the operator's role can be enhanced to be more active and accountable.

145

In an example from the wastewater treatment industry, the Hamilton Township Division of Water Pollution Control in New Jersey has implemented an enhanced SCADA system at its wastewater treatment facility. This system includes an electronic operator's log linked to the SCADA system with the following features:

- The electronic log takes inputs from the alarm output of the SCADA system, manual responses and entries by operators, manual entries by supervisors, and electronic mail. All entries in the log are date and time stamped, providing a detailed chronology of events and responses.

- Operators log in to the system and are identified as part of the log.

- SCADA system alarms are filtered, coded, and displayed with different responses required by the operators depending on alarm type. The bulk of alarms are for information only. Some require acknowledgment only, while others require acknowledgment and operator response in the form of a manual log entry. This facilitates the review of alarm and operator response procedures to evaluate the effectiveness and appropriateness of the actions taken.

- Supervisors may create must read log entries that appear for all operators. The system tracks who has read the must read entries.

- The electronic log is integrated with a plantwide e-mail system so that operators can conveniently extend their communication from the same interface.

The electronic operator's log is implemented as custom software, with information stored in conventional database files on the plant's network file server.

Example of Intranet and Document Management System

Using Internet technologies in a private Intranet may be one of the fastest, least expensive ways to improve access to information within any large, complex, or dispersed organization. The technology is particularly appealing because it can be implemented in a modest fashion, with immediate benefits and can be improved incrementally over time.

Initial Intranets implemented in most organizations have focused on providing online access to information, normally provided in printed form that requires frequent updating, such as employee benefits and policy information, internal phone directories, and basic information about the organization, its mission, teams, and projects. In second generation Intranets, limited information contained in human resources and financial databases has been made accessible through database gateways. These early implementations, which integrate the simple Intranet web browser interface with data from traditional applications, indicate how Intranets may become the central access point for information from many varied sources.

A moderately sized water commission in the greater Chicago metropolitan area is implementing an Intranet-based information access system (IAS) that includes traditional document imaging and management functions, and that provides a more complete information resource for operators and maintenance staff.

The first part of the system consists of an online operations and maintenance manual based on conventional Internet standard HTML (Hyper Text Markup Language) format. The HTML format provides easy access, better organization and a centralized place for updating procedures. The Intranet-based manual has a table of contents, full text search, hypertext links connecting related topics, and links to supporting material including drawings, manufacturers' equipment and procedure information, and construction contract submittals.

The second part of the system consists of a document imaging and management package. Reference documents available only in paper form are scanned, indexed with keywords for document type, source, and content, and then stored as images and data. Optical media are used for long-term permanent storage and archiving.

The third part of the integrated system is a database and image gateway that connects the HTML-based pages to the images in the document management system. The gateway allows

A Florida Intranet

West Coast Regional Water Supply Authority (WCR-SWA) is Florida's largest provider of wholesale water, serving Hillsborough, Pasco, and Pinellas counties and the cities of St. Petersburg, Tampa, and New Port Richey. It owns or manages 15 water supply facilities, producing about 42 billion gallons annually with an average of 116 mgd.

WCRWSA developed a prototype web-based Intranet application for using Microsoft's Visual InterDev, a high-end web application development tool. The application gives all staff access to more than seven years of rainfall and wellfield data collected from monitoring sites. The information will be used for more effective modeling and analysis and better planning.

The Intranet approach to information sharing has many benefits including ease-of-use through a simplified, common interface; accessibility from multiple locations at any time; sharing

common informa-
tion in an electronic
format for importing
into other software;
and a simplified
query process for spe-
cific data—resulting
in improved produc-
tivity and planning.
Intranets have
become more than
an enabling technol-
ogy, they have
become a sustaining
technology and an
integral part of the
utility enterprise.

these documents to be incorporated into the online manual as hypertext links. In this fashion, Intranet manual users can access procedures and supporting material from a single consistent interface.

Examples of Internet Web Sites

Organizations that need to provide public information and to serve a diverse customer base can benefit by establishing an Internet web site. The AWWA currently maintains a list of several hundred Water Utility Home Pages. A review of these sites indicates many are in initial stages of development.

The most common uses for web sites are to make available basic information about the water utility, its services, and customer service procedures. For example:

- The City of Los Angeles Water Services site (http://www.ladwp.com/water) provides a menu of information areas, including answers to frequently asked questions and a detailed description of how requests for service are handled.

- The Seattle Public Utilities web pages (http://www.pan.ci.seattle.wa.us/seattle/util/dw/watrinfo.htm) provide information on the system's water sources, distribution, and treatment; water consumption statistics; conservation advice; and links to other area sources of water information.

- The Henrico County Department of Public Utilities in Virginia web site (http://www.co.henrico.va.us:80/utility/) has similar basic information, but also provides an online form for customers to request information as an alternative to telephone-based customer service.

Beyond the water industry there are numerous Internet web sites that have provided an interactive, customer-focused experience designed to enhance customer intimacy and communication. For example, the Federal Express site, the most popular web site in the United States, gives customers access to real-time information on the status of their packages shipped via Federal

Express. In a similar fashion, water utilities could give customers "real-time" access to water usage, billing/accounts, and customer complaint information.

SUMMARY

This chapter on IT describes an approach for considering the opportunities available to water utilities to improve the efficiency of a wide variety of processes, functions, and activities. The correct application of IT is a critical element in achieving and managing a higher level of performance and improved efficiency by increasing productivity at all sizes of water utilities.

The best results are achieved by the application of IT to water utility functions that have been strategically planned and well thought out by the technology experts working with the staff that is expected to use it. Training staff in IT is critical to a successful application. The use of IT, including automation, is no substitute for making necessary improvement in the performance of a particular unit or function.

Given the critical supporting role of information technology, the challenge for water utilities is to select the right technologies at the right time, to realize the promised benefits, and to facilitate future improvements. By following a cyclical process of self-assessment (defining the organization's mission and vision), strategic and annual planning, implementation, and measurement, utilities can create an intelligent context for numerous technology decisions.

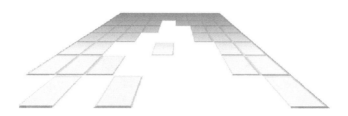

CHAPTER TEN

Creating an Energy Efficient Utility

Electrical energy is a critical resource to water utilities—both because of its fundamental role in water treatment and delivery, and because it represents a significant expense for most utilities. While many utilities are aware of the importance of electricity to their operations, a surprising number of utilities do not possess a detailed knowledge of facility and equipment energy use characteristics, and are not fully optimizing their use of energy according to current utility pricing practices. Under these circumstances, substantial major gains in efficiency and cost reductions may be achieved through investments in more efficient equipment, and through automated control systems that allow time of use to be matched to utility pricing schedules. These actions constitute the traditional approaches to utility energy managment, and are a major focus of this chapter.

At the same time, the electrical industry is in the midst of fundamental restructuring that will irrevocably change the way energy is sold, priced, and purchased. These changes will affect large energy users like water utilities during the next five years. The changes constitute both a threat and an opportunity. A threat because the rules are changing and adaptation is required to stay ahead, and an opportunity because there are ways to gain economic advantage even in the absence of investment in energy efficient equipment and controls. Deregulation is the other major focus of this chapter.

A thoughtful and balanced approach to both deregulation and the efficent use of energy together constitute the concept of an *energy efficient utility* and the subject of this chapter. Such a utility will adopt a strategic approach to optimizing its energy usage, resulting in significant cost savings while ensuring that its core mission of delivering water and protecting public health and safety will continue to be met without compromise.

Becoming energy efficient will require a new focus on time-of-day energy use and a new approach to planning and scheduling. This approach, under the direction of a skillful staff, will also manage the minute-to-minute operating systems, the longer range scheduling of equipment maintenance and construction, and the external purchasing of energy from power marketeers, brokers, and energy service providers. Energy efficient utility management of the future will also routinely incorporate alternative energy sources (e.g., natural gas and self-generation), load management flexibility (e.g., elevated storage) and possibly demand management incentives (e.g., time of use water rates) into their practices. The use of these emergent capabilities will ensure real-time flexibility for energy purchasing at minimum costs.

Before we describe the path to becoming an energy efficient utility, we need to spend some time gaining a more complete understanding of deregulation and its implications for energy purchasing and pricing.

UNDERSTANDING RESTRUCTURING OF THE ELECTRIC POWER INDUSTRY

The electric energy services business is a $250 billion industry. It is undergoing significant change, similar to other industries that have deregulated including the telecommunications, banking, trucking, and airline industries. However, the energy industry restructuring is likely to be the most dramatic of these due to the sheer size of the industry and the crucial role it plays in contributing to the nation's vitality.

Likely Changes in Purchasing Practices

Competitive purchasing of electric energy followed passage of the Energy Policy Act of 1992, which mandated wholesale "wheeling" of energy across the nation's transmission system, open transmission access for independent generators, and encouraged the development of the inexpensive gas-fired generation. States like California, Pennsylvania, New York, and Massachusetts are leading the way in designing a competitive market to replace the regulatory environment.

Restructuring requires "unbundling" of electric energy services, which breaks up generation, transmission, and distribution into separate functions. Those who are placing the industry on the slippery slope of deregulation believe restructuring will also create new services and markets that could expand the entire industry. No longer will there be only a single vertically integrated service provider, but rather several providers who offer unbundled services and others in a competitive market.

Electricity services will become similar to telephone services. That is, your local electric service provider will become the operator of the local electrical distribution system, providing you with a regulated service akin to local telephone service. Energy, like long distance phone service, will be unregulated and subject to competition and other forces of the marketplace.

Deregulation will require electric energy customers to decide how to "rebundle" the parts of the electric service. There will be many options for integrating electric services:

Remain a local electric utility customer. The local electric utility will purchase energy in the market and provide all of the related services for the water utility. Because the local electric utility will remain regulated, its rates will be controlled by the state's utility regulator.

Contract or commit to local, regional, or special interest partnerships. The water utility can join a larger group with similar interests that will leverage buying power and technology needs to gain a better deal. This will allow the water utility to obtain electric services competitively, although the rates will be based on the group's ability to manage and commit to energy load. To ensure they receive the best rates, water utilities should

153

review the capability and commitment of others in the group to shift load or otherwise meet the purchase commitments.

Do it yourself. This is for the water utility with skills and technology to compete for the purchase and management of energy. There may be some significant benefits of doing this, but the water utility should have control over the use of energy by maintaining optional energy sources.

Competing electric service providers are now emerging to provide a full range of electric energy services, moving from energy supply contracts to metering/billing and load management services.

Likely Changes in Pricing Practices

Energy costs may vary dramatically. Initially there may be transitional costs to allow for the change from a regulated to a deregulated industry along with separate charges for each function (e.g., generation and transmission). An example from California for the interim period between regulation and full deregulation has been estimated as follows (this is meant to be illustrative of the breakup and may not represent actual charges):

Components of an Average Electric Energy Invoice	
Energy Services (for generation)	25%
Transmission Services	5%
Distribution Services	25%
Stranded Investment Recovery	40%
Public Benefit	5%

The energy component of the utility's energy bill will depart most substantially from current circumstances because it is the only part of a customer's bill that is subject to the dynamics of supply and demand. It is possible that the cost of electricity could vary by a factor of 40 between off-peak hours and peak hours, as illustrated in Figure 10-1. On the other hand, pricing could more closely resemble current time-of-day price variations.

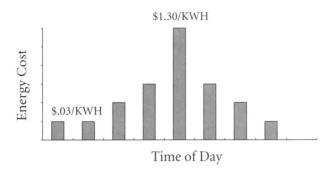

Figure 10-1   Example of electric prices varying with time of day

Responding to these fluctuations, whatever they may be, is where the opportunities for long run cost reduction exist. Those who use demand planning and scheduling to anticipate the frequency and magnitude of changes in electricity costs will avoid peak costs and take advantage of lower cost periods.

THE PATH TO BECOMING AN ENERGY EFFICIENT UTILITY

Your strategic approach to becoming an energy efficient utility will most likely include five key elements.

Reassessing your capital investment and operating practices related to energy using equipment.   Investigate the investments and practices in all utility facilities that require energy or affect energy purchase and use (e.g., efficient motors, starter devices, variable speed drivers, ozone systems, building HVAC and lighting, and water storage). Consider revising purchasing practices to include life cyle measures of costs and uncertainty in energy prices.

Exploring opportunities for automated approaches to system controls and decision-making.   Many control systems minimize energy costs while improving overall water system performance and cost without any additional investment in more efficient equipment. SCADA and Decision Support Systems provide effective system operations and scheduling and are very cost effective from a long run perspective.

Making energy a focal point in the utility. Include energy purchases and energy-related capital investments in the list of top issues brought to management. Perhaps you might incorporate energy investments into the utility's risk management framework and decision structure. Include energy costs in all activity-based reporting and analysis.

Preparing for energy purchasing in a restructured electric power industry. Begin to educate the organization about current changes and those expected in the near future. This allows the water utility to plan for its future contracts or commitments with local electric utilities and/or third party energy suppliers.

Adding staff expertise with energy experience. Begin to add the type of expertise capable of developing your strategy for purchasing electric energy and managing its use. This may be comparable to the stock market where the spot energy price fluctuation is real-time, requiring new information systems and skilled workers for trading and decision making.

MAKING A PLAN

To address the challenges and to capitalize on the opportunities created by deregulation, it is best to prepare an energy management plan to determine the best options to pursue and to assess the risks and benefits associated with those options. The creation of a plan will assist a utility in profiling its energy demand and, ultimately, in optimizing its energy use and service delivery. Figure 10–2 identifies the steps in creating a plan.

These strategic elements are most readily accomplished in several steps.

Step 1 – Perform an Energy Efficiency Assessment

Major power uses and definition of the power demand profiles for each use should be identified. I.e., when is power used during the day and how does this profile change with the day of the week, the season, or other factors. Reasons for the energy use should

156

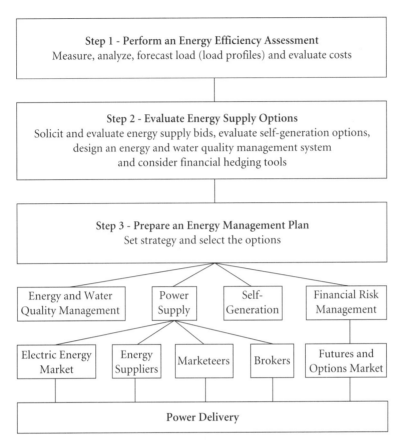

Figure 10-2 Creating a plan

then be determined to see if it is possible to shift the time of power use without impacting the water utility's performance.

This step is likely to be significant for many utilities that may not have any information other than that provided by the local electrical utility. Local electrical energy bills usually arrive monthly and may include only maximum demand and total kwh use. Figure 10–3 shows a typical water utility and the functions that are candidates for energy auditing.

- *Source*—well pumps, river pumps, or raw water transmission pumps

Figure 10-3 Typical water utility energy demands

- *Treatment plant*—ozone system, backwash/surface wash pumps, facility and grounds equipment, lighting, HVAC

- *Distribution system*—pumps

- *Administration/service offices*—facility lighting and HVAC

A more detailed assessment that may require system component audits (e.g., pumps and motors) is needed to fully establish an inventory of energy use within a system operation. Finally, the utility should compare energy use data and practices gained from this assessment with best practices in other utilities to develop performance measures. This comparison should be performed to identify key differences between the utility and the best in class. This assessment is likely to uncover recommendations for immediate action as well as establish a base of information for determining how and when to purchase energy in the future.

Step 2 – Evaluate Energy Supply Options

The assessment in the first step provides the basis for soliciting proposals from energy suppliers, power marketeers, and others to be your energy supplier. Your demand profile plus a projection of future requirements that may change must be developed. The level of precision in this load profile is important to the evaluation of the proposals. You should be prepared to identify the confidence level in your projections.

Step 3 – Prepare an Energy Management Plan

The utility's plan begins with the identification of its mission and goals, key performance measures related to electric energy, and its strategy regarding commitment to energy efficiency. The plan should not sacrifice customer service but should ensure it. The energy management plan will provide an ideal opportunity to review and update best practices to fit the new electrical energy purchasing process. These practices will include those discussed in this chapter, along with others tailored to the specific utility needs and, in the process, reveal necessary changes in the way the utility operates.

Based on these straightforward steps and the utility's commitment, an energy efficient competitive organization will emerge. A key to success is the staff's willingness to change the way the utility does business. Table 10-1 presents the paradigm shifts needed for competitive utilities to become energy efficient utilities as they prepare for the future.

Energy efficient utilities have effective operational practices. Developing a plan to prepare for the future and making the paradigm shifts identified in the table will take time. Some utilities have already initiated this transition. Some best practices are emerging and more will take shape as the electric industry restructures and the impacts on water utilities are realized and measured. Will water utilities that have not prepared for restructuring receive significant increases in their electric bill? Yes, and when this occurs changes will quickly develop in reaction to the sticker shock of having maintained the status quo. In the meantime, the following best practices are good candidates for water utilities to consider adopting:

Table 10-1 Becoming an energy efficient utility—paradigm shifts

FROM	TO
• Energy load unmanaged	• Energy load planned and committed
• Energy consumption decisions by many	• Energy commitment integrated
• Technology is an obstacle	• Technology is an enabler
• System operation by reaction	• System operation planned
• Organization discourages best performance	• Performance based organization

Load Profile. Energy load profiles of all energy components should be developed in at least hourly to half-hourly (or less) increments. Accounting for all combined energy uses provides an initial starting point to predict and control energy use and cost. The energy demand should include subcomponents of the demand for predicting and measuring performance. This includes audits of the components to verify efficiency, metering, and submetering or surrogate measures, and understanding alternative measures to energy use.

Load Management. How can utilities forecast and commit to controlling load? Load management requires the capability to monitor and control energy use in real time. Effective load management requires enhancements to the SCADA system and an organization with skilled workers. Best practices include a staff focused on overall system performance, using effective tools to balance energy costs with water quality and other service performance standards of the utility.

Development of Technology. Utilities need sophisticated tools and software that can meet long-term needs and be competitive yet affordable to many users. This will require water utilities to collaborate and engage in partnerships in the development of technology. A sophisticated information management system is needed to track and control energy use, plans, and expenses. Other innovative technologies might be developed by utilities working together and with the electricity industry.

Energy Purchasing. Utilities are considering purchase pools, third party contractors, and other procurement options. Best practices will vary based on the unique needs and capabilities of each utility. The general practice will be to balance energy management and use with load commitment, according to conditions in real time. Using sophisticated tools and staff will provide the opportunity to spot purchase electricity and take advantage of best pricing. Others who do not know how to predict and control usage or who cannot commit may need to have a third party or their existing service provider assume the risks of load commitment by balancing it over a large consumer base.

PURCHASING ENERGY IN A NEW ENERGY ENVIRONMENT

Practices to consider in planning for energy purchases in a deregulated environment depend on size of the utility. Large utilities may approach this challenge by developing a capability to manage power demand and energy purchasing. Smaller utilities may need to unite with other utilities or within their municipal structure to contract for services from energy service providers. The following optional structures of these arrangements are shown in Figure 10-4, where the ⚘ represents all of the components of a water utility, from source to tap including pumping, treatment, storage, and transmission. In the center of this symbol is an information and control system that links all of the components and provides the intelligence to plan and schedule energy purchases. Other symbols such as ❋ represent other energy users; e.g., airports, big buildings, and commercial facilities. The other structures are illustrative of options that are likely to develop for water utilities and others. An important feature in these structures is the need to manage the information according to energy need and scheduling. In the past this task has been performed by the local energy service provider.

ENERGY EFFICIENT UTILITY EVALUATION

How does your utility rate on the energy efficiency scale? Utilities can be ranked from excellent to innocent to provide an initial

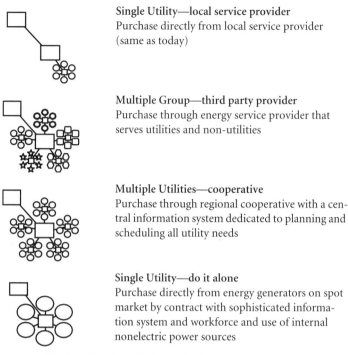

Single Utility—local service provider
Purchase directly from local service provider
(same as today)

Multiple Group—third party provider
Purchase through energy service provider that
serves utilities and non-utilities

Multiple Utilities—cooperative
Purchase through regional cooperative with a central information system dedicated to planning and
scheduling all utility needs

Single Utility—do it alone
Purchase directly from energy generators on spot
market by contract with sophisticated information system and workforce and use of internal
nonelectric power sources

Figure 10-4 Optional electrical energy purchasing structures and water utility energy sources

assessment of where the utility is relative to others. This ranking
also provides guidance to where the utility can improve.

Table 10-2 presents a description of factors by energy related
categories that distinguish the best in class from the ineffective.
This table can be used by utilities to perform a self test using the
rating sheet shown in Table 10-3, by rating individual categories
and scoring the one that best applies. The overall utility rating is
determined by totaling the points and referring to Table 10-4.
This rating process is greatly simplified but intended to provide a
guide in support of each utility preparing for restructuring.

Table 10-2 Energy Excellence Grid

	Strategy	Planning/Scheduling	Maintenance	Performance Measures	Information Technology	Workforce
Excellence	• Committed to planned Operations and Maintenance • Understand and establish leveraged position in competitive market for electricity • Develop and effectively use strategic partnerships • Provide information services and incentives to customers to achieve best service value • Use technology to enable workers to perform best practices efficiently	• Forecast energy demand and is almost always able to predict and control load within a few percent of actual on a 30-minute basis • Schedule facilities operation and maintenance to almost always effectively and efficiently meet demand • Energy buying and selling based on ability to commit to load	• Audit facility components as a system to maintain knowledge of system performance • Retrofits facility components by priority of rate of return on system-wide value	• System reliability exceeds expectations • Energy purchase and sales better than expectations • System-wide cost is lower than expected • Water quality to customer meets all regulatory requirements and is better than customer expectations	• Provides the following applications/systems capabilities and assumes leadership in the industry to promote their cost effective growth: - SCADA/Process Control Systems - Computerized Maintenance Management System - Energy/Water Quality Management System • Establishes and maintains standards to ensure effective and efficient sharing and integration of data, systems, and components	• Best employees who are flexible, knowledgeable and committed workers fully capable to exceed company expectations • Workers perform a wide range of tasks with broad skills • Motivations to excel with features such as pay for performance apply to most employees

table continued on next page

163

Table 10-2 Energy Excellence Grid (continued)

	Strategy	Planning/ Scheduling	Maintenance	Performance Measures	Information Technology	Workforce
Competence	• Most Operations and Maintenance activities are planned • Work with others to provide a leveraged position in a competitive market for electricity and/or maintain effective contract for some/all energy needs • Use some strategic partnerships • Use some technologies to enable workers to perform best practices efficiently	• Forecast energy demand and most of the time is able to predict and control load within a few percent of actual on a 30-minute basis • Schedule facilities operation and maintenance to effectively and efficiently meet demand most of the time • Energy buying and selling at key locations to commit to load	• Audit key facility components as a system to maintain knowledge of system performance • Retrofits key facility components by priority of rate of return on system-wide value	• System reliability meets expectations • Energy purchase and sales meet expectations • System-wide cost is as expected • Water quality to customer meets all regulatory requirements and most customer expectations	• Uses competitive technologies for information systems - SCADA/ Process Control Systems - Computerized Maintenance Management System - Energy/Water Quality Management System	• Highly qualified employees are flexible, knowledgeable and committed workers capable to meet many company expectations • Workers perform many tasks with broad skills • Motivations to excel with features such as pay for performance apply to many employees

Table 10-2 Energy Excellence Grid (continued)

	Strategy	Planning/ Scheduling	Maintenance	Performance Measures	Information Technology	Workforce
Understanding	• Some Operations and Maintenance activities are planned • Has contracts and/or partnerships for energy • Understands technology to enable workers to perform best practices efficiently	• Can forecast/ predict energy demand on an average daily basis for major energy uses • Has understanding of operating facilities to schedule effectively, efficiently meets demand and schedules major components • Has understanding of energy demand and average use	• Audit facility components as a system with some understanding of inputs on system performance • Retrofits facility components by priority of rate of return on system-wide value	• System reliability generally meets expectations • Energy purchase and sales generally meet expectations • System-wide cost is understood • Water quality to customer typically meets all regulatory requirements and customer expectations	• Understands information systems technologies and applies some such as: - SCADA/ Process Control Systems - Computerized Maintenance Management System - Energy /Water Quality Management System	• Flexible, knowledgeable workers usually meet company expectations • Motivations to excel with features such as pay for performance apply to some employees

table continued on next page

Table 10-2 Energy Excellence Grid (continued)

	Strategy	Planning/ Scheduling	Maintenance	Performance Measures	Information Technology	Workforce
Awareness	• Operations and Maintenance activities are occasionally planned • Has contracts and/or partnerships for some energy needs	• Typically knows daily water demand but may not know specific energy demand/use • Schedules some operating facilities and typically meets water demand but not aware of energy demand • Energy buying and selling based on system needs except for large components that are operated off peak rates	• Audit facility components as a system to maintain knowledge of system performance • Retrofits facility components by priority of rate of return on system-wide value	• System reliability sometimes questioned • Energy purchase and sales based on need and not expectations • Water quality to customer generally meets regulatory requirements	• Uses nonautomated technologies such as time clocks and remote start telemetry	• Workers generally only have moderate training, skills, or tools to make best use of worker productivity

Table 10-2   Energy Excellence Grid   (continued)

	Strategy	Planning/Scheduling	Maintenance	Performance Measures	Information Technology	Workforce
Innocence	• Reacts to system operation/ maintenance needs without consideration to energy costs • Not aware of the implications of deregulation on the utility	• Forecasts are daily average and within capabilities of components to react to instant changes • Schedules major operating facilities and typically meets water demand but not aware of energy demand • Energy buying and selling based on system needs except for large components that are operated off peak rates	• Audit facility components as a system to maintain knowledge of system performance • Retrofits facility components by priority of rate of return on system-wide value	• System reliability does not always meet expectations • Energy purchase and sales based on need • Water quality to customer usually meets regulatory requirements	• Uses nonautomated technologies such as time clocks	• Workers generally have only limited training, skills, or tools to make best use of worker productivity

Table 10-3   Energy efficiency self test rating sheet

Activity	Best in Class 5 Points	Effective 4 Points	Conventional 3 Points	Basic 2 Points	Ineffective 1 Point
Strategy					
Planning & Scheduling					
Maintenance					
Performance Measures					
Information Technology					
Workforce					
Total Score					

Table 10-4   Overall energy efficiency rating

Total Score	Category
26–30	Best in Class
21–25	Effective
16–20	Conventional
11–15	Basic
10–5	Ineffective

ENERGY AND WATER QUALITY MANAGEMENT SYSTEMS

Water utilities of the future will require new technology to manage water quality and to optimize energy consumption. To achieve the maximum benefit, water utilities will require an advanced automation system. A project developed by the AWWA Research Foundation is focused on developing this technology. It is called the Energy and Water Quality Management System, or EWQMS. It is described here to provide an example of the complex requirements anticipated to accommodate energy efficiency for the excellent utility of the future.

An EWQMS integrates technology, organization, and people to provide a systems approach to water operations and management.

EWQMS, a group of individual software programs that collectively provide information to individuals who use the computer systems to solve the utility's water quality and energy management problems, provides users with information to prepare plans for daily decision making. These plans are developed based on optimization and simulation techniques embedded in the software programs.

While these software programs are usually thought of as part of the Supervisory Control and Data Acquisition (SCADA) system (and do require data acquisition from the SCADA system), these programs are typically used by staff outside of operations. For example, operations planners, water quality engineers, energy use monitors, and planners use these programs to solve the least cost operating problem from the perspective of the entire water system while also complying with water quality constraints. AWWARF is leading the project to develop this software. The full development will include many building blocks as illustrated in Figure 10-5.

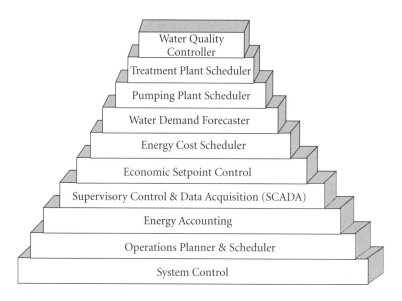

Figure 10-5   Energy Water Quality Management System

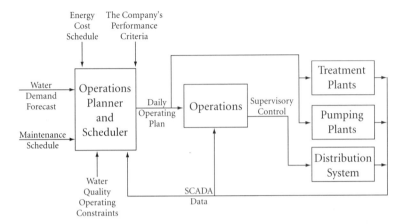

Figure 10-6 Functional overview of EWQMS

Building the EWQMS requires a building block approach, with basic blocks at the bottom and more sophisticated blocks at the top.

Figure 10-6 shows the functional overview of the EWQMS. The operations planner and scheduler develops the daily operating plan for the entire water system. The daily operating plan contains the production schedule for the treatment plants and prepares the pumping plant schedules. The operations planner and scheduler develops the daily operating plan using data from a maintenance scheduler, a water demand forecaster, equipment and facilities status information gathered by the SCADA system, input from an energy cost scheduler, and information about how the utility's management wants things done—the utility's performance criteria. Operations, the traditional function carried out by the operators, uses the daily plan received from the operations planner and scheduler. One of an operator's major functions is to compare the actual operation of the water system against the daily plan by monitoring information about the current operating status of the water system using the SCADA system. To help ensure the water system follows the plan, they use rate control values and other supervisory control capability (through the SCADA system) to guide the actual water system operation according to plan.

They also have responsibility to respond to emergencies that may supersede the daily plan.

The treatment and pumping plants are at the right of Figure 10-6. Treatment plant operators receive the daily operating plan directly from the operations planner and scheduler. When necessary, operators use SCADA to control the pumping plants and equipment in the distribution system to guide operation of the water system back to the plant. Both the operators and the operations planner and scheduler receive status information via the SCADA system.

Because of technology advances, all of this information transfer occurs instantly. A daily operating plan is prepared and distributed to the treatment plants, pumping plants, and to the operators. The operators monitor the entire water system to make sure the plan is followed. When the equipment strays, the operators use rate control valves and direct control of equipment in the distribution system to get it back on track.

SUMMARY

Increasingly, the electric bill will become a target for the water utility manager, or it will become a target for his/her critics on the utility's governing board and in the press and public. The effort required to reduce energy consumption and to negotiate advantageous rates will be substantially rewarded by lower operating costs and the respect awarded to cost-effective managers. Many water utilities have already responded wisely and aggressively to the challenges and opportunities of energy/energy-cost management. Utilities that have not yet followed suit are well-advised to do so.

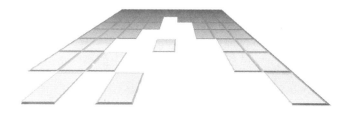

Tools for Change

Foreword To Part 3:
Tools For Change

Sometimes talk of strategic plans and quality management causes eyes to roll and stomachs to heave—if you have experienced more talk than action or if you have seen much energy expended in spinning wheels with no forward movement. Based on good experience, the approach proposed here looks at planning in a different light. Here you will see planning as a critical, active step in changing your organization's performance. When plans are not taken seriously they really don't matter. When employees operate as they always have, the plan on the shelf makes no difference. Here we talk about planning that everyone is going to have to live with.

We follow up on chapters 4 and 5 with rather detailed presentations on strategic thinking and planning, performance assessment methods, and procedures to improve your work processes.

Chapter 14 provides valuable insights into our innate fear of change and suggests techniques through which you can help employees cope with the unsettling experience of organizational change. Managers who are insensitive to the human dimensions of restructuring organizations are in for some surprise difficulties. You can't get the job done without taking employee fears seriously and responding helpfully. On the other hand, managers who are so keenly aware of employee anxiety and resistance that they are reluctant to change will also find help in this chapter.

The three other tools for change discussed in the following section concern specific changes in the way most utilities have conducted their business in the past. Organizations considering outsourcing their water system's operation and maintenance will find guidance on how to successfully manage those contracts in chapter 15. Similarly, chapter 16 offers advice on managed competitions. Chapter 17 explores several experience-based processes for public-private partnerships in Design Build Operate (DBO) and Build Own Operate Transfer (BOOT) arrangements. Most utilities are at a disadvantage in entering any of these contractual arrangements because they lack experience with them, while the would-be private partners across the table are quite savvy and experienced. We believe the advice given here will educate utilities to act in ways that protect the public interest.

These chapters deal directly with outsourcing certain utility functions. The authors emphasize that we are not recommending outsourcing to all utilities, but we do believe that this emerging practice should be examined with an open mind. For many utilities it can lead to substantial efficiencies and cost savings. The decision in each case is truly a local one and we trust these chapters will aid in your consideration of possible alternatives.

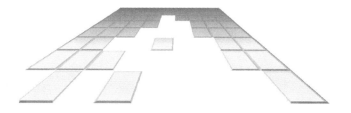

CHAPTER ELEVEN

Strategic Thinking and Planning

A modern organization must know where it wants to go and how to position itself in today's world. It then needs a plan and an organization structured to achieve the goals desired. Staying effective and relevant in the constantly changing times we face is an awesome challenge. The speed and comprehensiveness of change is revising organizational development thinking in the business and academic worlds, so it is not surprising that public sector organizations have difficulty staying current when businesses as strong and powerful as IBM and Apple stumble. Most thoughtful theorists of organizational structure address organizations comprehensively and emphasize the interaction of leadership, planning, customer attitudes, competent employees, appropriate controls, and systematic monitoring. When these aspects of an organization are tied together with coherence in a system, the organization is likely in productive alignment.

In this chapter we will talk briefly about the meaning of strategic thinking and how it relates to organizational development, then lay out the issues in designing a strategic plan to meet today's management problems, and finally, discuss how strategic thinking is tied to work process redesign to achieve a comprehensive self-improvement process. This chapter expands the concepts in chapters 3 and 4 to make them more useful tools for the utility manager. We also address ways to use planning to give coherence to an organization both in setting a vision and in laying out a road map. We also explore problems that will be encountered and ways to overcome them.

STRATEGIC THINKING

Strategic thinking means that an organization strives for relevance and focus, direction, and the capacity to achieve intended results. In order to achieve intended results it is essential to align the goals, functions, systems, and structure of an organization. The premise of this book is that organizations are facing change and that they must become competitive and customer oriented. In the recent past an organization could look to a simple mission such as "deliver clean water 24 hours a day," and organize around that goal. The goal was well suited to a command and control style of organization that had clear procedures and structures, where leadership could make a decision and see that it was carried out down the ranks. As the world becomes more complicated and organizations have to contend with multiple goals and a variety of clients and customers, the goals and organizational structure become more difficult to define and to manage.

New theories are beginning to emerge that discuss a "sense and response" style of management.[1] See Figure 11-1. New organizations that are trying to change have to think in terms of responsiveness and flexibility. Sense and response requires leadership that can lead conceptually, employees that can work responsively, and strong strategic systems that coordinate and align internal functions. Sense and response also requires new business behaviors that involve continual learning and response to constant change.

Strategic thinking is a way for organizations to begin dealing with their potential futures. Strategic thinking is a way to think and the concept of how to think. It is bigger than a plan. It means that the entire organization has engaged in thinking about and working

---

[1] Stephen Haeckel, "Adaptive Enterprise Design: The Sense and Response Model," *Planning Review*, May–June 1995.

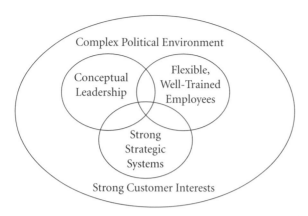

Figure 11-1 Sense and response management

on what it means to be responsive, creative, and competitive. It also means that the entire organization is committed to achieving corporately adopted goals that are designed to build a strategic intent that all employees understand. Only by having a strongly held corporate culture that understands and is committed to a set of principles and goals can employees be empowered in teams or as individuals to pursue and carry out a range of programs to be responsive to customers. If employees lack a strong sense of corporate goals and culture they will be empowered without clear direction or focus, which can be chaotic and anarchistic. Strategic thinking and strategic intent hold the organization together and direct people toward common goals.

Essential to developing strategic thinking in an organization are:

Leadership. Leadership in a creative organization must be highly conceptual and intuitive. Top leaders need to understand the direction of the business, pay careful attention to trends in society and the marketplace, and be able to translate the trends and ideas into actions that the organization should take. Leadership also requires enlisting and aligning all functions of the organization where cooperation, compliance, or teamwork is necessary to achieve the organization's mission. Leadership requires the ability to highly motivate the people who are critical to the implementation of the organization's strategy.

Planning system.   The planning system should be strategic and identify the vital goals of the organization and the appropriate methods of achieving them. The planning system should also identify critical success factors and include budgeting, accountability, and monitoring systems.

Employees.   Employees need to be competent, empowered, and highly flexible to be able to respond to customer needs in a customized way. Although it is necessary to have a core set of rules and regulations, the new organization needs an employee group that is not so bureaucratized by rules and regulations that it can no longer respond to perceived needs. Employees are treated as a critical resource and given appropriate training to keep their skills up to date and relevant.

Controls and oversight.   The control and oversight systems of a strategic organization will be coordinative and sensory, not command and control. A minimal amount of control and policies and procedures are necessary in any organization to keep it from being chaotic and to institutionalize goals so they are widely understood.

Systemic monitoring and communication.   In a strategic organization one must monitor the conditions and satisfaction of the community, measure performance, and respond with changes as needed. Internal and external communication must be viewed as critical components of a strategic organization.

In the private sector the Malcolm Baldrige Quality Award is given annually to businesses that represent the best in planning and performance. The Baldrige Award criteria are based on a framework that embodies a systems approach to analyzing an organization's performance. The criteria assume sound business planning; but the organization must go beyond planning to implementation in a systematic way that leads to results. Figure 11-2 depicts the connections between the criteria. The assumption is that all planning is focused on the customer and actions taken are based on information and analysis. If pursued aggressively, the criteria on which a contender for the award is rated (see sidebar, p. 180) should lead an organization to performance excellence.

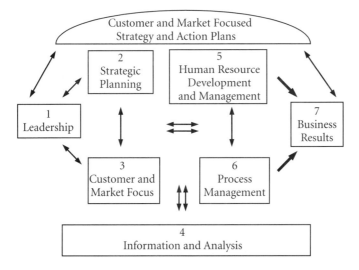

Figure 11-2 Baldrige Award criteria framework: a systems perspective[2]

STRATEGIC PLANNING PROCESS

Strategic planning is a tool that can be used to identify, define, and implement the strategic thinking of an organization. It can be used to establish a vision, a sense of direction, or a road map for where to go and the outline of a process to get there. It can be used as an organizational development tool to bring a group of people together around a goal, and as an accountability mechanism to measure the degree to which the organization is achieving its goals. Planning works only when employees are involved in developing the direction and when upper management is committed to act on and use the plans that are developed.

Challenges in the water industry demand best performances. In order to achieve them a leader needs to align employees in a direction that they understand and accept. Therefore one needs to set goals and to be able to measure them. The phrase "that which

---

[2] *The Malcolm Baldrige National Quality Award 1997 Criteria for Performance Excellence*, p. 42.

The Malcolm Baldridge National Quality Award is an annual award given to U.S. manufacturing companies, service companies, and small businesses to recognize performance excellence. The award promotes:

• *understanding of the requirements for performance excellence and competitiveness improvement; and*
• *sharing of information on successful performance strategies and the benefits derived from using these strategies.*

The award is based upon seven performance excellence categories:

• *The **Leadership** category examines senior leaders' personal leadership and involvement in creating and sustaining values, company directions, performance expectations, customer focus, and a leadership system that promotes performance excellence. Also examined is how the values and expectations are integrated into the company's leadership system*

gets measured gets done" contains truth and wisdom. However, it is important in an era of limited resources that an organization strategically chooses those activities it will measure and accomplish. These ideas relate to those in chapter 6 and design a process whereby an organization can tie together technology, organizational factors and operations and maintenance (O&M) issues. The organization of the future also needs to think "system" and think "technology" (see chapters 7, 8, and 9). Efficiency requires that systems be tied together both intellectually and technically. Therefore information systems and strategic thinking must be coordinated to affect results.

REASONS TO USE STRATEGIC PLANNING

Although planning can be a useful tool to assist an organization in identifying its goals and aligning systems to achieve those goals, it can also be a useful methodology for bringing employees together to set department goals and objectives. Strategic planning can be used to accomplish the following objectives:

Set a Vision

Employees working with management can establish the vision, mission, values, and goals of an organization. The vision statement should be lofty and far-reaching. It needs to be something that employees can aspire to and yet which challenges them to do the best work possible. Vision statements are long-term. Examples of far-reaching statements are: "We bring world class utility services to our community" (Seattle, Wash.) and "To be the best regional water and wastewater utility in the country where employees, customers, and the environment are valued" (Austin, Texas). Values can be drawn out of employee meetings and will set the tone of an organization. Examples include: respect, creativity, trust, diversity, productivity, and collaboration. Management can select values to encourage in the workplace and employees can set forth those aspects of the work environment that are important to them. As described in chapter 4, stakeholders other than employees should also be involved in establishing the organization's vision.

Goals can be developed in coordination with employees and should be those major directions that the organization must pursue to achieve its vision. Strategic plans should not select more than 6 to 12 major goals or the plan could lose its sense of priority (see Table 11-1). Each goal can have objectives related to it that are action oriented. Often the ideas that employees have for goals will actually be more action oriented objectives. Determining an organization's true goals can be a challenge and requires excellent leadership and facilitation.

Create a Road Map

The strategic plan is a road map for accomplishing the vision. The vision sets a future direction (15 to 20 years) and the goals are long term (5 to 10 years). Under each goal are several strategic objectives that can accomplish that goal. Each strategic objective should be accompanied by an action plan to carry it out. The action plans make up a road map of one to three years of activities that can be directed and managed by leadership. (An example of this planning structure is given in chapter 4.) Employees can see progress in their work and see the progress of the organization toward a goal and the desired future when a regular reporting system accompanies the plan. In order to measure progress toward a goal, a series of performance measures should be established. The performance measures can be both quantitative and qualitative and can be reported regularly so that progress is visible to employees and decision makers.

Prepare Action Plans

For each strategic objective within a goal, responsible parties can write action plans to accomplish those objectives. The action plans identify the actions to be taken, the outline of a scope of work, the people and partners involved in assuring that the action takes place, the products to be produced, due dates and persons responsible for completing the actions. Action plans can then be analyzed to determine resource needs and budget requirements. Action plans also give the operating agency the ability to write accountability contracts with senior managers or

including how the company continuously learns and improves and how it addresses its societal responsibilities and community involvement.

- *The **Strategic Planning** category examines how the company sets strategic directions and determines key action plans. Also examined is how the plans are translated into an effective performance management system.*
- *The **Customer and Market Focus** category examines how the company determines requirements and expectations of customers and markets. Also examined is how the company enhances relationships with customers and determines their satisfaction.*
- *The **Information and Analysis** category examines the management and effectiveness of the use of data and information to support key company processes and the company's performance management system.*

The Human Resource Development and Management category examines how the workforce is able to develop and utilize its full potential, aligned with the company's objectives. Also examined are the company's efforts to build and maintain an environment conducive to performance excellence, full participation, and personal and organizational growth.

The Process Management category examines important aspects of process management including customer-focused design, product and service delivery processes, support processes, and supplier and partnering processes involving all work units. The category examines how processes are designed, effectively managed, and improved to achieve better performance.

The Business Results category examines the company's performance and improvement in business areas such as customer satisfaction, financial and marketplace performance, human resources,

Table 11-1 Seattle Public Utilities goals and objectives [3]

SERVICES, ASSETS, AND INFRASTRUCTURE

GOAL: SPU delivers reliable, high quality, and cost-effective services; protects public health and safety; and maintains the service reliability of our infrastructure and assets.

OBJECTIVES:
- We maintain our infrastructure through regular and emergency maintenance to minimize future cost and service disruption and to protect private and public property.
- We accomplish projects on time and within budget.
- We plan for and provide clean, reliable quantities of water.
- We ensure that solid waste collection, transfer, and disposal services are integrated into a well-coordinated system that makes the most efficient use of resources.
- We plan for and provide systems for stormwater and wastewater.
- We ensure public health and safety:
  –untreated sewage and contaminated wastewater is isolated from the public;
  –solid waste is collected regularly and disposed of properly;
  –drinking water is safe;
  –opportunities for disposal of household hazardous waste are provided;
  –adequate water is available for fire suppression;
- Closed landfills are soundly maintained.
- We collaborate with our customers to solve problems.
- We maintain optimal resource levels, including staffing, equipment, and contracting services.
- We maintain and ensure technical expertise.
- We incorporate state-of-the-art technology, as appropriate.

CUSTOMER SATISFACTION

GOAL: SPU consistently meets or exceeds customers' expectations.

OBJECTIVES:
- We respond to our customers' needs in a timely manner.
- We provide accessible, professional, courteous service.
- We actively reach out to our customers to anticipate their needs.
- We deliver our services in a well-coordinated, high quality, and cost effective manner.

EMPLOYEES

GOAL: SPU employs and values a diverse, efficient, and productive workforce.

OBJECTIVES:
- We recruit and select a qualified and diverse workforce through a fair and open process.
- We compensate all employees fairly for work performed.
- We train all employees for their jobs and support and encourage their career development and job satisfaction.
- We are proud of a workforce where all employees are responsible and accountable for their performance.
- We provide a safe work environment and safe work practices are expected.
- We recognize and reward teams and individuals for outstanding performance.

Table 11-1 Seattle Public Utilities goals and objectives

- We actively seek the input and opinions of employees at all levels of the organization to inform decision making.

ENVIRONMENT

GOAL: SPU protects, sustains, and enhances environmental quality for current and future generations.

OBJECTIVES:

- We demonstrate responsible environmental practices in our everyday business activities.
- We manage all projects with a priority concern for the protection and sustainability of the environment (e.g., solving flooding naturally).
- We restore and preserve habitats that are threatened as a result of our activities.
- We effectively manage our liabilities to minimize negative environmental impact.
- We encourage the public to become responsible environmental stewards and wise users of our resources through cost-effective re-use, recycling, and other environmentally responsible practices.

COMMUNITY

GOAL: SPU reflects and strengthens the spirit and values of our community.

OBJECTIVES:

- We communicate clearly and regularly about our services and activities.
- We partner with communities and organizations on common issues to create integrated and innovative solutions.
- We involve individuals and organizations in decision making through collaborative processes.
- We develop and support integrated community education and outreach programs that further SPU's goals.
- We maintain long-term, positive relationships with our regional partners.

FINANCIAL AND ORGANIZATIONAL PERFORMANCE

GOAL: SPU is a responsible financial steward and improves productivity and performance each year.

OBJECTIVES:

- We provide clear, timely, and useful information to our customers and ourselves for managing our resources.
- We develop and implement quality, cost-conscious solutions.
- We have plans and tools in place to improve performance, manage risk, and act strategically.
- We set and meet financial targets that preserve and protect the rate payers' investment over the long-term and support City policies.
- We establish rates that reflect our financial, environmental, and customer satisfaction goals.
- We have work teams and work processes that successfully merge and integrate the utilities and our programs.

supplier and partner performance, and operational performance. Performance levels relative to competitors are also examined.

NOTE: Excerpted from The Malcolm Baldrige National Quality Award 1997 Criteria for Performance Excellence.

[3] Adapted from the Seattle Public Utilities Strategic Plan, 1997.

to do performance evaluations with employees to make sure they are working to accomplish department goals.

Establish Accountability Goals

Setting the direction of the organization and preparing action plans and performance measures allows a chief executive to write accountability contracts with lead directors and managers. In these contracts the executive can define expectations that are well established and measurable. The expectations can be tied to the organization's goals and be set forth in the action plans. In addition, the leader can establish expectations for ways that the directors/managers will work to achieve the values of the organization such as diversity, trust, or collaboration—all of which are hard to measure. The value of having an accountability mechanism tied to organizational goals and objectives is that in the process of setting it up the organization becomes aligned with the direction of the plan. Alignment—making sure that all employees are going in the same direction—can be hard to achieve in mid-to large-sized organizations.

Develop A Performance-Oriented Organization

Developing a method of measuring goals and objectives is critical to the success of implementing a strategic plan. Performance measures are difficult to write and are sometimes difficult to quantify, but they are the mechanism for making progress visible. It is important to develop both quantitative and qualitative methods of measuring performance. Qualitative measures (such as employee satisfaction or customer satisfaction) may need to be based on surveys or focus groups. Quantitative measures are difficult to select because they should be indicators of what drives organizational performance and not just something easy to measure. The ability to measure performance is critical in defining changing levels of progress and efficiency.

ELEMENTS OF A STRATEGIC PLAN

Developing a strategic plan is a process in which a number of strategic actions are tied together in a systematic way. Although plans

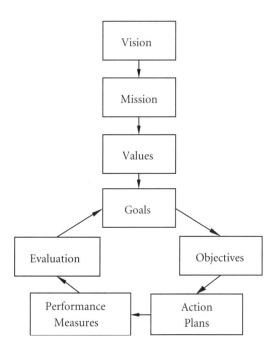

Figure 11-3   Strategic planning process

can certainly differ, Figure 11-3 diagrams steps to set up a plan that goes from establishing a vision to identifying performance measures. A final step would be to reassess your performance to determine whether goals should be changed.

Let's discuss what is meant by each step.

Vision.   The vision is a long-range (20 years) statement about what the organization wants to be in the future. It should be conceptual, demand excellence, and be a "reach" for employees to attain (see examples in the section on Reasons To Use Strategic Planning).

Mission.   The mission is a short statement about the work and purpose of the organization or its reason for being. The mission should be stated in general and conceptual words and must be clear to both the organization and its employees. Examples include: "It is our commitment to deliver quality water and dependable wastewater services of outstanding value to our customers, while earning a fair return on our shareholder invest-ment" (Florida Water Services); "Denver Water will provide our customers with service through responsible and creative steward-ship of the assets we manage. We will do this with a productive and diverse workforce. We will actively participate in and be a responsible member of the water community" (Denver Water).

Values. Values are statements about the organization's pri-orities and how to enhance them in the quality of work life. They are principles that motivate the organization and the way it oper-ates. Examples include: respect, environmental stewardship, innovation, collaboration, and diversity. Both the mission state-ment and the values of an organization must be simple and repeated often.

Goals. Goals are long-range statements of desired states which, when taken together and achieved, fulfill the long-range vision for the organization. An example of this type of goal is "We consistently meet or exceed customers' expectations."

Strategic objectives. Strategic objectives are end states that must be reached to accomplish the goals. They are more specific and action oriented than a goal, but have a strategic quality; i.e., describe how, why, or when a component of the goal will be achieved. For example, "We respond to our customers' needs in a timely manner."

Action plans. Each strategic objective needs to have one or two action plans to accomplish it (Table 11-2). The action plan is very specific and includes a statement of the problem/project, what will be done, what the product will be, when it is due, who the partners are on the project (strategic effort), and who is responsible. The assembly of action plans makes it possible to develop a budget or resource/personnel list for accomplishing the goals of the organization. The set of action plans will most likely be updated regularly and are often kept separate from the vision-ary parts of the strategic plan.

Performance Measures. Performance measures are ways to measure progress toward accomplishment of the goals and actions. Once a list of performance measures is established for the plan, reports can be generated quarterly or semiannually to report on the progress toward achieving the plan. The report can be a useful tool to communicate to both employees and decision

Table 11-2 Goals, objectives, and strategic efforts[4]

CUSTOMER SATISFACTION
GOAL: SPU consistently meets or exceeds customers' expectations.
OBJECTIVES:
- We respond to our customers' needs in a timely manner.
- We provide accessible, professional, courteous service.
- We actively reach out to our customers to anticipate their needs.
- We deliver our services in a well-coordinated, high quality, and cost effective manner.

1997–1998 STRATEGIC EFFORTS:
- Establish systems, structures, and procedures to provide one stop, integrated, seamless customer service.
- Develop automated and online systems that increase access to reliable and timely customer information.
- Develop and present proposals to increase effectiveness of staff serving the public.
- Develop service level agreements for the most common services provided within SPU and to other City departments.
- Reassign decision making authority and responsibility to the appropriate levels in key areas to meet customer needs.
- Create a system that provides timely and ongoing customer feedback and response.
- Identify and implement improvements that will increase customer confidence in the accuracy of charges for service.
- Develop a functional directory of the organization including department, division, and project descriptions and people to contact for answers to customer questions.

makers about the organization and the degree to which the utility is working on its goals.

Evaluation. All plans need to be regularly updated and revised. The utility needs to decide whether it will update objectives and action plans annually or biennially. The longer term goals will be reviewed less frequently. Every process needs to include an assessment phase that leads to regular review.

Strategic intent. Once a plan has been written with a vision, goals, and objectives, the leadership should be encouraged to develop a statement of strategic intent. The strategic intent is a short statement that summarizes quickly and intuitively what the

---

[4] Ibid.

In the mid-1980s a new course of action was developed for the Des Moines Water Works Public Utility (Case Study 1.5). The utility developed a new strategic plan. A team of directors defined the components of the plan and a process and timeline for developing each of those components. It was then reviewed in focus groups of individuals at all levels of the organization. The vision included a comprehensive review of many aspects of operations.

First, existing policies and procedures were reviewed and evaluated to improve efficiency and eliminate wasteful practices. Second, the team evaluated the true costs of various functions. Third, the team observed the technologies and methods used by private operators at other facilities in order to learn how such practices could be used to improve their own facilities. Fourth, the team sought new areas of growth for developing and marketing a range of services beyond the metropolitan area. Last,

purpose of the plan is. It is a statement that employees can keep in mind to guide them in day-to-day activity. It should be meaningful, action oriented, and colorful. Some examples: Act as one, think world class; Knock-your-socks-off performance; Think number one, be number one; and Great service, great win.

DEVELOPING A STRATEGIC PLAN

The Approach

Although it is not necessary, most organizations will likely hire a consultant to help develop the plan. It can be done with excellent results by organizational leadership. If one does hire a consultant, he or she will bring along a particular approach to planning. Be sure it meshes with your own. There are many ways to accomplish a plan. However, it is important to think about the roles of leadership, stakeholders, employees, and communication.

Role of leadership. Leadership in an organization must be committed to the idea of a plan and must see its value. If the leadership is not involved and delegates the task to staff or a consultant, employees will sense the distance and will not want to participate because they think their involvement will not produce results. Leadership must also be deeply involved in setting the vision and goals. Leaders must be conceptual and intuitive to foresee the issues that the organization will be facing. In some organizations goals are set by leadership alone; however, employee involvement in all aspects of the process tends to lead to a stronger result.

Role of employees. Commitment to carrying out a plan is stronger if employees have been involved in its development. It takes time and some employees find the process tedious and unrewarding. Hiring a facilitator for employee involvement in developing concepts is often a good idea. A facilitator can more easily clarify what can often be muddy ideas without employees feeling that leadership is trying to put words in their mouths.

Role of stakeholders. Every organization has a series of stakeholders and must determine what role they will play in the development of a plan. The plan will be stronger if an organization takes the time to reach out to customers and clients as well as

188

to decision makers to gather ideas for the future and to identify gaps in performance and future market niches. At a minimum the goals and objectives of an organization should be shared with decision makers (elected officials or board of directors) early in the process before an organization completes a plan and discovers that for some reason it has not accommodated the issues and concerns of its board.

the skills and flexibility of the department's employees were evaluated with the goals of maximizing the flexibility of staff.

Role of communication. A strong internal communication program is important to keep employees and interested stakeholders engaged in the process. It is also important to have a plan and schedule and to keep it. A brisk schedule that keeps people moving is easier to manage than one that is protracted and leaves employees wondering if anything is happening. Communication can include information from a weekly newsletter, an internal web site, a hotline, regular briefings of the leadership team (down several layers) and scheduled meetings with employees to report on drafts of materials as they emerge. Allowing reasonable time for comment on drafts assists the employees with a sense of ownership.

Use of Assessment

If the organization takes time to conduct a self-assessment before launching a plan, the information gained will improve the plan. The purpose of an assessment is to gain a sense of the strengths and weaknesses of the organization so that future goals and strategies can be geared to meet perceived needs and to correct perceived weaknesses. An assessment can also be useful in determining the market niche in which future initiatives may be successful. Market niche analysis is less important in an organization that is perceived as a monopoly, but the more that utilities can mirror private sector organizations the stronger and less vulnerable they will become to competition from the private sector.

Assessments can be accomplished by doing customer and employee surveys and by using focus groups. The information will clarify that gained in general assessments and will help develop clearer ideas about preferred futures for the organization. Another form of assessment that should be used to analyze the current state of the organization is benchmarking studies or analyses that have been done, such as for field operations, where best practices and strategies are being assessed for water maintenance activities.

Some effort should be done in the planning process for a futures assessment of the direction that the industry is going. The futures assessment should include technology, customer interests, economics, and business and organizational trends.

The AWWA has published a self-assessment program for water utilities that was developed by utility personnel working with consultants. It involves asking a sampling of employees from all levels within a utility to complete a rather comprehensive assessment instrument. The assessment can be accomplished in a short time frame. For information on cost, scheduling, and materials, contact the QualServe Program Manager at AWWA headquarters in Denver, Colo. (303-794-7711).

If the impartial opinions of qualified outside reviewers would be helpful in refining your self-assessment or in convincing your board and staff of the need for improvements, AWWA has a follow-on Peer Review service. Trained volunteers from comparably sized utilities visit your utility for three or four days and develop a report on your strengths and functional areas that would benefit by an improvement program.

Using a Consultant

Writing a plan can be a difficult, time consuming, and sometimes tedious job (editing and rewriting work that comes from employee consensus can be difficult). If leadership is distracted with other job responsibilities and does not have a strong, skilled employee to lead the planning process, it is wise to get a consultant. The ideal consultant is a consultant coach, one who assists leadership in developing a plan that they are committed to. If a plan is the consultant's plan and not leadership's plan, employees will know that and the plan will not garner respect or interest or inspire action. Consultants can also be used as facilitators for employee discussion groups either in the assessment phase or in developing strategic actions, or both.

Achieving Employee Ownership

Achieving employee buy-in is a combination of strategies: using employees effectively in the development of the plan, having a strong communication plan, and thinking of ways to market the plan so employees understand it when it is complete.

Employee involvement in development requires preparation of a process that includes employees in present performance assessment, development of goals, and strategic actions. Communication throughout the process should be transparent and easily accessible. It needs to include a strong monitoring process that is visible and shows what progress is being made.

A strong communications plan requires that multiple types of media are used to broadcast the process and to regularly describe progress in accomplishing results. Every organization needs to have a strong internal communications program to inform employees. The media might include e-mail, written newsletters and documents, a web site and regular employee meetings.

When the plan is completed it should be distributed broadly, perhaps with a video attached that explains its purpose and importance. A regular method of reporting on and updating it should be developed so that employees are aware of the changes occurring and of the progress being made.

ISSUES TO BE MANAGED

Every planning process will have its own set of problems and quirks, some of which cannot be anticipated. The following are some of the most common that should be avoided if possible. Any significant problem along the way should be handled immediatly so that it does not undermine the effectiveness of months of work.

Employee Involvement

Employee involvement is important, but it is also a potential problem to the process. The problem can be either too much or too little. Too much employee involvement can slow down the process or lead to paralysis as someone tries to negotiate among employee interests for what should go into the plan. Some

When Cincinnati Water Works (Case Study 1.7) did their strategic business plan, they learned the following lessons:

- *The development of a strategic business plan is a time consuming process that cannot be done quickly.*
- *It is essential that all levels of the organization be involved in the plan's development in order to gain the necessary shared acceptance and support for the strategies.*
- *Active support and patience from top management is vital as the organization works through the consensus-building process.*
- *Top management must be committed to following through with implementation of the plan once it is developed.*

employees prefer to be involved in processes rather than to do their own work. Such an employee has an interest in expanding the process just to stay involved in it. Sometimes employee interests are in conflict with each other and then there must be a decision maker who will make the final calls. It is important for the plan to emerge as a good balance of the interests of employees and the interests of leadership.

Too little employee involvement can lead to the plan being seen as a tool of management with no ownership or engagement by employees. Such a plan is not likely to lead to change or be the road map to the future it was designed to be. The process of developing a plan can also be a change management process in which employees learn the goals of the organization and how it is going to change. They begin to understand that values may have changed and that behavior and actions have to change to reach a new future. If employees are not involved, the plan can be a missed opportunity.

Unending Editing

Someone involved in the process of preparing the strategic plan itself must be a clear and concise writer with a willingness to say no. If the leader of the process accommodates every request and interest the plan will lead to unending editing or such a garble of vagueness that it will have no value or meaning. Writing clear, concise, future-oriented prose with an action orientation is a difficult job and requires an excellent writer. Be certain one is on the team.

Completing the Process

The best plan is a completed one. Planning processes have a tendency to go on and on in an effort to please many interests. It is critical to have a due date, firm deadlines (even with some planned slippage), and a task master to rule the process who will insist on products by deadlines. Plan writing is hard work because it requires considerable clarity of mind; at the same time it is abstract and conceptual. Many employees will put off writing tasks for the plan, and some incentive must be developed to keep people on time and on task.

Sitting on the Shelf

Next to an uncompleted plan the worst plan is one that sits on a shelf unused and unreferenced. Employees get demoralized by processes that seem to be the "management fad of the month" and ones in which they invest time and energy that do not lead to any change. Then they will not volunteer the next time you ask them to be involved in a process. In order for the plan to be used there must be management/leadership buy-in and an accountability/reporting mechanism whereby employees see that progress is being made and where managers have performance evaluations tied to completion of tasks on the list.

Making it Relevant

To be appreciated and valued, the plan must seem relevant to employees. Employees, especially field workers, have a strong sense of authenticity. If the plan does not ring true to their sense of the real needs of the department it will not gain respect. For many employees the action plans and the specific actions that will be taken to implement them are the most important part because the actions are concrete, relevant and something that they will be involved in. If the plan is all vision and goals with no implementing actions it will not be seen as a relevant document. Much of the task in making it relevant is in communicating what it is and what it is for because the printed document is likely to include more general goals while action plans are likely to be in an accompanying document. It is important for employees to know when the document comes out that there are action plans that support it.

Short, Simple, and Readable

Strategic plans should be short and clear with pictures and colorful text that convey the right "feeling" to employees. The writing should be good, simple, and readable. Ideally, pictures should show a range of different employees doing tasks in the organization. The document should be attractive and something that employees wish to pick up. In order to keep a document short and simple it is often necessary to keep action plans and performance measures in some other document.

Strategic planning is a tool that can be used in an organization to manage change. It can be used to set forth the vision and strategic direction of an organization. It can also be used to tie employees into the mission and need for change as well as into a new set of values to guide the agency to a new future. The vision and goals can be long-term and bound to a set of strategic actions that are shorter term and tied into the accountability processes of the organization. By working together to design the future for the organization, employees and leadership can come to new understandings of directions that will make their organization more efficient and effective and less prone to criticism from stakeholders or competitors.

SMALL UTILITIES

For medium-sized or small utilities the concepts in this chapter are valid; the process for implementing a plan will be different. Smaller utilities with fewer staff members are more likely to want a consultant coach to assist with the process and to do the primary job of writing a plan. It then becomes especially important for the consultant to involve leadership in plan development and to use employees to assure plan ownership. Smaller utilities can also "borrow" ideas and draft plans from other larger utilities to reap the benefits of customer surveys and assessments. Often change concepts or customer preferences sweep from larger urban areas to smaller urban centers, so a customer survey from a large city can be a predictor of customer preferences for the future.

Employee involvement is especially important in smaller utilities in which employees are often multitasked and are both the thinkers and the doers in the organization. In these situations employee ownership of a strategic plan is crucial to its successful implementation.

COMPREHENSIVE IMPROVEMENT PROCESS

The next chapter contains information on performance assessments, work process improvements, and quality management. These are the systems that need to be added to the strategic organization to achieve goals in the strategic plan. Efficient work process flows are complimented by technology and organizational design to produce high performance outcomes. The comprehensive improvement process must be continuous so that as market and business forces change the internal systems can also change. If employees can customize products and services to the needs of customers then even the employees will be making demands on the organization to change in response to perceived needs from either customers or the market.

SUMMARY

As stated in the beginning of this chapter, strategic thinking is the glue that holds all the other organizational activities together. It is more than a plan—it is a way of life in the organization. The plan becomes a tool to help leadership define the most important coordinating values and principles necessary to keep the entire organization aligned to a particular goal or direction. The process is not easy and there is more than one way to get to a goal. What is critical is the commitment to the process and the belief that strategic thinking is essential for a utility to remain relevant in the changing business world.

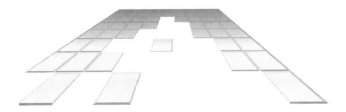

Assessment Tools

Gap analysis is an assessment of where you are compared to where you want to be. In chapter 4 we establish that there are three complementary perspectives from which to approach the assessment of your utility: the customer perspective, the internal perspective of the utility, and the external perspective. This latter view can include your peers, industry leaders, and other stakeholders such as regulatory agencies.

This chapter presents some assessment tools to help in your analysis and to introduce you to some of the work currently being undertaken in our industry and by others. Some specific methods of assessing internal capabilities are provided and tools are included to aid in decisions about outsourcing as an approach for certain services or operations.

Topics covered in this chapter include:

- Assuring alignment with your customers by methods including surveys

- Assessing internal capabilities and performance including evaluation of core competencies and contracting opportunities

- Uncovering span of control issues

- Developing performance measures with a framework and examples

- Comparing your performance to others by using metrics, work process evaluations, and other techniques

ASSURING ALIGNMENT WITH YOUR CUSTOMERS

Capturing the Voice of the Customer; How Important Is It To Water Utilities?

In competitive industries understanding what customers want and providing it is an integral and obvious part of business. Without that focus you are apt to lose your customers to your competition. Therefore, the best performing companies continuously listen and respond to their customers. For water utilities, there has been less competition for customers and, as a result, less necessity to listen to them.

Chapter 1 describes a future in which utility consolidations and threats to public water utilities will increase competition for customers. This competition will increase the need for improved performance. Only your customers can tell you what they want and how well they think you are providing it. Capturing their voice is an integral part of assessing your performance.

The voice of the customer for utility organizations represents more than consumers or bill payers. Therefore, any assessment of the customer must first identify and then acknowledge the various voices that impact utility performance. Answering the following questions helps us understand and use information from a range of customer voices.

Which of these voices guides our investment in service/product improvement?	- customers - governance and decision makers - advocacy groups - media - employees - industry standards - industry performance - technology developments - regulations - community sentiment - other

How do we listen?	- reactive listening - proactive listening - systematic listening
What do we do with what we hear?	- investigate - capture and organize - look for patterns of change - disseminate - problem solve - support decisions

Formal surveys conducted either in house or by one of the many private companies that provide these services are just one way to listen to the customer. On a daily basis there are other routine ways that your organization has contact with customers and through which you can elicit feedback. For example, meter readers and customer service representatives speak to customers every day. Each phone call is an opportunity to ask questions and to obtain feedback. Community and public processes are means of listening and interacting with the public, as are citizen advisory groups.

Surveying Your Customers

Customer surveys provide the most obvious means of determining if utility programs and priorities reflect the interests of the community being served. They can also communicate if messages or information are being heard and understood, whether customers are satisfied with performance, and if there is customer preference for certain actions to be taken by the utility. As indicated, the customers can refer to a range of voices—from employees to decision makers to the public at large. Several types of surveys are used.

Multi-utility survey. These surveys are one way to gather customer survey information, compare across utilities, and take advantage of survey instruments that have been developed and targeted at utility interests. These surveys allow you to compare customer satisfaction levels for water quality and other service dimensions against those of other utilities.

Considerable research has been performed or is in progress through AWWARF, member utilities, and consultants. Currently,

AWWARF Project 359, "Developing Benchmarking, Clearing-house, and Customer Satisfaction Measures," includes a component devoted to standardization of a customer survey instrument. Its intent is to provide a tool to track customer service and satisfaction data over time and to facilitate interutility comparisons of this data. The project includes the development of standardized survey tools and support for their use.

Targeted or project specific survey. A second type of survey is focused on a specific project or issue. This survey could have the purpose of educating your customer by simply asking questions, gathering information, or obtaining feedback from them or of marketing your project or issue. This could include surveys at community meetings, door hangers asking customer opinion about work being performed in their neighborhood, a survey targeting all or part of the customer base on an issue or project being considered by decision makers, or an internal employee survey attempting to obtain suggestions or feedback on utility or communication issues. Determining customer preferences can often help speed up a project process or help assure community acceptance. Most utilities require professional assistance in developing these surveys, but this should not stop utility managers from reaching out to communicate with customers in a variety of ways.

Service or product survey. Surveys may also address certain service elements of the utility or identify attributes of the service that are important to customers. Customer focus groups are often used for this purpose. Attributes that are identified through this process can be used to develop a larger customer survey to measure performance in areas that have been identified as important. Professional assistance in designing and conducting focus groups may be necessary.

ASSESSING INTERNAL CAPABILITIES AND PERFORMANCE

Managers are challenged to objectively evaluate the utility's capabilities and performance within the organization. This section identifies several perspectives and tools that can help assess strengths, weaknesses, and strategies for your utility. It introduces methods of evaluating core competencies and of identifying contracting opportunities.

Evaluating Core Competencies

Many utilities are confronting the challenge of self-improvement by first assessing which services, if any, they could contract to a private entity. Decisions of this magnitude tend to be driven by factors unique to the circumstances involved. However, several tools are useful in assisting utilities and decision makers with this issue. Some questions must be asked: Which functions should our utility forces perform for us to be competitive and efficient? Which functions are vital to our essential mission? What are the essential core services and skills that should be resident in our workforce for us to be a high performing organization? Which are the ancillary or support services that may be best suited for the private sector to provide? Which strategies are in place or are needed to address contracting issues with employees and unions? What work do we currently perform well? Can we attract and maintain necessary resources or technical skills in our utility?

Figure 12-1 is a quadrant matrix in which utility functions can be arrayed. For functions performed in each utility, this tool may assist managers, employees, and utility boards in initially identifying candidate functions for contracting. A sample, representative list of utility functions is displayed. While functional placement may vary from utility to utility, the matrix should help stimulate thinking about contracting issues. Consideration of each utility service or function is often required. In this regard utilities can learn from the experiences of others who have begun this assessment or have developed evaluation tools that can be used to identify any services that could be contracted out.

In 1989 the Colorado State Auditor's Office developed a privatization assessment workbook (a series of questions and a scoring methodology) to determine the feasibility of contracting out particular services. The workbook identifies nine categories of questions important to outsourcing decisions:[1]

---

[1] "Evaluating Service Contracting," *MIS Report* 25, no. 3, (March 1993).

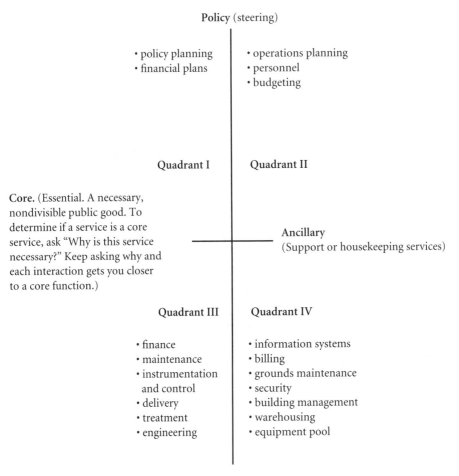

Quadrant I: Core, Policy
These services and activities are the last to move into the marketplace. They can typically be done with a small number of people.

Quadrant II: Ancillary, Policy
These activities such as personnel and budgeting are carried out by the administration. Probably not suitable for competition or outsourcing.

Policy (steering)

· policy planning
· financial plans

· operations planning
· personnel
· budgeting

Quadrant I   Quadrant II

Core. (Essential. A necessary, nondivisible public good. To determine if a service is a core service, ask "Why is this service necessary?" Keep asking why and each interaction gets you closer to a core function.)

Ancillary
(Support or housekeeping services)

Quadrant III   Quadrant IV

· finance
· maintenance
· instrumentation
  and control
· delivery
· treatment
· engineering

· information systems
· billing
· grounds maintenance
· security
· building management
· warehousing
· equipment pool

Quadrant III: Core, Implementation
These functions may be candidates for market competition although government may have a role to play.

Quadrant IV: Ancillary, Implementation
These functions may be strong candidates for market competition.

Source: Adapted from Husock, Howard. *Organizing Competition in Indianapolis: Mayor Stephen Goldsmith and the Quest for Lower Costs.* 1994. The Case Program of the John F. Kennedy School of Government. Harvard University, Cambridge, MA.

Figure 12-1   Identifying contracting opportunities

1. The strength of the private market for the service

2. Political resistance

3. Impact on service quality

4. Impact on public employees

5. Legal barriers

6. Risk to the government

7. Resource efficiency

8. Importance of in-house control over the service

9. Cost efficiency

In 1995 the City of Portland Bureau of Waterworks developed its own set of criteria for determining which of its services were appropriate for public-private competition. Portland groups these criteria into four broad areas for consideration.[2] The first is the service itself. What type of service is it and is it ready to be contracted out? The second is the internal market surrounding the service. Is the agency willing and able to contract out the service? The third is the external market associated with the service. Is the private (or nonprofit) sector willing and able to provide the service or good? The fourth is the political and legal environment surrounding the service. (Portland calls this category "other considerations"). Is it legal to contract out this service and is there political support to do so? Let's integrate the Portland and the Colorado State Auditor's Office frameworks to discuss other issues that are important to consider in making in-house versus outsourcing decisions.

Is the service itself a good candidate for outsourcing? The first step to identifying good candidates for outsourcing is categorizing services by type. Certain service types (e.g., ancillary, hard, and new) are better candidates for contracting out than others. Questions to ask of a service or function include: Is it a support service? Are the desired outcomes of the service easily defined?

---

[2] "Selecting Services for Public Private Competition," *MIS Report* 28, no. 3 (March 1996).

Are its outcomes required for the successful delivery of another service?

On a basic level, services or functions can be categorized as either internal or external. Internal services include those that the utility provides to its internal customers to ensure that it can function. Internal services include personnel, training, information services, finance, accounting and other administrative functions. External services are those that an agency provides to outside customers. Some of both internal and external services are candidates for outsourcing. There are nationwide private sector providers and precedents for both.

Other service characteristics provide more guidance as to which services are ready to be contracted out. Ancillary services (versus core services) is one such characterization. Ancillary services—support or housekeeping services—are better suited to contracting out because control is not as important for these services.[3] They involve fewer policies. In addition, more contractors are available for these services.[4] Although core services, those essential to the mission of the organization are sometimes contracted out. They provide additional challenges and involve more risk. Should the contractor fail to live up to the contract for a core service, the mission of the organization (e.g., the health and welfare of the public) could be jeopardized.

One of the challenges of outsourcing is developing and monitoring the contract. To ensure a successful contract an agency must define exactly what it wants to buy. Services for which this is possible, "hard services" such as vehicle towing, water main installation, or street service repair, are therefore easier to contract out. Services that are not as quantifiable leave more to the discretion of the contractor and leave their successful delivery at

---

[3] For the City of Phoenix, the tradeoff between cost and control is the basis of their outsourcing decision. Services for which control is more important than cost are not contracted out. Services for which cost is the primary concern are contracted out if the private provider is less expensive. *Municipal Service Delivery*. Kirkland, Wash.: Municipal Research and Services Center, July 1993, p. 19.

[4] *MIS Report* 28, no. 3 (March 1996): p. 3.

greater risk. If the contracting agency cannot define what it wants to buy, it will be hard pressed to find it in the marketplace.

The public sector can avoid a disadvantage of contracting out—loss of information—by contracting out independent or stand-alone services. The services of an agency are not always independent from one another. The delivery of one may depend on the product of another. For example, in most utilities the billing system cannot assess customers accurately without accurate meter reads. Although service outcomes that are required for the delivery of another are often contracted out (including meter reading), the contracting out of stand-alone services is less complicated. The information generated by the outcomes of stand-alone services has less bearing on the other operations of an agency. Their success or failure is contained in the service itself.

New services also provide a good opportunity for contracting out. The agency and its employees are not yet invested in new work. Workers then will probably feel less threatened by outsourcing new work rather than that which they currently perform. Contracting out new work may also allow the agency to avoid large investments in training or equipment required for the new endeavor. Seattle's Tolt Treatment Plant Design Build Operate (DBO) project is such an example.

Temporary, seasonal, or peak workload situations are also good candidates for outsourcing. One of the benefits of outsourcing is the labor flexibility of the private sector. This flexibility is especially advantageous for seasonal work or work that requires specialized skills not necessary on a day-to-day basis. The agency may also save large investments in specialized training and equipment by hiring the private sector for this work. In addition, by outsourcing, the agency may avoid costly overtime and preserve its own staff and training resources for daily operations.

Does the utility possess sufficient political will and skills to outsource? Opposition to outsourcing is a barrier to its success. It is difficult to develop adequate contracts without political and organizational support. Identifying functions and services for which there is little resistance can be an important step to successful outsourcing. These might include services for which there are few current employees involved or impacted.

Agencies with contract writing and monitoring experience or capabilities will be more successful with outsourcing. Of course, experience is only gained by actually contracting out. Agencies have to start somewhere. Recognizing where there are trained resources—staff available to write the contracts as well as to monitor performance—will increase the chances for successful contracting.

A final internal consideration is the current delivery of the service. How well does the agency currently perform this function? Services that are rated below average may be good candidates for outsourcing. Contracting out these functions or services allows the agency to focus its resources on core competencies and the services it performs well.

Does capability exist in the marketplace? It is important to consider the external market for contracting. Are there private sector providers that currently provide or are able to provide the service in question? Does the private sector have the equipment and skills necessary to do the job? Finally, do other cities successfully contract out for similar services?

Services for which there are not established private sector providers are difficult to contract out. It is usually not wise to hire beginner vendors. Start-up costs and lack of experience in the private sector can be prohibitive. Conversely, services for which there are multiple private sector providers may be good candidates for contracting out. From a cost and efficiency standpoint it is important that there be multiple providers of the service. It is the competition among them that often brings down the price and improves the service.

Another external market consideration is whether or not the private sector is willing to contract with the utility. Some contracts may be so small that it is not worth the private sector's time to compete for it. Multiyear contracts are more attractive to both the contracting agency and contractor. The agency then has time to work out the kinks of the contract before soliciting new bids, and the private sector has a contract of sufficient size (length) to make its investment worthwhile.

A final external market consideration is whether or not other cities are contracting out similar services. The experience of other cities can provide valuable insight into the private provision of

Table 12-1   Good candidates for outsourcing

- Certain internal and external services
- Services not essential to the mission of the organization (ancillary services)
- Hard services (services with easily definable outcomes)
- Stand-alone services and services for which there are few consequences if stopped
- Seasonal services
- Services that require specialized skills not necessary on a daily basis
- New services
- Services that require specialized equipment or operators
- Services for which there are multiple interested and capable contractors
- Services that do not draw opposition from, involve, or displace a large number of utility employees
- Services with in-house delivery problems
- Services where cost is more important than control over the service
- Services contracted out by other cities or utilities

Source: Adapted from *MIS Report*, Volume 28, Number 3, March 1996 and Volume 25, Number 3, March 1993.

the service and the development of the contract. Other cities can provide useful information on vendor quality or performance. It can be reassuring to know what has worked and what has not. Table 12-1 summarizes some of these considerations.

Is it legal to contract out?   Does the legal framework exist for the private sector to provide a public service or function? It is important to consider the legal constraints. While legislation can be amended, it is easier to contract out those services with no legal barriers.

UNCOVERING THE SPAN OF CONTROL ISSUES

Part of any internal assessment should include a careful analysis of organizational characteristics and efficiency, including management and decision systems. Utilities today are concentrating on flattening organizations, developing matrix approaches to accomplishing projects, and breaking across the "silos" or barriers that fragment the organization.

Research and tools have emerged to evaluate to what degree organizations are hierarchical in nature or have too many layers of supervision. Consistent with principles of delegation and accountability at lower levels of the organization, utilities are not only decreasing layers within the organization, but also increasing the number of direct reports to supervisors. This tends to reduce overhead costs and to apply labor more directly to productive work.

Determining the appropriate span of control, or the number of staff reporting to one manager, is not easy. There is not a universally accepted standard. The federal government recommends that the span of control ratio be 7:1 and management expert and author Tom Peters recommends up to 25 to 75 workers for every one supervisor.[5] In addition, there is not necessarily an appropriate standard within any one organization because the nature of the work influences the span of control. To calculate the span of control you must determine who to include in your analysis. Counting both full-time and part-time workers but not temporary workers is one suggestion. In addition you need to define the layers of management. Who is a supervisor? This could be anyone with responsibility to conduct performance evaluations.[6] If you choose to compare your span of control to other utilities, be sure to consider the parameters used in their calculations.

DEVELOPING PERFORMANCE MEASURES

Addressing the performance gap depends on knowing how you compare to desired performance goals. To do this utilities should measure performance, particularly in those areas where improvement is needed. One useful model for developing internal performance measures is called the balanced scorecard concept.

Devised by Robert Kaplan and David Norton of the Harvard Business School, the balanced scorecard method results in a set of measures intended to give top managers a fast but comprehensive view of the business (see Figure 12-2). It includes financial measures that tell the results of actions already taken, and includes operational measures on customer satisfaction, internal

---

[5] From Red Tape to Results, *National Performance Review,* 1993.

[6] *Ratio of Staff to Managers in City Government,* Office of the City Auditor, City of Seattle, January 1996.

Financial Perspective
(How do we look to shareholders?)

Goals	Measures
Survive	Cash Flow
Succeed	Sales, Growth, & Operating Income
Prosper	Increased Market Share and ROE

Customer Perspective
(How do we look to customers?)

Goals	Measures
New Products	Percent Sales from New Products
Customer Partnership	Number of Cooperative Efforts
Responsive	On Time Delivery

Balanced Scorecard Goals and Measures

Internal Business Perspective
(What must we do to excel?)

Goals	Measures
Technology Capability	Capacity Versus Competition
Excellence	Cycle Time, Unit Cost, Throughput
New Products	Actual Introduction Versus Plan

Innovation and Learning Perspective
(Can we continue to improve and create value?)

Goals	Measures
Technology Leadership	Time to Develop Next Generation
Time to Market	New Product Introduction Versus Competition
Process Improvement	Performance of Indicators Versus Competitive Benchmarks

Source: Adapted from R. Kaplan and D. Norton, "Using the Balanced Scorecard as a Strategic Management System," *Harvard Business Review,* January-February 1996, 76.

Figure 12-2 The balanced scorecard goals and measures

processes, and innovation and improvement activities. The balanced scorecard allows a manager to look at the business from four important perspectives:

1. How do customers see us? (customer perspective)

2. What must we excel at? (internal perspective)

3. Can we continue to improve and create value? (innovation and learning perspective)

4. How do we look to our shareholders? (financial perspective)

From an internal assessment point of view, performance measures can apply to the most detailed functions and operations. All parts of the organization, and likewise all employees, should ask "How do I know if I am doing a good job?" A family of performance indicators can be developed for water utility functions. These can assist in answering the performance question, but only if clear goals and strategies exist and if accurate and timely reporting of data is produced. Considerable expertise is available in utilities and in the marketplace to assist in this assessment activity. Typical categories of a balanced set of measures include:

- Customer satisfaction

- Quality of work life (employee satisfaction, safety performance, and absenteeism)

- Resource utilization (equipment, materials, and people)

- Service or timeliness (speed, performance versus schedule)

- Quantity (output, productivity, and efficiency)

- Quality

- Cost

The measures you choose should directly relate to the goals and objectives you establish for your utility. By connecting measures to goals you can demonstrate progress toward achieving goals. Table 12-2 presents potential measures from the seven listed categories within a framework of suggested goals for a utility.

Table 12-2 Organizational performance measures

Services, Assets, and Infrastructure	Customer Satisfaction	Employees
The utility delivers reliable, high quality, and cost effective services; protects public health and safety; and maintains the service reliability of our infrastructure and assets.	The utility consistently meets or exceeds customers' expectations.	The utility employs and values a diverse, efficient, and productive workforce.
Reliability 1. Number of unplanned water service interruptions **Adequacy** 2. Number of fire hydrants out of service for repair **Quality** 3. Number of validated quality complaints 4. Number of regulatory violations **Infrastructure** 5. Percent of CIP completed versus planned 6. Performance to scope, schedule, and budget 7. Capital investment as a percent of replacement asset value	**External Customer Satisfaction** 1. *Percent of customers whose expectations are met or exceeded in terms of:* • *quality of service (courteous, professional, accessible)* • *quality of product (value, reliability, adequacy, taste, odor)* 2. An index of quantitative measures of customer satisfaction • percent of correspondence answered within 7 days • percent of calls answered within 60 seconds • percent of total billings appealed • percent of total billings with corrections to prior billing periods • percent of meters that can't be read **Internal Customer Satisfaction** 1. Percent of internal customers whose expectations are met or exceeded 2. An index of quantitative measures of internal service quality	**Employee Satisfaction** 1. *Percent of employees satisfied* • *job satisfaction* • *career development* • *job training* • *performance recognition* • *tools to do the job* • *quality of work life* **Quality of Work Life** (quantitative measures) 1. Percent of sick leave used 2. Number of grievances **Safety** 3. Number of incidents per 100,000 person-hours 4. Average amount of worker's compensation paid per year per employee 5. Number of days lost to on-the-job injuries **Organizational Capability** 6. Median number of training hours per employee 7. Average number of years of experience **Diversity** 8. Affirmative action profile

table continued on next page

Table 12-2   Organizational performance measures (continued)

Environment	Community	Financial and Organizational Performance
The utility protects, sustains, and enhances environmental quality for current and future generations.	The utility reflects and strengthens the spirit and values of our community.	The utility is a responsible financial steward and improves productivity and performance each year.
Conservation/Sustainability 1. Percent reduction in water demand per residential customer Environmental Stewardship 2. Percent of envirostar site awards	Community Outreach I 1. Percent who think the utility communicates clearly and regularly about programs and activities in their neighborhood or community 2. Percent who find our community programs valuable Community Outreach II (quantitative measures of community outreach) 3. Program participation rate (performance to plan) 4. Number of outreach efforts	Financial Stewardship 1. Debt coverage 2. Bond rating 3. Net income Efficiency 4. Overhead rate 5. Operating expense per million gallons of water delivered Sales and Revenue 6. Total revenues compared to rate study projections 7. Total volumes compared to rate study projections

ASSESSING YOUR PERFORMANCE AGAINST OTHERS

Chapter 4 describes a framework for identifying performance gaps and comparing yourself with other comparable utilities. This section presents a brief background and several tools for the practitioner to use for metric and process benchmarking. Metric benchmarking is a quantitative, comparative assessment that enables utilities to track internal performance over time and to compare this performance against that of other similar utilities. Process benchmarking involves identifying work procedures to be improved through step-by-step process mapping and then locating external examples of excellence in these process elements for setting standards and for possible emulation.[7] As previously indicated, the water industry is still in the formative stage in developing cross-utility indicators of performance, so tools are not always full of data or comparisons.

---

[7] "Performance Benchmarking for Water Utilities," American Water Works Association Research Foundation, 1996.

Emergence of Performance Standard in the Water Industry

In competitive markets performance comparisons are continually being made between firms and their competitors and between firms and the industry as a whole in terms of costs, performance, quality, customer satisfaction, and ultimately in terms of sales and profitability.

Water utilities have been slow to develop a set of standards for the industry. Perhaps this is explained by the fact that 70 percent of the industry is publicly owned and operated. In the public arena, performance imperatives are more closely tied to a political process that centers on rates and budgets. If rates do not increase markedly on a year to year basis, the assumption is often made that the utility is performing adequately.

This approach to public utility management is no longer adequate because of increasing rate payer expectations for better services and lower costs, and because of the potential for competition by private utility service providers. Accordingly, leaders in the water industry in the United States, Canada, and abroad have begun to develop industry-wide performance measures and standards. These efforts, which have been conducted thus far in terms of benchmarking, are being undertaken under several auspices.

An initial major effort at benchmarking was conducted in the United Kingdom. Although it was plagued with many of the difficulties that arise in numeric comparisons across utilities, it ultimately led to notions of "yardstick competition" and to the performance measures that are in use today by United Kingdom's Office of Water Regulation (OFWAT).

A second effort is an AWWARF project aimed at developing an industry-wide clearinghouse for benchmarking in the water industry. This effort includes development of a model for water utilities for process benchmarking, the creation of high level performance indicators, and coordination with WATERSTATS, AWWA's industry-wide water utility database. Efforts are also underway to link the water industry benchmarking standards to the American Quality and Productivity Center in Houston.

Another significant effort has been the West Coast Benchmarking Study in the United States. This effort involves the collaboration of 15 peer utilities in western states. The first goal of this effort is to develop a common business model to facilitate process comparisons across the utilities. By using this model the group has so far focused on maintenance processes and on generating both quantitative and qualitative data.

Making Metric Comparisons Among Utilities

WaterStats is a data base maintained and updated by AWWA. It includes quantitative information and explanatory factors for U.S. water utilities of all sizes and characteristics. Data collection and definitional limitations make the use of the data imperfect. However, with adequate data filtering and selection of appropriate comparisons, the information can be assembled for assessment.

The City of Austin Water and Wastewater Department (WWD) began a self-assessment effort in 1992 that included metric benchmarking (Case Study 2.13). In 1997 25 utilities responded to a survey intended to collect raw data and identify industry standards. Austin WWD's continuing work has led to process improvements and workforce reductions among supervisory ranks. They have learned about the challenges of comparing data across utilities, confidentiality issues, and the need to combine metric benchmarking with process benchmarking and other self-assessment efforts.

A close look at performance data (metrics) will generate a large number of questions that require investigation with potential benchmarking partners. Are the same definitions used? What is the quality or accuracy of the data? Is it verifiable or is it an estimate or a guess? Does it reflect comparable accounting assumptions and are the same activities reported? Are the utility characteristics similar: wholesale versus retail; age and size of system; ground water versus surface water; filtered or unfiltered; gravity or pumped supply; and so on. While no two systems are identical, segmentation by utility type and characteristics, including within particular utility functions, can assist in comparing performance.

Using Process Benchmarking to Achieve Performance Improvement

The desire to improve performance by learning from and comparing to others is leading to greater interest in process benchmarking. The American Productivity and Quality Center's (APQC's) International Benchmarking Clearinghouse (IBC) and Arthur Andersen and Co. developed the Process Classification Framework to encourage organizations to "see their activities from a cross-industry process viewpoint instead of from a narrow functional viewpoint." The framework provides a common language that forces organizations to think outside their functional boxes and to map their businesses processes. The framework includes 13 operating and management support processes. The 13 process categories

Table 12-3 IBC Benchmarking Clearinghouse Process Framework

Operating Processes
- Understand markets and customers
- Develop vision and strategy
- Design products and services
- Market and sell
- Produce and deliver for manufacturing
- Produce and deliver for service
- Invoice and service customers

Management and Support Processes
- Develop and manage human resources
- Manage information resources
- Manage financial and physical resources
- Execute environmental management proofread
- Manage external relationships
- Manage improvement and change

Source: International Benchmarking Clearinghouse Process Classification Framework, 1995.

are listed in Table 12-3 and have been used in an adapted form by the WCWUBG and AWWARF Project 359. This classification framework allows for comparisons not only between water utilities, but also with partners outside the industry.

Business Practice Scaling is a process benchmarking tool that can be used to evaluate and improve specific business areas. This approach uses a series of questions to give a comparison of a utility's practices with the practices of best in class utilities. The logic of this approach is that a utility can find ways to improve performance for certain functions by comparing to the best in class utility.

The Business Practice Scaling Matrix, shown in Table 12-4, is divided into five categories: frequency, emphasis, formality, systems, and results. The appropriate category should be selected when answering business practice questions. In Table 12-5 some questions aimed at the maintenance function are presented. These questions are arranged by the appropriate categories of Table 12-4. This example was used in the West Coast Benchmarking Study and is included here as an example.

Table 12-4   Business practice scaling matrix

Percent Used Category	0–5%	6–25%	26–50%	51–75%	75–95%	96–100%
Frequency	Never do this	Done on an as needed basis for critical programs and activities	Sometimes used, numerous programs and activities	Typically do this, on many programs	Often do this, omitted only in exceptional circumstances	Always do this; standard operating procedure
Emphasis	Not emphasized	Receives modest emphasis; some efforts underway	Moderately emphasized; attempt to adhere to	Generally emphasized; done and checked	Strongly emphasized, standard we measured and recorded by	Heavily emphasized, one of the principles business is based on
Formality	No formal process	Done informally only; ad hoc procedures	Semiformal process; some routine procedures exist	Formal process exists, modestly documented, good but still evolving	Formal documented process, well tested and well followed	Strict formal process exists; well documented, no deviation
Systems	No system	Manual system exists or plans for automated system in place	Automated system exists, meets basic user needs	Good system in place, widely available, meets nearly all user needs	Strong system in place, meets nearly all user needs	State of the art system in place, meets all user needs
Results	No results	Minimal results, long way to go	Some results still below expectations	Some results still below expectations	Excellent results, still some room to improve	Unparalleled results, a total success

Source: Paralez, L. Rose Enterprises, Inc. Feb. '97 Preliminary Maintenance Practices Report for the West Coast Water Utilities Benchmarking Group, Ogden, UT.

Table 12-5   Maintenance questions applicable to scaling matrix

Frequency
1. *To what extent do your maintenance and operations practices include the following:*
   a.   *Bar code labor reporting*
   b.   *Activity based cost tracking*
   c.   *SPC techniques for tending and prediction*

Emphasis
2. *To what extent do you use the following planning practices:*
   a.   *Technology performance trends are charted and evaluated against life cycle, risk and potential*
   b.   *Publish a technical plan*
   c.   *Patent strategies*
   d.   *Participate in testing programs*
   (*If there is little attention to new technology—materials, methods, and hardware—then preventive maintenance may be higher because of lagging behind in use of more reliable methods and materials.*)

3. *To what extent do you update preventive maintenance planning based on:*
   a.   *Manufacturer standards/quotes*
   b.   *Actual performance data and trending analysis*
   c.   *Simulation*
   d.   *Best guess, experience, or intuition*

4. *To what extent do you emphasize the following performance indicators in evaluating operations and maintenance performance:*
   a.   *Schedule adherence*
   b.   *Labor productivity*
   c.   *Cost of missed maintenance*
   d.   *Cycle time*
   e.   *Equipment utilization*
   f.   *Budget performance*

5. *To what extent are the following practices used to handle field failures:*
   a.   *Failed part/system undergoes laboratory testing*
   b.   *Accurate and detailed failure history is maintained*
   c.   *Failure data is online and accessible to all operation/maintenance crews*

Formality
6. *To what extent do operations crews/teams participate in the design of preventive maintenance programs? Are operations and maintenance activities scheduled and managed using an integrated, event driven master schedule for the entire utility?*

7. *To what extent does the maintenance planning process formally incorporate the following:*
   a.   *Set quantitative goals and allocate resources to achieve them*
   b.   *Monitor/report internal progress towards goals*
   c.   *Monitor trends and adjust targets*
   d.   *Employ a formal process to get input from operations and maintenance crews/teams*

table continued on next page

Table 12-5   Maintenance questions applicable to scaling
matrix (continued)

8. *To what extent are the following elements automated?*
    a.   Work instructions
    b.   Material requirements for maintenance activities (kiting)
    c.   Quality and test requirements (specifications)
    d.   Priorities for maintenance work scheduling

9. *To what extent do you have a computerized system for scheduling and tracking preventive maintenance?*

Results

10. *To what extent do your operations emphasize preventive maintenance as evidenced by:*
    a.   Percent of (valves, meters, water mains, and other supplies—list each separately) on a preventive maintenance program
    b.   Percent of (valves, meters, water mains, and other supplies—list each separately) overdue for service

11. *To what extent are operations and maintenance activities scheduled and managed using an integrated, event driven master schedule for the entire utility?*

Source: Ibid

---

*A*WWA *is adapting its QUALSERVE program of self-assessment and peer review for use by smaller systems. The self-assessment instrument is a valuable and informative first step in identifying key opportunities for improvement. Peer review is a particularly efficient way to get expert advice and counsel that can reinforce and even guide the thinking of your stakeholders and advisers (For information contact the QUALSERVE program manager at AWWA.)*

Obtaining Third Party Assessments

QUALSERVE peer review.   The AWWA peer review program provides utilities with an opportunity to get an expert third party review of their organization. The peer review process is intended to help organizations highlight excellent utility performance, to identify business area(s) needing priority attention and to develop an agenda for improvement. The reviewers are volunteers with the experience and expertise to gather and analyze information and provide suggestions for improvement. As listed in Table 12-6, the program is a comprehensive review that covers 14 performance categories identified for the program, or can be a more specific review that covers a subset of these categories. Reviewed utilities receive a written report from the reviewers that includes strengths, weaknesses, opportunities, and barriers.

Other audits.   For water utilities there are typically a number of audit opportunities available that can be leveraged as assessment tools. While sometimes considered adversarial, forward thinking utility managers often use these audits to identify both performance gaps as well as opportunities to use recommendations and

Table 12-6 AWWA QUALSERVE peer review performance categories

1. Leadership and organization
2. Water resources planning, development, and operations
3. Water treatment
4. Water quality management
5. System management and operations: maintenance management
6. Capital improvement program
7. Financial and fiscal management
8. Business and strategic planning
9. Emergency response
10. Customer service
11. Customer account management
12. Human resources
13. Government, business, and community relations
14. Information systems

Source: AWWA QUALSERVE program

findings to support needed improvements. Included here are regulatory, environmental, safety, financial, engineering, and management audits.

SUMMARY

Knowing where you really stand and where you would like to be in any function or activity is the only sound framework for launching a program of performance improvement. Assessing this gap is necessary to define the specific elements, features, and processes to be changed in the utility. This assessment can be achieved by determining the extent of alignment with customers, understanding internal capabilities and performance, and comparing against others, both in terms of metrics and work practices. Although our experience is limited to well developed, industry-wide assessment tools, utilities have many resources for examining performance in various functional areas. The fact that many utilities are applying a variety of methods and devising new ones demonstrates that the industry is recognizing the practical value of measuring performance.

Following recommendations of an audit, the Los Angeles Department of Water and Power (DWP) began a work management improvement process in its water operations division (Case Study 1.1). This ambitious effort led to reduced crew sizes, warehousing improvements, and training initiatives. Its workforce was reduced by 20 percent while productivity increased by 30 percent. The audit generated a specific directive for DWP to move forward on benchmarking and performance measurement. DWP is now focusing on process benchmarking and redesign in conjunction with the West Coast Benchmarking Group.

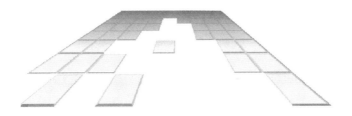

CHAPTER THIRTEEN

Quality Management and Work Process Improvements

The best management practices in any organization are those that keep the people, processes, and larger systems focused on the vision of the organization and on the short- and long-term needs of customers and stakeholders. These practices should also align decision making with this vision and with customer needs. As utilities seek to expand and enhance the role of employees and to create more flexible and efficient processes, the command and control management systems in traditional organizations must give way to the more difficult management practices guided by vision, values, and customer needs.

This chapter introduces the utility manager to a Total Quality Management (TQM) approach to self-assessment and improvement. First it describes some basics of quality management; next it provides an outline and some useful strategies for improving work processes; and finally, it gives some pointers on managing the improvement program that have proven effective for utilities and other organizations.

INTRODUCING QUALITY MANAGEMENT

A large number of water utilities have introduced quality management or TQM programs in their organizations. First and foremost

"*Total Quality Management is a term initially coined in 1985 by the Naval Air Systems Command to describe its Japanese-style management approach to quality improvement. Since then, TQM has taken on many meanings. Simply put, TQM is a management approach to long-term success through customer satisfaction. TQM is based on the participation of all members of an organization in improving processes, products, services and the culture they work in. TQM benefits all organization members and society." From Quality Progress magazine, a publication of the American Society for Quality Control.*

these programs connect the organization's work processes to the customer. Our customers experience the results of our services and operations. One way to improve results is to improve the processes that create those results. Utilities are challenging themselves to meet or exceed customer expectations.

Considering Quality as a Measure of Customer Satisfaction

It is useful to think of quality in terms of the following characteristics or customer perspectives.[1]

Quality is:	
Transcendent	It possesses innate excellence. The customer knows when it exists. They can tell if water tastes good, if water flows from the tap, if timely and courteous service is provided.
User based	It is fit for consumption and satisfies customer expectations and preferences. The customer drinks tap water; bottled water is not a necessity.
Standards based	The product conforms to necessary requirements. Facilities are designed to standards; water quality meets regulatory standards; environmental, safety, personnel, maintenance, and supply standards are met or exceeded.
Value based	The product gives the customer the greatest value for the lowest price. The utility is efficient and provides a high standard of service at a bargain rate.
Product based	Quality can be distinguished in terms of differences in quality of some desired ingredient. Fluoride is added to the water; levels of water quality protection are achieved; levels of water pressure are provided to customers; certain information is provided with customer bills.

Managing for Quality

Effective quality programs require developing and maintaining the organizational capacity to constantly improve quality. A framework for understanding quality management involves planning your quality management program, designing and using tools that

---

[1] Adapted from Ron Zemke, "A Bluffer's Guide to TQM," *Training,* April 1993, 50.

help assess performance, and focusing on key work processes that promise the greatest potential for improvement.[2]

Planning for quality.   The utility must first figure out who the customers are they are planning to serve. This could include retail and wholesale customers, consumers of water, and stakeholders such as regulators, governing bodies, employees, unions, citizen and community groups, and other agencies.

The utility should next determine the needs of those customers by developing a means to define service attributes of importance. The quality management process designs and employs techniques to capture the voice of the customer, translates that voice into a quality planning effort, and engages in planning features that respond to the customer's needs.

Finally, the quality planning process is not complete until the plan has effectively been transferred into operations.

Designing and using assessment tools.   For quality management to be effective in the operations of an organization, the organization must be able to use specific tools and techniques. The tools may include performance measures, benchmarking, or peer review, and be aimed at evaluating actual operating performance to improve or create customer satisfaction. The use of proper tools includes comparing performance against targeted levels and acting to improve performance gaps. Chapter 12 discusses the tools that can be used.

Focusing on key work processes.   The third facet of managing for quality is aimed at attaining new and higher levels of performance. This effort should focus on identifying performance gaps by using metric or work process comparisons with peers. In the water utility industry there are many potential opportunities for making improvements, some of which are discussed in this book. Improving performance can often involve stripping away constraints that define current processes; eliminating productivity limitations driven by acceptance of traditional union rules or traditional utility structures; minimizing limitations caused by a lack of investment in technology enhancements (automated meter

---

[2] Juran, J. M. and Frank M. Gryna. *Juran's Quality Control Handbook*. New York: McGraw Hill, 1988.

reading, treatment operations, or management information systems); and changing practices that lead to wasted capacity (e.g., laboratories using only one-third of available operating hours).

IMPROVING WORK PROCESSES

The fact that utility work processes deliver outputs, services, or products is inherent to the challenges of quality improvement. It is impossible to separate the performance of an organization or individual from the performance of the system they work in. In fact, experts tell us that greater leverage for overall improvement can be gained by focusing on improving the work processes.[3]

Redesigning work processes in the water utility industry will challenge new ways of thinking in the organizations. Traditional demands for bureaucratic control have lead to the creation of fairly rigid and inflexible processes, aligned in a hierarchy under strong, vertical, chains of command. The environment in which most utilities now find themselves demands flexibility, innovation, and cross-functional thinking and action. Managers must know whether or not a process works, and what dynamic forces drive the process in order to know whether it will result in the desired outputs. The goal is to have value-added work processes that result in delivering value as defined by the customers of the process in question.

All functions within a water utility have the potential to be improved by examining specific work processes. Automation of meter reading, customer billing, mapping, drafting, management reporting, surveying, supply management, laboratory information, and water quality monitoring are but a few examples of processes in which huge improvements have been made by many utilities.

Four components are common to process improvement approaches used in utilities: understanding the work process, analyzing the work process, redesigning the process, and measuring

---

[3] Deming, Dr. W. Edwards. *Out of the Crisis.* MIT, Center for Advanced Engineering Study, Cambridge, MA, 1986.
Harrington, H. James. *Business Process Improvements.* McGraw Hill, New York, 1991.
White, Thomas E. and Fischer, Layna, *New Tools for New Times: The Workflow Paradigm*, Future Strategies, Lighthouse Point, Florida, 1994.

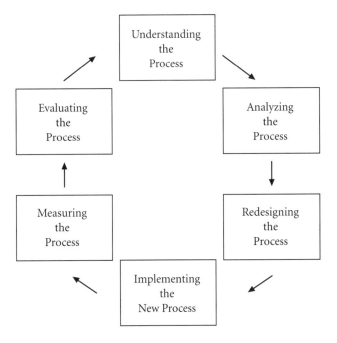

Figure 13-1 Improving the work process

the new work process performance. Figure 13-1 demonstrates these and two additional components of process improvement.

Understanding the Work Process

Understanding a work process is best accomplished by process mapping or flow charting, usually conducted with the help of a trained facilitator and normally involving those who work in the process being mapped. This starts with diagramming what actually happens in a work process including all activities and decision points. Participants define a beginning and end point for the process being studied through use of input/output models. The use of labor, materials, equipment, and work methods are documented along with associated time intervals. The group also defines how that process connects to its larger work system.

Analyzing the Process

Once the work process has been flow charted or described in a way that users can understand, an analysis is conducted to search for the following types of features. These represent strong improvement opportunities for efficiency gains or customer responsiveness.

Redundancies. *Improvement can be achieved from parts of a process that are repetitive or conducted by more than one person, such as multiple reviews, sign-offs, signatures, or inspections.* An analysis of redundancies might reveal ways to automate reviews, to group redundant activities earlier or later in a process, or to eliminate unnecessary steps. Typical examples of redundancy in a utility might be unnecessary layers of approval for personnel, contracting, or work order activities; inspection teams that are used for construction or cross-connection activities that do the same work as a project manager; or work crews that are too large because individual members are over specialized and not allowed to perform multiple skills in completing work.

Bottlenecks or Constraints. Doing time analysis of work-flow through a process can reveal bottlenecks that cause operational inefficiencies. It is always important to understand where in a process bottlenecks exist, since improving workflow below a bottleneck is not likely to result in much improvement in efficiency. Looking for backlogs is one way to identify possible bottlenecks and constraints. Purchasing, timesheet processing, accounts payable and receivable, material requisitions, customer service orders, laboratory analysis, as-built engineering drawings, and water system maintenance activities are examples in water utilities where backlogs are often found. Work process teams and process benchmarking can help a utility define better ways to perform these activities.

Complexity. The most complex work processes often present streamlining opportunities. The presence of too many decision points in a process is an indicator that perhaps a process is being made too complex. Any control feature in a process, such as reviews and approvals, is likely to slow a process and create complexities that contribute to poor process performance. Complexity also contributes to error when there are numerous potential actions that might be selected with limited immediate feedback on what works best. *Creating standards is a way to streamline many utility operations.* These standards could include standard customer charges for service work, use of standard prepackaged materials for system maintenance activities, creation of standard engineering plans and specifications, and standard work methods for field operations, preventive maintenance, and so forth.

Feedback loops. The identification of feedback loops is probably the most powerful of all process analysis steps. *Utilities need to see the effects or results of their actions and service delivery.* When effects are far removed from actions, it is difficult to assess performance. In charting processes utilities often identify initial actions that cause the most problems later in the process. Shortening feedback loops and closing information gaps within a process is a powerful improvement technique. Customer surveys, water quality monitoring, Supervisory Control and Data Acquisition (SCADA) systems, customer call frequency, field inspections, and auditing are a few ways that information tools are useful in water utilities as feedback mechanisms.

Technology enhancement. It is also important to evaluate whether or not existing technology is being fully exploited, and whether or not the addition of technology would be worth the investment to improve work processes. A menu of opportunities is provided in chapter 9. Bar coding, GIS technology, automated meter reading, and SCADA technology are examples where significant improvements to cycle times have resulted from data storage opportunities available through technology.

Alignment. *Results from utility work processes need to be measured so alignment can be assured with the strategic vision, mission, goals, and values of the organization.* Are customer needs being met? Are there steps in processes that do not contribute sufficient value to warrant the costs involved? Are there entire activities that are not worth doing? Is organizational performance improving?

Skills assessment. Sometimes it is necessary to evaluate individuals in the utility to make sure they have the skills needed to carry out their tasks efficiently and well. Further training may be part of your improvement solution. Employees, particularly those performing the work tasks, are usually the best source for identifying skill deficiencies. *Investing in employee training and development is critical to efficient and effective work processes.* Utility self-assessments, strategic plans, information technology plans, and employee performance evaluations are also good starting points for skill assessment.

Redesigning the work process

Process redesign is the next step to correct weaknesses and problems uncovered during work process analysis. It is also an opportunity to develop improvement and development plans for products, services, and individuals. *Redesign affords an opportunity to benchmark with comparable utilities.* It helps to develop new workflow scenarios and to test them on a pilot basis.

Measuring Work Process Performance

An essential element of redesign work is determining how well the work process is performing. Measuring and managing quality during the transformation of a process is your route to building quality into water utility work. Success here must be measured against the system's guiding principles and against the standards set by the reward and measurement system you have established.

What comes out of an organization is measured at the output stage. For the water utility industry there are fairly clear expectations with regard to product, but less definition with regard to the quality of service outputs, quality of financial performance, quality of maintenance operations, quality of work life, and so forth. Yet it is possible to set standards for these "softer" processes. Having standards enables everyone to measure success on the same terms.

The management and measurement of quality depends on understanding customer satisfaction. Although this process model can look linear and sequential, it is in reality a circle where customers on the receiving end of your system are also often present in the planning system. If they are not satisfied with your performance or outputs, they are able to create a new set of expectations and criteria for you to manage. Surveys and other methods are discussed in chapter 12 as methods of assessing customer satisfaction.

Organizations typically select a family of measures designed to reflect process performance. These include, but are not limited to:

- *Cost measures* such as cost per million gallons of water sold, performance within budget, and ratios of input costs to output value

- *Customer satisfaction* or customer success measures that are usually qualitative, but can sometimes be quantitative, such as percent positive on a satisfaction survey; quantity of problems, concerns, and complaints; or quality impact of failures, errors, and omissions

- *Resource utilization* measures that include features such as number of labor hours per unit of output; labor or equipment utilization and capacity; materials lost (to waste, rework, or error); and energy use

- *Quantity measures* such as total production, number of items produced, decisions made, or services rendered

- *Quality measures,* which can be related to water quality, safety, or environmental standards

- *Service or timeliness measures* relating to cycle time, duration of tasks, delays, or labor hours

- *Quality of work life,* or employee satisfaction measures that are reflected in safety, performance, absenteeism, grievances, tardiness, percent positive on an employee attitude survey, or corrective actions

MANAGING YOUR IMPROVEMENT PROGRAM

The principles of quality management and work process improvement may seem elusive or vague to utility personnel who are attempting to implement them for the first time. Here are a few practical tips, drawn from the experience of a number of organizations that may help ensure success.

Irvine Ranch Water District (IRWD) has implemented a comprehensive Total Quality Management program (Case Study 1.6). Objectives are to implement a continuous improvement program, address internal and external customer needs, and improve management and work processes. IRWD created a quality council to oversee the program, trained all employees, created cross-functional process teams, established a performance measurement system, and used a three year action plan to implement the program.

Teams were formed to work on a wide range of priorities in their utility: customer service reporting, engineering project management, energy management, employee ridesharing, customer satisfaction, and others. IRWD has learned that top level management support, prioritization, employee participation, and continual program updating are factors for success.

Choose Winners First

Utilities carry out their work through hundreds of processes. Based on your self-assessment, start with high priority processes that are likely to improve quickly as candidates for the first performance improvement efforts. Visible progress and success in your initial efforts will build staff confidence in the improvement process.

Select the Right Employees

The process review team must include people involved in the process and the major stakeholders. The team also needs either a consultant/facilitator or a trained, enthusiastic, highly competent staff team leader. As much as possible, the team members should have positive attitudes and be committed to what the utility is trying to do.

Give Your Teams Training

Sending a team off to do a job without training them in the process and showing them its value to the organization and to themselves is folly. Invest time and resources in good preparatory training; the team will then be eager to apply their new skills to the task at hand.

Communicate with the Entire Staff

All employees need to be well informed about what you are setting out to accomplish and why it is necessary for the utility and for them. They need to understand that they will share in shaping the utility's future. They also need to be informed of progress and results as your program moves forward. Take every opportunity to reinforce your seriousness and the importance of their help in achieving the organization's new goals and vision.

Support Your Process Teams

To the extent possible, provide the resources needed to ensure that teams will be able to do their jobs. This includes allowing time for meetings and follow-up assignments; providing information from other sources and utilities; gathering information on potential technical and electronic aids; giving the opportunity to attend appropriate professional meetings; and providing special consulting or professional assistance when they encounter barriers.

Be Realistic About Time

Process improvement activities and their larger organizational improvement programs take time—lots of time. You want to maintain forward movement, but understand that you are entering an ongoing process, an open-ended process involving years of working in a new way. You want some early victories and successes along the way but do not frustrate yourself or your staff by short-term thinking.

Stay Involved and Interested

Arrange for feedback and reports from process teams along the way. Both management and employees should be kept informed. Demonstrate your interest and support. Praise contributions and progress and successes. Confirm by your actions and behavior that the work of the teams is at the center of your interest, and that you also want it to be at the center of theirs.

Trust Your Teams

Experience has proven that empowered, committed employees can recommend improvements not even imagined by management. Encourage risk taking, uninhibited thinking, bold innovations and open-minded creativity. Teams need to understand that they are not searching for a solution that you might have in mind. They have been commissioned to discover or invent new and imaginative ways of doing things more efficiently and more cost effectively than the utility has ever known before. They will embrace the challenge and respond to it.

SUMMARY

Process improvements require much time, effort, and creativity. There is a wealth of experience to use within your organization.

Utility staffs, given the opportunity to learn from their experience, can help lead your utility to a new level of high quality performance. Look within the organization for those work processes that you believe can yield large returns. Provide strong leadership and commitment to the quality improvement process and involve employees in identifying and making internal improvement. Measuring performance and addressing customer needs and satisfaction should be the constant focus of your quality management program.

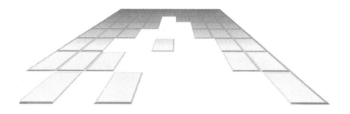

CHAPTER FOURTEEN

People and Change

The water industry has been very stable. Water is supplied to customers 24 hours a day, 365 days a year with reliable pressure and reliable quality. This type of stability has led to an industry that has historically been immune to change. Stability and reliability have been excellent qualities and characteristics for employees. However, with changes now affecting the water industry, employees are being asked to become more flexible and creative in problem solving. The change will engender fear in some employees used to stability. The fear will have an impact on their attitudes, postures, and behaviors. In many cases the employees will not realize that they are reacting to fear. They will see circumstances around them change; they will not like the situation and think proposals are wrong; then they will resist the proposed changes.

Some managers don't want to deal with employees who are reacting to change. The manager's own fear of change will keep him or her from being bold and from making the changes the organization needs to respond to today's conditions. This chapter is about change and fear of change, and gives managers and leaders some strategies to deal with the problems they will encounter as they try to turn their organizations in a new direction. There is not one simple way to handle problems that will arise; there is no formula for success. Each situation is different and must be addressed with a combination of insight, skill, sympathy, and energy.

Once organizational change has been introduced it is important to nurture and maintain the organization in a learning process so it continues to accept ongoing change. Maintaining a nurturing environment in a time of change requires a strong team-based orientation that is flexible and able to respond to new demands. It also requires strong training and development programs that support employees and provide new skills to match new opportunities. A combination of teams and a strong training program will lead to an organization that has a learning environment where continuous change is expected, planned for, and valued.

This chapter also addresses issues of communicating and working with labor unions for those utilities that are organized. Unions are a strong influence on employees. They can be collaborative partners by identifying issues that will cause concern to employees and by helping to explain management's objectives to employees. However, they can become a source of continuous discontent if relations break down between labor and management. It is important to work with unions and to have a solid basis for communication with them so that small incidents don't get blown into significant problems.

Because of the breadth of the material covered in this chapter, it cannot be thorough and comprehensive. Entire textbooks and seminars address aspects of issues relating to employees and change. This chapter identifies the main problems and issues that a manager must be aware of and sensitive to. Once a problem is encountered, a manager is advised to review literature or to seek further assistance. The reference section at the end of the book identifies some leading books and thoughts on management. The American Water Works Association (AWWA) also offers programs and seminars on aspects of change management.

THE CHALLENGES OF CHANGE

What causes change? Change is all around us. It can come from a simple variation, such as a new director or manager of an organization who has new ideas about how to do things. It can be brought about by significant organizational restructuring, such as a merger or an effort to privatize. It can come because a special

234

interest group is challenging a basic program of the utility. Sometimes change comes because the utility decides to implement a new technology or process that affects current operations. This book is all about the types of changes that might occur and the reasons for them.

In most cases the reasons are logical and wise, and the change being implemented will, after a few bumps, lead to a new structure that will be better for the organization as a whole. The situation may also be better for individual employees; however, they often do not see it that way initially. They see that their lives are changing. They have new work to do, a different shift, a new manager, a different office or workplace. Their reporting systems have changed, their colleagues and peers may shift locations, and the water cooler may even move. The challenge facing the supervisor, manager, or director is to help employees see the opportunities embedded in the changes and to help them embrace the change as a better way of doing business. It is important to work with those employees who cannot make the shift. They can be moved to different work, or as a last resort, moved out of the organization in a humane way. Management will be watched and judged by all employees on how respectfully they deal with the problems they encounter.

Which theory of change management is right? Some creative managers will look to the theories of change management for advice and counsel. The literature will certainly guide and assist, but each change situation is different enough that most managers will have to customize their response to the needs of the organization. William Bridges' book, *Managing Transitions: Making the Most of Change*, sets forth a number of checklists that are useful guides to moving through a process. The idea that change is a process of transition with a beginning, a middle, and an end is a useful concept for thinking about the stages that different people and sections of an organization might be in at any one time. Following are problems to keep in mind when determining your own approach to change management:

People change at different speeds. Employees in an organization and even in a group will accept change differently. Some people (those who probably rearrange the furniture in their living room every six months) get exhilarated by the idea of doing

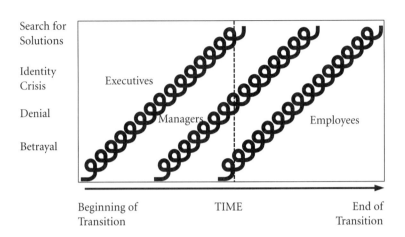

Source: Adapted from *Business as Unusual*, Pritchett & Associates, Inc., 1991.

Figure 14-1   Perceived transition

something differently. Others like stability and have lived in the same neighborhood or house for decades. One of the difficulties of change management is that reactions go through an organization in waves. Some employees will accept new ideas and respond readily; others more slowly. Just as the first group is overcoming its resistance and is ready to move forward, someone else will just be realizing that the change is real and that the boss really means it. Resistance may then arise at that point. (See Figure 14-1.) One important strategy discussed later in this chapter is to repeat messages frequently because the time of change acceptance is varied.

Employees at different levels hear different messages. The leader of an organization usually starts with a leadership team at the highest level to begin to define a change. It is expected that the top level of directors will carry the message down to managers, and managers to supervisors, who will carry the messages to employees. The problem with this system is the delayed reaction caused by the cascading effect. It is important for leadership to try to get messages out to all levels of the organization at about the same time. Although it is not realistic to think that change will be symmetrical throughout the organization, the cascading system of information delivery means that the complexity of change will be enhanced because employees at all levels are at such different points in the process. Messages should be sent out centrally to all employees at one time so that information is consistent and appropriately timed.

Public employees have not faced much change. For most public sector employees change such as reorganization, downsizing, a merger, or privatization is a change of major proportions. Most cities and water utilities have been immune to significant reversals. Most public employees believe that they will have a job until they retire and feel quite safe from the types of changes that have become commonplace in the private sector. That kind of stability engenders the thought that "it will not happen to me" and leads to a strong resistance to proposed change. There is a common feeling that employees have been there "longer than the boss and will still be there when he or she is gone." Such an attitude leads to a belief that an employee can out wait the proposed idea and see if it will go away. Employment stability in the public sector is a kind of "tenure" that can lead to deep-seated feelings of resistance and denial among employees and that may require strong action by leaders to overcome. Employee trust in "tenure" has also contributed to lower productivity and efficiency in quasi-monopolistic utilities.

Employees need to understand the benefits of change. Often managers describe the need for change in terms of a problem that needs to be fixed. Employees interpret this type of description as a criticism of their work and tend to become immediately defensive of their performance and productivity. Hence, they defend the status quo and current way of doing business. In order to motivate employees to make a change, it is critical for them to understand how they will benefit from the new way of doing business. Utility management must make an effective case for the expected benefits of a change and be clear with employees about the new opportunities it creates. One way to build a climate for positive attitudes toward change is to emphasize the strengths and skills of the workforce and to develop justification for new approaches as enhancements to an existing well-developed program.

THE CHANGE PROCESS

Most theorists agree that there are stages to change that are much like the stages in the process of death and dying. People will move through denial, anger, fear, and bargaining until they reach an ultimate stage of acceptance. Once they have reached acceptance there may still be a period of slow recovery before an organization is back to full productivity. The stages of change mean that productivity will likely decrease during the transition, and managers should expect that. The desired change should be one where there will be significant improvement in productivity so that the results of going through the change process are worthwhile.

Employee behaviors will be different during a change. Some will be excited, responsive, and optimistic about the outcome. These are the employees that should be given special tasks and leadership roles where possible. By their enthusiasm they will bring others along. They will also participate in a creative, dynamic way that will bring new ideas to the organization. However, other employees will express resistance in less productive ways; some will express anger and outright disbelief, some will be silently hostile and express discontent indirectly, and others will simply be quietly unhappy and less productive. It is important to understand that all these behaviors are normal. With the right listening and encouragement, and by responding to employee concerns, much of the unhappiness can be overcome so that they will return to being productive and energetic with the new vision in mind. Management should expect that it can take three to five years to adjust to a new work culture following a major change.

Psychologists tell us that the fear of change is in large part due to fear of failure. People know they can do their job the old way, but they are not so confident about doing new tasks and making new judgments. They hate to give up the familiar, comfortable patterns that have worked for them. Letting go of the tried and true can be profoundly frightening. Stepping into the unknown can be risky. A person needs time to get used to the idea of changing.

As they support employees, directors and managers also have to know that there is a point to draw the line to determine what is expected in the work environment. Productivity is ultimately the bottom line, and employees have to decide that they want to be part of the new organization and are willing to make the changes that performance improvement demands.

STRATEGIES FOR WORKING WITH AND OVERCOMING FEAR OF CHANGE

The literature of change management is full of ideas and examples of ways to overcome resistance to change. The following are the main issues that a manager will have to deal with and some suggestions for strategies that have been effective. Again, it is important to emphasize that these are just a few approaches and a manager should study the literature or seek guidance from experts on problems that seem extremely serious and likely to undermine proposed changes.

Communication

Excellent communication is critical to the success of any change strategy. It is almost impossible to have too much communication. Employees will want to know about proposed changes as soon as they are being discussed and everyone wants to hear news at the same time. When key decisions are being communicated it is often important to have lead managers and labor union officials hear the news slightly in advance of line employees so that they can interpret the announcement correctly. Communication should take place in as many ways as possible: print material, e-mail, newsletters, hotlines, and meetings. Employees have different listening and learning styles. If the change is major and significant, the general manager of the organization should have meetings with as many employees as possible to explain the vision and concept of the change. It is important for employees to have the vision communicated directly from the top, from the person who will be responsible for directing its implementation. The general manager should have both enthusiasm and passion for the change and should be able to explain it in a way that is convincing and believable. Throughout the period of transition leadership should hold meetings with employees to answer questions, to describe the process and report on progress, to explain decisions that have been made, and to encourage participation in the process.

A good communication strategy requires that information be widely available, accurate, and timely. A periodic newsletter, while useful as a regular source of information that people can count on and expect to receive, will not be sufficient. Rumors will fly and it is important to have a method of instant updating to inform people in a timely manner of new developments. If one is relying on a computer system for current new information, it is important to have a system for downloading that information into hard copy that is posted for field employees who may not have regular or easy access to computers.

Communicating with employees requires listening skills so that utility leaders hear the issues being raised and respond to them appropriately. It is important to acknowledge and respect employee concerns about the loss of the "old way" and not to push them too quickly into the new vision. Time spent understanding the feelings of loss attached to the old way of doing things will help many people move on more quickly and rapidly to a new way of doing business.

Utility managers should identify colorful phrases and quotes that may characterize the change required. Images will often be remembered longer than just words. Quotes and images help those people who learn visually and help convey messages more quickly if employees begin to identify action with a colorful phrase or image.

Employee Involvement

Successful change management requires a considerable amount of employee involvement. The difficulty is balancing involvement with moving ahead with the program. If so many staff are involved that the process of change is slowed, dysfunctional complaining can retard movement toward goals. If, in fact, certain decisions have been made about the direction to be taken, it is important to communicate and move toward that direction and not spend needless time getting input if it is not likely to be used. Employees are very sensitive to whether or not their involvement is useful and appreciated. If they believe that management is making all the decisions and ignoring their advice, they will resent and criticize the process.

Management should decide up front who will be involved in the reorganization, what decisions employees will have input on, and how that input will be taken. It is often useful to have a consultant assist with gathering input from employees about their experiences. It is easier for a consultant or facilitator to guarantee anonymity to employees who fear that any negative comment could lead to retaliation by management, and it is important to hear employees' worst fears. Management should be clear with employees about which decisions they are seeking input on and which ones might be delegated to employees. Confusion about the difference between management seeking input, while retaining decision making control, and management actually delegating decision-making can lead to misunderstanding.

Use of Outside Experts

Outside experts can be useful and helpful in carrying out a change, but they should not be in charge. They should be coaches and advisers to leadership. The burden of articulating and desiring the change is one that utility managers cannot transfer to a consultant. However, the consultant can be instrumental in advising how to implement change. The roles for experts in change management are advising leadership on organizational development activities, gathering employee input, organizing meetings, assisting in resolving issues and conflict, and team building around new organizational themes and processes. Consultants will often be more skilled at hearing and understanding employee concerns than some managers. They can provide valuable additional staff assistance by being responsible for devising plans for overcoming problems. Often managers themselves are struggling to understand and appreciate the change, and will have difficulty hearing and solving employee concerns.

Once a change has moved into an implementation phase, consultants can do a valuable service by working with restructured teams to help them understand the new demands of their changed roles. After determining the appropriate change structure, moving quickly is important and consultants can act as staff

An excerpt from a letter from David Ragar, Director of Cincinnati Water Works, on the topic of employee involvement in developing a strategic business plan:

"*Within the Cincinnati Water Works the selection of who served on those teams was a source of much debate. Initially top management in consultation with others including the union selected employees to serve. But when the first draft of the vision, mission, and values was presented to the entire organization for comments many complained that it was a plan developed by the same selected few. So a third group of employees was added to the process. This third group was made up of employees elected by the organization to represent them. Generally it was a ratio of one employee elected for every ten employees. Employees of similar job classifications elected one of their own to serve on this group.*

"*The lesson learned and probably true for most organizations is that if top*

241

management and the union leadership select individuals to participate in the planning process, it will not have the broad support and acceptance that is needed for it to succeed. Some sort of process needs to be developed that allows the employees to have input on their terms to the process."

advisers to managers who are having difficulty with resistance from an employee group.

Visible Actions

If the leadership team has taken the time to learn about employee concerns and the types of activities they believe will bring about responsive change management, it is important for leadership to make their actions visible. Demonstrate your responsiveness! As activities get completed, information must be transmitted to employees so that they know progress is being made. Again, it is important to transmit the information by as many media as possible. Managers should also continue meeting with employees after major changes have taken place so they will hear about progress in the organization, have opportunities to get their questions answered, and make contributions to the change. Progress and accomplishments by work process teams and other subgroups should be publicized and celebrated. Use all opportunities to point out how the organization is moving toward its goals and its vision of what it wants to be.

Concern for Employees

If the reorganization or change is going to require downsizing, that process should be done humanely with concern for employees in mind and with appropriate information to labor unions if they are active. If it is possible to downsize by using attrition and some type of early separation process, where employees are offered early retirement options or options to leave with some type of severance, there will be less stress than in a layoff situation. Layoffs should be avoided if alternative methods of downsizing can be developed. Public sector entities have the reputation of taking care of employees and tools are available such as hiring freezes, retraining opportunities, and transferring employees to other departments or organizations to assist in downsizing situations. However, if the bottom line is efficiency there may be times when employees must be laid off. In those cases, early warning and provision of types of retraining, counseling, and retooling programs will alleviate tensions.

MAINTAINING A NURTURING ENVIRONMENT IN A NEW ORGANIZATION

Today's organizations will be facing considerable change for the foreseeable future. In order to keep employees connected and challenged but not overstressed, organizations need to find a healthy balance between continuous learning and increasing expectations of employees. Human resources (personnel) is a critical resource to a utility. Once it has selected an employee, continuing to train, retrain, and upgrade the person's skills represents an investment in human capital. Other ways to develop a culture of continuous learning include using teams to accomplish tasks and supporting employees.

Teams as an Integral Element of Management

Teams are beginning to be used extensively in the utility industry as utilities restructure for more efficient operations. Teams have been used widely in sales and manufacturing industries and their value as an important element of organizational change is becoming commonplace in other businesses. Teams can be structured in a number of different ways to achieve business objectives. The concept is to bring together employees with diverse skills and talents to deal with problems that require a multidisciplinary approach. The range of skills desired depends on the nature of the problem to be solved. Teamwork can energize employees while providing them with an excellent learning experience through working with other people in the organization. Additionally, it is beneficial to address problems from a multidisciplinary approach.

Teams can be used for evaluation, solving management problems, designing solutions to problems, running projects, or assuring completion on a production schedule. They can be ad hoc in that they are established to work on a particular time-limited problem. They can exist for the life of a project, or they can be semi-permanent until a series of problems has been solved. Chapters 3 and 4 address the use of a team as the "oversight group or steering committee" for the development of the strategic planning process. Examples of other successful uses of teams are:

A project team. Set up to manage a major capital project through the process of bidding and construction. The team might include engineers, a financial analyst, a planner, a community relations specialist, a lawyer, and a community representative. The team could develop the RFQ/RFP, run the evaluation process, monitor the implementation of the contract, communicate with the community, and develop financial reports.

An organizational support team. Set up to work across an organization on an issue of importance to the department. The team might be established to develop a program for recognition and rewards within the organization, to improve communication, to further diversity objectives, or to develop a department training program. It should be culturally diverse and include representatives of each division in the organization and an employee from each hierarchical level. The team could design and implement the program and communicate it to employees.

A process redesign team. Set up to restructure a system within a division. An operations work process redesign team might consist of line workers, a supervisor, a planner, a financial analyst, and a safety specialist. The team could be responsible for identifying a process that needs improving, analyzing the steps in the process, finding a faster way to complete a task, determining whether time and money is saved, and deciding how to schedule crews to make the change happen.

A strategic policy team. Set up to determine a strategy for solving a regional supply or transmission problem. The team might consist of strategic planners, project engineers, a financial analyst, a community or intergovernmental representative, an environmental analyst, a representative of another jurisdiction, and a lawyer. The team could be established to determine a method of sharing resources across a regional watershed, to determine the engineering requirements to make it work, to negotiate a contract with another jurisdiction, and to communicate with the community on the progress of the project.

Structuring Teams for Maximum Effectiveness

In order for teams to be effective, everyone in the organization has to understand that they are designed to achieve results. They must be empowered to make decisions and to carry out projects or tasks. Some of the key ingredients that lead to successful team designs are:

Strong leaders. A strong team requires members who are designated leaders and viewed as influential by their peers. In a period of change, successful teams include some of the best and most intelligent employees because leadership needs the teams to succeed. When institutional leaders are picked to be part of special teams, then employees who want to succeed should be selected for this kind of special purpose work.

Clear mission. A successful team needs to have a clearly understood mission and objective. The mission is sanctioned by leadership and is broadly known so that team members can have influence within the organization. If all employees know that the team has the authority to carry out a special project, they are more likely to want to share information and work with the team in a cooperative fashion. The team's latitude to make decisions should be clear.

Self-management capabilities. The team should be made up of self-starters who are entrepreneurial in their approach but who understand the organization's mission and values so they can work with other team members in a united purpose. There is a delicate balance between teams of independent self-starters and those who have the maturity and judgment to know when to consult leadership on the direction they are going. Developing skills of self-management within the team will be beneficial for the organization because the skills the individuals develop will prepare them well for potential leadership positions in the department.

Performance measures. The team needs to understand the value of performance measures and targets so that it works to achieve measurable outcomes. If some of the first teams created in the organization use performance measures to track their work, it will be easier to implement measures in other aspects of the utility.

Coordination with other teams. Specialty teams must work with other teams in the organization and be supportive of the theory of team management. An ethic of coordination and cooperation can be promoted by teams to help the entire organization adopt cooperative values. Seeing such values demonstrated by example will improve the quality of cooperation throughout the workforce.

Empowering Teams

In order for teams to be successful they need to be empowered by leadership. The concept of team management should be integrated into the organizational way of working. Enough teams should be used that teams are not unusual, but a common way to organize resources for problem solving. Teams need to have a clear mission and boundaries so that they understand their sphere of influence and can operate within it freely. Within the organization, managers need to develop the general ethic of team management so that the older hierarchical style of leadership is slowly broken down and employees begin to accept the empowered entrepreneurial and more egalitarian style of interaction. Teams need to have both management support and sufficient training to be successful at what they do. Resources in the organization can be set aside to support team development.

As the team concept grows in the organization, the role of managers will need to change from command-and-control to a support and facilitation role. Managers will be cooperating with teams and giving them support and encouragement but not telling them what to do or how to do it. The decision making style in the organization will, therefore, also have to change from a top-down to a collaborative approach. It is important to solicit the input from all the employees on the team. Although decision making on a particular project or task is delegated to the team for a specific project, it is important for the team to gain skills of communicating upward with management so that there are no surprises about the direction they are taking. Empowerment requires employees who are taking action to communicate laterally as well as up and down the chain.

Team based decision making in an organization can lead to a significant change in ethic in which creative, entrepreneurial skill and collaboration with peers and cross-department communication are valued and rewarded. Many old line organizations have built "silo mentalities" in which communication is up and down a division of an organization. It does not go across the divisions and sections of the organization so there is not broad, general understanding or communication.

CREATING A LEARNING ORGANIZATION

Much of today's most progressive management literature talks about creating a "learning organization." Although defined differently by different authors, a learning organization is one where change is accepted as a constant and is seen as a positive improvement in the work ethic. The change is managed and planned so that the environment is not chaotic for employees, but employees are expected to be constantly looking for ways to improve work processes. For a learning organization to be a nurturing environment, it is essential for the leadership team to make a strong commitment to training and skill development for employees. As the work environment changes, leadership should support employee development and value them so that they will be ready to take on the new tasks and assignments expected of them.

As utilities cut their budgets, training is often the first to go. It, after all, takes time away from "real" work. This is very much a shortsighted strategy. Leaders need to budget for training because training and development are vital to an effective learning organization. A learning organization is likely to be competitive and efficient, because it is constantly scanning the horizon to find better, more efficient ways of getting work done.

There are numerous opportunities for training and development that leadership can take advantage of. These opportunities should not be reserved for top management. Staff from all levels of the organization should be sent to other utilities to learn new ways of doing work, and to training sessions and seminars such as those developed by AWWA. These include specialty conferences, workshops, and certification programs. Staff should also be encouraged to pursue continuing education units (CEUs). The change of pace found in these opportunities and by interacting with peers in other jurisdictions often provides a powerful way to recharge employees.

In 1995 the Colorado Springs Water Resources Department (WRD) Treatment Services Division (Case Study 2.2) began an optimization program aimed at improving productivity and being competitive with private companies. Management understood that employees could contribute significantly to optimization efforts if they were given more participative, broader roles. At the same time management was committed to protecting the jobs and financial security of employees. With these two guiding principles, a cultural change process began. Additional training was provided on communications, conflict resolution, and other teamwork skills. They believe that some of the lessons learned in the process were that cultural change requires educating employees, obtaining commitment, and embracing new philosophies. By substantially involving employees and facilitating reengineering, new behaviors and routines were adopted. A skill based compensation program supports concepts such as workforce flexibility, which are needed for productivity improvements.

Training and Development

Competitive organizations of the future must make a commitment to training and development for employees. Investment in human resources is among the most important investments a utility can make. Employees who understand and are excited about the culture of the organization they belong to can make a significant difference in the quality of the work done. Energized and empowered employees are the best representatives to the public of the services offered. Once an employee is hired and trained it is usually more efficient to continue that training and bring the employee to a new level of success rather than to hire someone new and begin the training process all over again. Even with a commitment to training and maintaining employees, organizations of the future will be faced with significant turnover and with the need to have diverse employee groups that reflect the customers they serve. An excellent training and career development program is one of the best ways to maintain stability of employment. Following are some of the types of training that a modern organization should offer to be able to maintain a learning organization. If an organization is small, its training needs can be met by sending employees out to AWWA and other existing programs.

Leadership development. Leadership development should be a program offered to the highest level of leadership in the organization. It is the way to instill common values and a common cultural ethic among the leadership group. Depending on how large the organization is, it might be offered every year or every other year. It would include topics such as decision making, organizational development, diversity, communication and listening skills, strategic policy development, community outreach, team development, and other subjects of the group's choice. The utility director might even teach some sections.

Supervision and management. Supervision and management training should be offered to managers and first line supervisors. It should include basic skills of management as well as cultural values of the organization. Subjects might include: diversity, personnel, decision making, team building, communication and listening skills, basic supervision, community outreach and other subjects that the leadership team decides are critical to first line supervisors. In smaller organizations this category of training may be combined with leadership development.

Safety and certification training. Depending on the work being done in the utility, there are a series of courses that are basically required to ensure that the staff is properly qualified. These courses can be offered in house if the organization is large enough, or can be scheduled in local community colleges or through labor unions. They might include a wide range of topics such as basic safety training, CPR, operator training, handling hazardous materials, and other technical, job related training.

Apprenticeship programs. In most large utilities it is worthwhile to develop an apprenticeship program to prepare water or wastewater workers with the skills necessary to do their jobs. In smaller utilities these skills can be taught in union programs. On-the-job training usually takes too long and provides an inconsistent training program because it depends on the quality and knowledge of the crew chief or mentor involved. Today's worker often comes to the job market inadequately prepared in language and math skills or without specialized technical training, such as in field maintenance practices. These skills can be improved through a creative apprenticeship program. With skill development comes improved self-image and self-confidence. Often labor unions will be partners with utilities in providing apprenticeship programs if the utility is not large enough to organize one of its own.

Skill development. Some skill development training courses can be organized centrally but delivered at various locations to make them convenient for employees. A special concern is computer skills. As organizations become more and more technologically complex, they will have to find convenient, cost efficient ways to keep employee skills at a high level. It is often more cost efficient to offer internal training courses, video training, or self-tutoring courses on computer skills to employees than to send them to another organization for training. Professional training for senior technical specialists is also important to keep them abreast of developments in their fields. This includes personnel in customer service, training, and financial management, as well as laboratory and water quality specialists, and managers of water resources and distribution systems.

Mentoring, bridging, career development. Many employees today are asking organizations to help them with career development. In many cases the most ambitious employees (the ones you want to keep) want to upgrade their skills and be ready for a new job after a period of time. One way to upgrade skills and to encourage employees is to provide mentors or buddies that will help them with specialized skill development, or to develop "bridge" or training positions that they can fill while they are learning a higher level of skill. Cross-training employees to do a number of jobs is another wise investment for utilities. If labor unions are active in the utility, it is important to get their buy-off on cross-training programs. Career development is a complicated process and requires a sensitive and creative training department (or training officer if the utility is small), but is worth the investment if an organization is able to upgrade and maintain its best employees.

The preceding is a simple list of the types of training that the modern organization needs for its employees. A well-orchestrated array of courses will guarantee that employees will begin to reach their potential. They will also be positive and grateful to the organization for assisting them in skill development. Ultimately some type of skill-based pay may even be developed by the most forward thinking organizations. Perhaps utilities can develop broad classification systems that permit employees with higher skill levels to be paid more and asked to do more specialized work. Then it will pay off for utilities to develop skill development programs. Keeping "the best and the brightest" calls for meaningful commitments by the utility.

Continuous Learning

The organization that prides itself on continuous learning will be one that has a strong training and development program for employees and also one that has a well-developed work process redesign program. Continuous learning requires that an organization has the skills to develop itself and to continually question the direction it is going. It must be willing to rethink old systems and to develop new ones. The result of continuous learning will be continuous change. However, if the organization has well-trained employees who believe in the cultural norms of the organization, then they will be comfortable with the level and frequency of change and will be able to cope with it. In order to establish both continuous learning and a nurturing environment in the workplace, the leadership of the organization must be committed to employee welfare and development. Only in situations where employees feel they are valued and supported will they give their best efforts.

An important element in creating the culture of continuous learning is providing opportunities for staff members to meet their counterparts in other utilities whenever possible through opportunities such as meetings of professional associations, benchmarking activities, and professional-technical conferences.

LABOR UNIONS

No discussion of employee issues could be complete without talking about labor unions and the role they play in organizational development. In some cities and states unions play a small or insignificant role. But in most large utilities management must maintain a relationship with anywhere from one to 15 different labor organizations. Needless to say, the more unions that are active in an employee organization the more complex the management environment is. The best managed utilities will look at labor as a collaborative partner and a friend. There are situations where for some reason labor has taken an extremely negative view toward management. In those cases partnerships are difficult to impossible, but those are also cases in which management should spend some time assessing the situation and trying to get at the root of the confrontation for the mutual benefit of both interests. Sometimes the key to changing the relationship with unions is to move toward collaborative interest based bargaining rather than the more traditional position based bargaining. Focusing on interests rather than positions allows both sides to seek common ground when solving a problem.

The Role of Labor

In most utilities labor plays a role of both protecting employees and preserving stability for them. Labor can be effective in communicating the goals of management to employees and the needs and desires of employees to management. However, in times of great change the union usually takes the role of protecting employees against potential job instability. Unions have become quite aggressive in situations where privatization, managed competition, or downsizing is proposed. In those cases the union will tend to support employee desires for job security.

A number of the utilities where downsizing and managed competitions have been successful were either in areas with right to work laws (no unions) or where management has worked closely with unions to develop creative solutions. It helps when the unions can see that job security for employees means working closely with utility leadership. If utility leadership has been able to prove that the utility is dedicated to customer service and that customers are requiring efficiency, then unions and employees will be more likely to work with leadership to attain that goal. In cities where there appears to be interest in downsizing, unions often make a strong pitch for guarantees of no layoffs. Some cities will not want to make the commitment to a no layoff policy but will be willing to work with employees to assure downsizing by attrition, voluntary separation agreements, or some type of retraining or rehiring policy.

For labor unions, change can be a threat to their own stability, which is dependent on strong membership drives and stable funding bases. Changes that result in layoffs will reduce union membership and destabilize the union funding structure thereby threatening their way of operation. Management must emphasize that change requires a partnership; that addressing the need to be more efficient is a common concern and one that both organizations must face together. Neither labor nor management can put their heads in the sand when addressing issues of competition. Both must work together to find cooperative solutions to the problems.

Keys to Interaction

The most significant key to positive interaction with labor unions is to have good communication. Smaller utilities can often have a better relationship with labor than large ones because communication with labor is more informal and individuals interact more regularly. Unions want to demonstrate leadership and assistance to their employee members. In order to do that they must know what is going on and must be able to explain policies and events to members. In times of significant change the need for good communication gets more important. Usually union leaders want to hear about a proposal or a change before it is announced to employees so that when those first calls begin to arrive in their offices, they are already informed and know how to respond. Like most politicians (and unions are a political interest group), they do not want surprises. Some of the best ways to maintain quality communication with unions are the following:

Labor management committees. Most utility directors will have a labor management committee that he or she meets with regularly to discuss issues of concern to either side. If there are only a few unions involved in the organization, the committee may be a meeting scheduled in the director's office. If there are many unions involved, then it is often a good idea to have coleaders from both sides and have leadership from the organization meet with all the interested unions. The regularly scheduled meeting agenda can be prepared by the coleaders from both sides and be a sign of respect and egalitarian collaboration with union leadership. The ground rules and mission of the committee should be established early in the process so that it defines what issues can be discussed and it is clear that this is not a bargaining process.

Employee Involvement Committees (EICs). Employee involvement committees can be established with the support of unions to solve issues that exist within the organization that are best solved at the lowest level. These may be issues that will ultimately require some type of bargaining or change of conditions of work. The union representatives can see that the goal of the committee is to find a solution that is in the best interest of employees and the employees are the best people to make decisions about how to change the conditions of work. Under such controlled situations unions are often willing to enter a process with management to let employees make recommendations about ways to change working conditions. Some types of work process redesign efforts will require this type of cooperation and collaboration with the unions. EICs are similar to other teams that may be established in the utility. They are different in that they are formed around an issue that is definitely of concern to a union and may impact working conditions or other bargainable rights.

Job security commitments. Depending on the level of cooperation between labor and management and on the level of anxiety within the unions, the unions may make aggressive demands for job security guarantees in times of significant change. Although this is understandable given the risks that unions might be facing, management often does not want to make that type of guarantee because it may be facing politically driven change conditions that will require downsizing. This situation can lead to arguments and high levels of tension, or it can be worked through carefully by defining the interests of both parties and by making strong efforts to provide some type of predictable job security for employees through voluntary separations or retraining.

Interest based bargaining. Some unions are moving toward interest based bargaining in which each side identifies interests rather than positions on issues and then works in a collaborative way to find win-win solutions. Although more time consuming and less predictable than traditional position based bargaining processes, interest based bargaining has a higher likelihood of achieving results that are more creative, strong, and lasting.

Recalcitrant unions. Occasionally a utility manager will get into a disagreement with a union, or a union will decide to oppose utility policy aggressively. There are no simple solutions for this situation. It is like any other political or stakeholder stalemate. Utility leadership has to try to work the problem out. That can be done by using mediators, interest based discussions, increased communications or some type of political go-between. Managers will not win all fights and will have to compromise in some way.

As long as employees rely on unions to assist them in understanding management issues and to protect them against unpredictable behaviors, it will be a management responsibility to build collaborative partnerships with unions. One can take the attitude that this is a bother or one can recognize the power that unions have to benefit organizations and employees.

SPECIALIZED PROBLEMS

The modern utility faces a number of specialized problems in dealing with change management and in increasing utility efficiency. Some of these are the result of changes in the world around us; some are the result of competitive forces putting pressure on the utility to change. Although only briefly identified, each needs some attention paid to it.

- **Diversity:** The face of the modern utility will need to look like the customers we serve as we pay more attention to customer interests. Even utility leadership must change to reflect the multiracial nature of our society. In order to have diversity in leadership we need to recruit and train an excellent, diverse staff in all aspects of utility work.

- **Temporary employees:** As utilities become more competitive, they begin to differentiate between core staff and staff needed for seasonal or temporary work peaks. The private sector uses temporary employees to meet peak work loads, and public utilities need to establish staffing patterns that are efficient, meet work needs, and that are fair to employees.

- **Unique shifts:** Privatized utilities in Europe are gaining efficiencies by using technology to permit unattended facilities and one person shifts. As use of technology increases in public utilities, labor saving efficiencies will be evaluated and implemented. Issues that must be addressed relate to safety and labor requirements as an organization proposes decreasing shift size.

- **Problems with middle management:** Often the source of resistance to change in a utility is middle management. Wedged in between first line supervisors and upper management, middle managers often fear for their jobs and do not get enough direct communication with leadership. As organizations downsize, middle management has become a target for reduction. Utility managers need to make middle management jobs relevant. They need to increase the span of control or increase the percentage of time a manager is doing "real work" rather than just coordinating or passing paper. As a team concept grows within a utility, middle managers will become more project focused.

- **Management by Walking Around (MBWA):** Leadership needs to spend a higher percentage of time being directly visible and involved with employees. MBWA helps break the old chain of command communication system in the hierarchical structure of management and gives the utility manager direct access to more employees. When employees actually interact with leadership they feel more involved with the organization; they cease to feel anonymous; and they are likely to increase their output and productivity.

SUMMARY

It is not possible to deal comprehensively in one short chapter with all the issues that management faces with employees and change. The very nature of change is that it is unpredictable. Modern theories even extol the virtues of chaos saying that it keeps an organization alert and makes it more creative. The thesis in this book is that change is inevitable, but that it can be managed in a thoughtful way. Leaders in our utility organizations have the skills and talents to be creative and thoughtful about how to approach the problems they face.

Keys to facing the future successfully will be to create a learning environment in which employees are comfortable with change, feel respected, are trained, developed, and nurtured, and are supported by their unions in a collaborative relationship with management. The future will most likely be marked by teams of people working in collaborative partnerships to accomplish objectives that they believe in and have had an opportunity to define. Organizational improvement programs aimed at competitiveness and efficiencies are prime opportunities to introduce a new and more productive corporate culture.

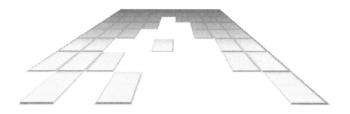

CHAPTER FIFTEEN

Thoughtful Contracting Processes

Contracting services to augment staff is becoming an increasingly popular practice in both industry and government. Water utilities have used contractual arrangements for a broad spectrum of services, from legal counsel and engineering services to groundskeeping, payroll, and billing and collection services. Most common are contract services for engineering, design, and construction of facilities and systems.

This chapter describes complete, detailed, thoughtful contracting processes that are suitable for significant contracts. The same principles also apply when smaller organizations consider smaller contracts, although in those cases the contracting work can often be streamlined, usually through the use of qualified consultants. The selection of experienced and knowledgeable contracting experts can be the determining factor in the success of a contracting activity.

Utilities routinely perform the majority of their own operation and maintenance (O&M) activities through direct employment and daily operational management. In recent years, some utilities have been contracting out all or a portion of O&M for their facilities and systems, sometimes including capital improvements as part of an integrated project. The methodology for contracting engineering and construction services has been well established. For example, the format of construction bid documents that include a contract, drawings, and detailed specifications is a widely accepted national practice. The practice of contracting for O&M services, however, with or without capital improvements, is not well developed. Contracting approaches often vary from utility to utility and from state to state. The combination of design build operate and design build alone is not common or well developed.

This chapter deals primarily with the contracting process for O&M services, which may include capital improvements and related services. Such contracts provide for the contractor or service provider to staff and operate the facility to meet drinking water specifications and to achieve regulatory compliance, with financial terms established in a service agreement that includes penalties for noncompliance. The contract (or service agreement) between the utility and the contractor contains the terms and conditions of the engagement. The service agreement outlines the responsibilities for permitting (if included), compliance issues, routine operations, repair and replacement issues, capital improvements (if any), payment provisions, and a host of other technical, legal, and financial requirements. The service agreement is the end result of a procurement process defined in this chapter that outlines the methodology for a thoughtful contracting process, ranging from the initial steps through execution and implementation of a service agreement.

The various types of public/private partnerships for the design, construction, operation, and ownership of water and wastewater facilities are depicted in Table 15-1.

Of the options listed in Table 15-1, the use of contract operations is the most common. This chapter describes the procurement of multiple year and multimillion dollar services for contract operations, which may or may not include capital improvements. Although contract operations have been used to depict the contracting process, this process can also be used for the range of public/private partnerships listed in the table.

The contracting process is systematic and methodical, and is applicable for medium- to large-scale projects. Contracting for O&M services for small facilities with relatively small operating budgets may not need to utilize as complex a process as provided

Table 15-1 Types of contract relationships

Type	Service Provided by Contractor
Contract Operations	O&M of a facility or system for a specified period, with or without capital improvements
Concession	Full contracted services including customer bill collection
Design-Build (Turn-Key)	Design build services under a single contract
Design-Build-Operate (DBO)	DBO services under a single contract; usually long-term
Build-Own-Operate-Transfer (BOOT)	Financing with ownership retained for a specified period of time
Asset Sale	Purchase of a system or facility

in this chapter. Such services can be procured using an abbreviated competitive process, or even a competitive bid process, similar to what is typically used for construction contracts. In many cases it has been treated as a negotiated, single source procurement or even as a professional service.

For the purposes of this chapter, terms "contractor," "service provider" and "contract operator" are used to refer to the party, private or public, that is providing the contracted services.

CONTRACT OPERATIONS OF PUBLIC TREATMENT FACILITIES

The current concept of private sector O&M of public treatment facilities was introduced in the United States 25 years ago in Burlingame, Calif., by Envirotech Operations Services, soon followed by Fairfield-Suisun, Calif., and Great Falls, Mont. These early contracts were characterized by three- to five-year contract terms, guaranteed costs, and assumption of risk for treatment plant performance. Ownership, permits, pretreatment, billing and collection, capital replacement, upgrades and expansions, and related utility management functions remained with the public sector. Despite the benefits of such early arrangements, the use of contract O&M did not gain widespread popularity.

Even after the Tax Reform Act of 1986 and subsequent IRS rulings, operating contracts for publicly owned facilities with outstanding tax exempt municipal debt continued to remain limited, with certain exceptions, to a maximum term of five years. The 1986 tax law changes also eliminated the tax advantages that had created some design, build, own, and operate arrangements for private sector provision of the facilities and their operation. Chandler, Ariz., and Scottsdale, Ariz., are two examples of such programs from pre-1986 (refer to Case Studies 5.1 and 5.2 in appendix A).

The past decade, however, has seen a change as an increasing trend emerged involving contract operations of municipal water and wastewater treatment works. By the mid-1990s there were nearly 1,000 municipal water and wastewater facilities using three- to five-year contract O&M services. Most of these contracts, however, were for smaller facilities less than 20 mgd in capacity. Because of the dollar amounts of these smaller and short-term contracts, the procurement and contracting processes were not complicated. These contract operations have been largely with private firms, although some contract operations have been between governmental agencies.

NEW LEGISLATION

In January 1997, the IRS issued Revenue Procedure 97-13 (Rev. Proc. 97-13), which permits contract periods of up to 20 years under specific conditions. With the use of Rev. Proc. 97-13, extended term (10- to 20-year) contract O&M services for facilities with outstanding tax exempt financing can be utilized. In addition, capital investment by the service provider is possible. Price adjustment methodologies that are accurate and fair can be developed over the extended contract term but should be carefully considered for the longer 10- to 20-year term. A further, more complete discussion of Rev. Proc. 97-13 is included in chapter 19.

The new procedure has prompted utilities to rethink the contracting approaches historically used for smaller three- to five-year, strictly O&M contracts, particularly regarding the size, complexity, and term of contracts; as evidenced by the high renewal rate and trend to take advantage of the recent IRS rules. Contract operations are likely to continue and to accelerate as a service provision alternative for water utility managers.

STARTING THE PROJECT

Recognizing the complexities of longer contracts that may include capital improvements, the person who contracts services needs to consider:

- Building a qualified project team

- Developing a strategic approach

- Allocation of risks

- Determining site-specific contracting alternatives and approaches

- Market availability

- Conducting the solicitation process

- Negotiating of a comprehensive service agreement

- Conducting performance monitoring

The following sections focus on each of these elements and how they must come together to develop a successful service agreement.

BUILDING A QUALIFIED PROJECT TEAM

A cohesive and capable project team is a prerequisite for successful project development. The project team should include the interests of all of the stakeholders, representatives of the policy-makers and elected officials, utility management and staff, and specialized expertise. As larger and more complex projects are considered for service contracts, expert specialized help becomes more critical for project success.

Policy-Makers and Elected Officials

For the best results there should be a partnership among the policy-makers and the utility management and its staff to achieve common objectives. The goal is to obtain a service provider to improve operations, to enhance performance, and to reduce or control costs. This may remove your governmental agency from responsibility for day-to-day operations of a facility or function. The idea of obtaining a service provider is often initiated at the level of a policy-maker or elected official, but can be initiated by the utility management staff.

Depending on the size and structure of a system, your policy-makers and elected officials may or may not be involved with the daily activities of the utility. Regardless of the level of involvement, the leadership must be clearly committed to the goals and must understand the risks before commencing the process. Such leadership and guidance are essential for your project team to develop a thoughtful contracting process.

For a successful competition, the utility must be viewed as fair and open during all phases of the solicitation process. The contracting community gauges the commitment of a utility's leadership before expending significant efforts in tracking the potential business opportunity and in responding to solicitation documents. Once your decision is made to procure a service provider, it is essential to manage communications and keep the process as apolitical as possible. This will attract the best quality firms and minimize potential protests from firms not selected.

Some utilities and municipalities have taken a stance against lobbying efforts by service providers. In some instances, informal contacts with officials, a practice which does not follow the solicitation documents guidelines, may result in disqualification of the respondent. These potential conflict of interest issues, including several utility approaches, are further discussed in chapter 20.

If possible, it is advantageous for your policy-makers to participate in the goal setting process with the utility's management staff. The procurement process can then be developed and implemented to accomplish these goals. Some governing boards and councils have removed themselves completely from the process, except for formal actions required by their local ordinances. However, resolutions may be adopted to begin the solicitation process, to issue solicitation documents, to endorse a short list, and to approve a negotiated contract. Sometimes representatives of boards or councils serve on the solicitation project team.

Utility Management and Staff

The solicitation process is best conducted with the direct involvement of several agency departments. Table 15-2 outlines roles that may be assigned to each respective department of a utility.

Table 15-2 Possible roles and divisions of a contract operation solicitation project team

Department or Function	Possible Roles
Manager, Superintendent or Executive Director (or designee)	• Link between policy-makers and utility staff • Establish project teams (solicitation, evaluation, and selection) • Key decisions • Overall program management • Coordination with other departments • Conduct contract negotiations
Engineering	• Define contract limits • Develop performance specifications • Prepare background information • Develop solicitation documents • Assist with evaluation, selection, and negotiation process
Operations	• Define operating conditions • Status of system(s) for contract operations • Input on repair and replacement plan • Input on transition plan (plant and employees)
Financial	• Establish measurement and payment provisions • Determine any financial barriers (i.e., bond covenants and grants) • Determine financial qualifications (credit worthiness) of respondents • Establish financial goals • Assist with life cycle cost analysis
Legal	• Evaluate state and local statutes for procurement • Interpret tax laws • Develop service contracts • Evaluate business and legal submissions • Conduct contract negotiations
Human Resources	• Personnel considerations and compensation • Union relations

A core project team should be formed from each department and charged with managing the process. Each project team member is assigned specific responsibilities.

Specialized Expertise

The eventual contractual arrangement between your utility and the contract operator will be embodied in a service agreement between the two parties. While many larger utilities have procurement management, legal, financial, and technical expertise in house, it may be advantageous to augment the utility's procurement team with outside experts who have assisted with solicitation and negotiation services elsewhere.

Respondents to solicitation documents may include an experienced group of technical, legal, and financial experts who are familiar with contracting techniques. It is important that your utility have similar capabilities to balance the expertise brought by the service provider.

Utilities can enlist outside experts for procurement management, legal and financial advice, or work products. Properly directed, they provide considerable added value. Furthermore, the costs of such services are a fraction of the overall contract amount and pay for themselves in cost savings. Consultants' expertise may include:

- Specialized knowledge and interpretation of present and future regulatory impacts and water treatment requirements

- Assessment of alternative approaches for service provision

- Assistance with developing and implementing specific procurement strategies, including a fair allocation of risks

- Development and management of the RFQ/RFP processes

- Analysis of resulting rate structures, cost of service, and benchmarking

- Assistance with developing contract terms and conditions

- Assistance with developing performance-based specifications

- Assistance with evaluating existing plant conditions and future capital improvement programs

The United States Conference of Mayors Urban Water Institute provides a forum to assist local government in exploring privatization and other public-private partnership approaches.

DEVELOPING A STRATEGIC APPROACH

Once your project team has been formed, it should develop a strategic approach to the procurement process. One way to develop the project strategy is to conduct a meeting to address the major issues. This strategy session is desirable at the beginning to set the framework for subsequent activities. The following major issues should be addressed:

- Project philosophy

- Roles and responsibilities

- Bid versus proposal

- Work to be included

- Market strategies

- Length of contracting period

- Risk and control sharing

- Confidentiality

- The evaluation team

- Ethics

Project Philosophy and Goals

Underlying the procurement process is the philosophy of your procurement and the desired results. The project goals need to be established early in the process. If the goals are well defined, the procurement process can be properly aligned with them. Goals can vary from utility to utility, and typically include several or all of the following:

- Long-term stabilization of rates and cost reductions

- Access to new technologies and experience of a service provider

- Implementation of capital improvements

- Divestment of utility operations

- Access to private capital

- Reduction in labor force

- Compliance with regulatory requirements

- Single source of responsibility for design, construction, and operation services

- Solicitation of alternative and creative approaches

- Training opportunities for personnel

The goals and philosophies dictate the contents of your solicitation documents. For example, if private capital is desired, you must solicit firms with the financial strength to secure such capital. If capital improvements are desired, the respondent will be required to demonstrate design and construction experience.

Roles and Responsibilities

Typically a procurement process includes a person who coordinates the involvement of all project team members. This person will likely need to be involved nearly full-time during critical portions of the project. In order to maximize the use of the project team, including outside consultants or advisers, roles need to be clearly defined. For example, assignments are typically made for the following responsibilities:

- The contact person for the potential respondents

- Gathering all background data on the facilities and systems

- Soliciting input from policy-makers

- Interpreting legal requirements

- Document production and quality control activities

- Tours of the facility and site

- Communicating with existing employees

- Preparation of press releases and addressing public comments

- Relations with regulatory agencies

An outline and/or table of contents of the solicitation documents should be developed prior to an initial strategy session. A sample outline is included in appendix B. At the strategy session assignments can be made to individuals for completing each portion of the documents.

Bids versus Proposal

A basic element of the strategic approach is to define the processes that you will use to solicit a contractor and the processes and mechanisms that you will use to make a selection. At the outset you should determine whether the primary vehicle for selecting the contractor will be a price bid or a proposal covering several factors. In most instances the latter is used and selection is based on a number of factors identified in the RFP, only one of which is cost.

Nonfinancial factors used to evaluate the technical quality of a proposal may include:

- Qualifications and experience of staff promised to your project

- Reference facilities

- Technical approach for O&M

- Quality of capital improvements (if any are proposed for the project)

- Ability to meet regulatory requirements

Financial factors used to evaluate the monetary aspects of a proposal may include:

- Cost effectiveness (life cycle cost analysis)

- Financial strength of the respondents

- Compensation package to existing employees

In those cases where cost alone is the deciding factor, the RFQ must be carefully crafted, assuring that all entities permitted to submit a bid meet specified minimum criteria. The lowbid approach does not recognize value associated with the service provider and proposed staff experience, skills, service, and other enhancements above the minimum criteria.

Work To Be Included

The work desired can be divided into two major categories: the scope of project and the type of the services to be provided.

The scope of the project refers to the physical limits of responsibility such as whether the work is limited to a particular facility site and facility, or if it encompasses the collection and distribution systems, pump stations, reservoirs, and so on. The interface points between utility and contractor responsibilities must be clearly established to prevent later dispute. At this point it is useful to develop an inventory of all facilities and systems included as part of the contractor's responsibility.

The types of services to be provided are the activities that the contractor will provide, and may include:

- Facility operations

- Routine maintenance of equipment

- Major repair and replacement of equipment

- Capital improvements

- Project financing

- Laboratory services

- Permitting responsibilities

- Billings and collections

- Customer service provisions

- Meter reading

The strategic approach includes a framework for the scope of the work and the types of services that ultimately form the basis of the service agreement. Since the ultimate responsibility for the protection of public health, regulatory compliance, rates, risk exposure, and other interests of the public and customers remain with the utility, the desired level of service must be carefully defined.

Market Strategies

It is often advantageous for RFQ and RFP or solicitation documents establish only the performance requirements and leave the method of achieving the performance up to the service provider proposers. This encourages creativity and gives the proposers flexibility to utilize their own experience and capabilities to the maximum. Utilities are accustomed to using carefully crafted, detailed specifications for contracting construction work, and there is often a tendency to use a similar approach in developing the solicitation documents for O&M services. However, a well-developed, performance based RFP can better serve the utility's interests by providing for:

- A guarantee of the finished water quality that must be produced by the facility without specifying the detailed method of achieving the required results

- The level of quality of a capital improvement, along with the desired output, without specifying the exact size and type of a given component

- The frequency of sampling and testing without specifying the procedure for collecting samples

- The required hours of operation or production requirements without specifying how to conduct maintenance

While the competitive process can be enhanced through the use of a performance based RFP, other techniques can provide additional benefits. These procedures include:

- Uniform and consistent information to all proposers

- Alternatives in addition to the base proposals

- Communication of clear rules for the procurement process to all respondents

- Maintaining competitiveness through the receipt of final proposals *and* negotiation of final terms and conditions

Length of Contracting Period

Because limited information exists on long-term contracts, it is difficult to evaluate the benefits and risks of the length of the contract, and there are pros and cons associated with the various options. A longer contracting period offers the following potential advantages:

- The service provider can spread the investment in servicing and contracting the project, as well as start-up and shut-down costs, over a longer period of time, thereby reducing annual service fees.

- Any capital investment costs that benefit operations can be spread over a longer period of time.

- The contract operator is encouraged to implement new technologies and systems since they can be amortized over a longer period.

- The larger dollar amount enhances competitive interest for the longer term contract.

- It lessens additional cost to all parties to enter into a new service contract.

The primary disadvantages to the public utility in entering a service contract of longer than three to five years are:

- There is limited history of long-term contracts.

- Public perception may be negative if they believe there is a benefit to periodic competitive procurement.

- Changing conditions, including changing regulations and the need for change in capacity, may require renegotiation of the service fee in a single source, noncompetitive environment.

- It is difficult to develop a price adjustment mechanism that projects time-cost performance over the longer term that is fair to both parties.

Based on the limited experiences to date, a reprocurement after a five-year term usually results in savings. This has been the experience in Fairfield-Suisun, Calif.; Houston, Texas; Oklahoma City, Okla.; and Lynn, Mass. On longer term contracts it is advisable for the utility to have the right to terminate at its sole discretion and without penalty at specified intervals.

Risk and Control Sharing

One of the most important elements of contracting services is the allocation of risk and control. It is not good to transfer all risks to the contractor, but on the other hand the benefits of the contracting process are reduced if the utility maintains all the risks. A thoughtful contracting process identifies all the elements of risk and includes an appropriate risk allocation in the solicitation documents. Risk should be assigned to the party best able to assume and manage it. The eventual agreed upon risk allocation between the utility and the service provider should be contained in the terms and conditions of the service agreement.

A useful practice is to develop a risk matrix that forms the basis for risk allocation. This matrix can be developed early in the process, typically as part of the initial strategy session. Tables 15-3a, 15-3b, 15-3c, and 15-3d illustrate a risk matrix developed during the strategy session for the Seattle DBO project for a 120 mgd water filtration plant. The Seattle risk matrix was incorporated into the RFQ, and respondents were encouraged to provide comments on the risk allocation. This risk matrix served as the foundation for the subsequent RFP and service agreement preparation. The Seattle

Table 15-3a Risk matrix developed for Seattle DBO project

Risk	Allocation	Remarks
Design		
• Technology Selection	Contractor	City reviews designs through an established review procedure in Service Agreement.
• Technology Obsolescence	Contractor/ City	Contractor is responsible for selecting technology that is proven, will be permitted by agencies, and will meet performance guarantees. Contractor is responsible for technology obsolescence, except for change in law, unforeseen circumstances, and unspecified conditions for raw water and water demand.
• Unforeseen Preexisting Site Conditions	City	Risks for change in law, unforeseen circumstances, and preexisting site conditions are the City's risks.
Construction/ Commissioning		
• Construction Period	Contractor	City monitors construction and tests to determine compliance with service agreement.
• Acceptance Test	Contractor	Service agreement specifies guaranteed construction period after fulfillment of conditions precedent. Notice to proceed given after conditions precedent satisfied.
• Payment	City/ Contractor	Facility not deemed suitable for commercial operation until test is passed. Retest principles outlined in service agreement.
		Construction payments based on drawdown and milestone schedule in service agreement. City is responsible for payment when milestones are met by contractor.

project included design build services on an undisturbed, undeveloped site in addition to O&M services. Therefore, the risk matrix depicted in the tables is more comprehensive than would be normally required for a contract operations arrangement without capital improvements. The thought process, however, provides helpful insight for any contract you are considering.

Confidentiality

Unlike the public opening of construction bids, you may need to keep proposals confidential until after award of a contract. This is particularly true when you use a qualification and merit-based

Table 15-3b Risk matrix developed for Seattle DBO project

Risk	Allocation	Remarks
Operations and Maintenance		
• Payment	City	City monitors performance via review of records and reports. City may conduct periodic inspections.
• Preventive Maintenance	Contractor	Monthly service fee paid with a fixed and variable component consistent with tax laws and forms of financing (i.e., pass-through costs, the only variable component). Monthly reports typically accompany invoices.
• Repairs and Replacements	Contractor/City	Standard of care provisions and contractual obligations requiring proactive preventative maintenance program.
• Capital Improvements	Contractor/City	Contractor is responsible for all repairs and replacements to meet performance requirements, except for certain major improvements where the City may be responsible for costs.
		Contractor is responsible for all capital improvements required to meet performance requirements, except for certain major improvements where the City may be responsible for costs. City is responsible for capital improvements as a result of changes to performance standards. Renegotiation principles are included in the service agreement.
Supply of Raw Water		
• Infrastructure (e.g., pipelines, reservoirs, etc.)	City	City is responsible for supplying water to facility site at interface point. Contractor assumes responsibility at the interface point.
• Quantity	City/ Contractor	Specified range of flows based on historical data is provided in the service agreement. Contractor assumes risk for flows within the specified range. City provides relief for flows outside of the range. Contractual provisions included for contractor to justify adjustments to service fees for flows outside of specified ranges.
• Quality	City/ Contractor	Specified ranges of quality based on historical data are provided in service agreement. Contractor assumes risk for quality within the specified range. City provides relief for raw water quality (additional payment or reduction in treatment rate) outside of range.

Table 15-3c   Risk matrix developed for Seattle DBO project

Risk	Allocation	Remarks
Plant Performance		
• Quality (without change in law)	Contractor	Contractor is responsible for supply of specified water quality. Contractual provisions for the need to shut down facility if raw water quality prohibits ability to meet standards.
• Quality (with change in law)	City	City is responsible for costs associated with upgrading and operating facility to meet new standards. Renegotiation principles are included in service agreement.
• Quantity and Flow	Contractor/ City	Contractor is responsible for flows within specified range. Contractual provisions for delivery of water quantities requested by City outside of specified ranges.
• Infrastructure for Transmission	City	City responsible for installing and maintaining transmission and distribution systems for specified and requested flows.
Environmental/Permitting		
Additional Environ- mental Review	Contractor	Contractor is responsible for complying with mitigation in existing final EIS and to prepare supplemental EIS/ addenda if needed.
• Existing	Contractor	Reporting by contractor to regulatory agencies and the City. City to monitor contractor's performance.
• Change in Law	City/ Contractor	Typically allocated to the City. Limited risk can be allocated to contractor (i.e., dollar limit). Renegotiation principles are included in service agreement.
• Permitting	Contractor/ City	Contractor secures most permits. City may be copermittee. Securing permits typically undertaken as part of conditions precedent in service agreement.

selection process followed by competitive negotiations instead of a low bid process. Confidentiality helps maximize competition, and does not compromise your negotiating position. In addition, especially for projects involving designs, respondents may want to minimize the distribution of innovative designs and technologies that may provide a competitive advantage or be proprietary. Your ability to keep the selection process confidential will vary depending on state and local ordinances. To preserve the integrity of the

Table 15-3d Risk matrix developed for Seattle DBO project

Risk	Allocation	Remarks
Other Factors		
Financing	City	City responsible for financing project as part of conditions precedent.
• Escalation of Costs— Construction	Contractor/ City	Contractor hold price until a specified calendar date. Thereafter, price escalates at a percentage of a specified index (i.e., CPI, ENR, etc.)
• Escalation of Costs— Operation	City	Service fee escalates annually at a percentage of a specified index (i.e., CPI). Certain pass-through costs are allowed.
• Taxation	Contractor	All taxes (i.e., income tax) are contractor's responsibility.
• Natural Disaster	Contractor/ City	Insurance; renegotiation principles; force majeure provisions. City has responsible risk for amounts above uninsured portions.
Industrial Relations		
• Prevailing Wage Rates/ Force Majeure	Contractor	Contractor's choice whether or not to pay prevailing wages. Contractor's risk if initial choice not to pay such rates is incorrect.
• Strikes	Contractor/ City	For local strikes against the facility, contractor assumes risk. For national strikes, city assumes risk.

process, the need to keep information confidential must be conveyed to all project participants. Once the final selection is made and a contract is executed, it is unlikely that a governmental agency can keep information confidential.

The Evaluation Team

The evaluation team should be formed early in the process and typically consists of members from the project team. It is the evaluation team's responsibility to recommend the order of ranking of proposers for contract negotiations. Depending on the utility this recommendation may or may not be binding on the final decision maker.

The evaluation team should consist of unbiased individuals who are committed to an open, fair evaluation of proposers in accordance with the criteria established in the RFQ (to develop a short list) and the RFP (to rank proposers for negotiations and ultimately recommend a single proposer for contract). The evaluation team might consist of as few as five individuals or as many as 10 or more. It can draw technical support from the entire project team or from other designated technical, legal, and financial experts.

Ethics

Effective management of ethical issues can determine success of a contracting process, a fact that must be recognized in developing a prudent strategic approach. Some issues, such as communication, confidentiality, and selection of outside experts are often within your control. Other issues arise outside your influence, including potential conflicts of interest and attempts at political influence on decisions and on the process itself. These matters are discussed further in chapter 20.

DETERMINING SITE-SPECIFIC CONTRACTING ALTERNATIVES AND APPROACHES

Many of the earlier and short-term O&M contracts were developed using a simple bid process. In some cases contracts are awarded on a sole source and noncompetitive basis. Over time, and as the contract operations business has matured, the elements of a successful contracting process have emerged. The approach that emerged considers a multiple step competitive procurement process with three major elements.

- Issuance of RFQ and subsequent receipt and evaluation of a Statement of Qualifications from contractors

- Issuance of RFP and subsequent receipt of proposals; followed by evaluation and ranking

- Contract negotiation and contracting

In certain states, a two step RFQ/RFP process is not allowed, and requests for both qualifications and proposals must be a single document. For convenience, however, the RFQ and RFP

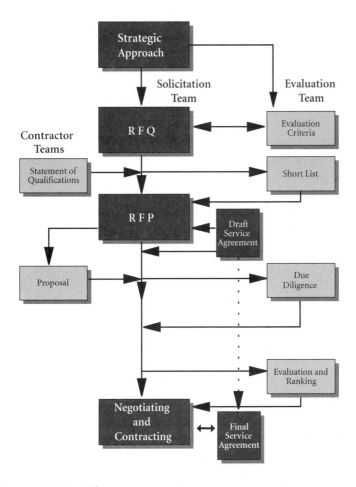

Figure 15-1 The procurement process

discussions of this chapter assume that such documents are issued separately.

Figure 15-1 illustrates the sequence of a thorough procurement process that includes RFQ, RFP, and negotiating and contracting, along with typical actions of the service provider and the evaluation team. This comprehensive process includes the elements usually involved for larger-scale projects. State and local legislation, individual desires of the utility, and the size and complexity of the project may result in project specific modifications of the diagram. Figure 15-1 will assist you in visualizing the subsequent discussions in this chapter, and also provides the framework for determining contracting alternatives and approaches.

MARKET AVAILABILITY

As with any business transaction, there are buyers and sellers. In this case the buyer is your utility or municipality seeking services from a service provider. The seller is the service provider of contract operations and related services. As the buyer you need to know as much as possible about the available firms and entities and the services they can provide. As the sellers, service providers want to fully understand your philosophies and goals so they can focus their efforts on meeting your goals and securing the contract. Because of the competitive process required of most utilities, and the inherent need for equitable distribution of information, information must be disseminated properly. It is your role to establish a sufficient exchange of information without compromising the integrity of the procurement process.

Understanding the types of sector providers available is useful since it helps you understand how to best exchange information between your utility and the service providers. Service providers are emerging from various business segments, some of which have been historically associated with the water and wastewater industries, including:

Environmental Engineering Consulting/Design Firms

Many regional and national engineering firms have provided comprehensive planning, design, start-up and testing, and operations consulting to the water utility community for decades. Recognizing a potential market opportunity, some firms have created operating divisions or other entities to provide contract operations and related services. Their previous experience with traditional environmental engineering consulting services provides them with many of the tools necessary to form divisions to provide these services. Their expertise is especially applicable where capital projects are involved.

International Firms

While the use of contract operations is relatively limited in the United States, these practices are prevalent in many other parts of the world. With the increase of private sector service opportunities in the United States and Canada, several international firms have entered the North American market either directly or via affiliated companies. This international presence has stimulated the market.

Full-Service Companies from Other Industries

Many large-scale companies and conglomerates have been providing full-service design, build, and operate services to the solid waste management and the electrical power production industries. These companies have developed divisions or affiliates to provide contract operations and related services in the water field.

Equipment Suppliers and Construction Contractors

Traditional construction contractors and equipment suppliers have recognized the potential to enter the contract operations market as an alternate way of providing goods and services. These entities have joined forces with other specialists to provide contract operations, especially where equipment needs and construction services are involved.

Other Contractors

The private sector has many small and large contractors who specialize in such work as paving, electrical, instrumentation, excavation, and plumbing.

Investor Owned and Public Utilities

Many long established, investor owned water companies and some public utilities have recognized that their operating experience makes them well suited to pursue opportunities for contract

Public-Public Contract Operations

Regional public sector water utilities in New Jersey are engaged in a legislative and legal battle with investor owned water companies that may, when the dust settles, reshape the privatization of municipal water systems in the state.

In New Jersey, nonprofit and for-profit utilities have gone head to head over the past two years competing for the operation of municipal water utilities. In large measure, the motivation for the public sector response came from the signing of the New Jersey Water Supply Public-Private Contracting Act in May 1995.

Among those participating in these activities have been the North Jersey District Water Supply Commission, the East Orange Water Commission, and the Passaic Valley Water Commission. All have signed, alone or in combination with one another, contracts of 20 years or longer for the operation and

operations. Such companies sometimes join a team of other specialists in order to provide comprehensive services.

The nature of the services sought will dictate the type of firms or consortiums that will respond to solicitation documents. For example, if the contract operations include substantial capital improvements, the team will likely include experienced engineering firms and constructors.

Because of the trend to enter into long-term contracts with more extensive services, the market is expanding to include firms that can provide engineering and construction services in addition to routine operations. It is likely that over time the current contract operators will be aligned both internally and via mergers, acquisitions, and spinoffs to meet market demands.

Contract operators do not have an operations staff ready to provide operating services. Rather, operations staff is usually developed by hiring some or all existing employees and others as needed from within the community. Contract operators will usually provide the appropriate management personnel and other operations staff to train and direct the locally hired personnel. When responding to solicitation documents, some service providers often lack the necessary certified operators on hand, and must hire such operators once they are awarded a contract.

Creating Market Interest in Your Project

With the increasing number of contracts available, service providers can be selective in choosing the best projects to pursue. A professionally managed procurement process, with complete support from elected officials and management personnel, will draw the greatest interest from a wide range of respondents. Any preconceived notions about the preferred service provider will deter competition and interest.

Careful dissemination of information through formal resolutions, results of feasibility studies, requests for expressions of interest, and other presolicitation activities will create interest in the market. The use of printed media to provide information to the marketplace can be very effective, provided that the information is balanced and useful. While it is advantageous to maximize interest in the project, it is important not to create interest by inaccurate statements. The use of the Internet and the World Wide Web is another good way to disseminate information, and was successfully used by the City of Seattle for the Tolt River DBO procurement process.

Soliciting Input from Potential Contractors

Service providers are usually willing to share information with you to shape the project and to increase their competitive position. Input is useful during the early stages of the project, and can take the form of formal presentations to municipal officials and your staff, submission of information packages, sample contracts from other localities where services have been provided, or other written materials.

Information from potential service providers can be received prior to initiating a formal solicitation process or after the process has begun. While information or input received beforehand can be relatively informal and unstructured, once the solicitation process begins (i.e., publishing RFQs, RFPs, and suggested service agreement), input should be accepted only within the context and guidelines of your established procurement process.

The RFQ document is usually the first official document of a formal solicitation process, and its contents are discussed later in this chapter. It is worthwhile to note, however, that the RFQ can be used to solicit comments on the project concepts, which can be received as part of the statement of qualifications submitted. Helpful comments can also be generated by issuing a draft RFP for review and comment. It is common to have presubmission meetings with the potential respondents after issuance of each of these documents. Some utilities prefer not to have individual meetings with potential respondents, while others consider such meetings to be useful.

Once a formal solicitation process begins, input should be requested in written form, with written responses provided to all interested parties. Failure to manage information in a formalized fashion can lead to allegations of favoritism toward one or more contractors and possibly to future litigation.

maintenance of municipal water utilities.

Thomas Golodik, "Public Sector Competition Contract Operations for Municipal Water Systems," presented at the 1997 American Water Works Association Annual Conference in Atlanta, Ga.

SOLICITATION DOCUMENTS—REQUEST FOR QUALIFICATIONS (RFQ)

The most widely used solicitation documents are the Request for Qualifications (RFQ) and Request for Proposals (RFP). Earlier in this chapter, you were introduced to some of the elements of both documents. The sections that follow provide more detailed information on how to prepare them and what to include.

Request for Qualifications (RFQ)

The RFQ is the first official document issued in the procurement process. Its purpose is to invite all interested parties to submit their qualifications to provide services sought by the utility. Typically, the RFQ is in the range of 10 to 20 pages, and may include:

- An overview of the project, including the sponsoring utility's overall goals

- A brief description of the types of services sought

- An outline of the procurement process and schedule

- A preliminary statement of contract principles, if available

- A description of the technical and financial information that must be submitted to demonstrate the qualifications of respondents

- A description of the evaluation and selection criteria

Appendix B provides a more detailed outline of an RFQ. There are several features of an RFQ that are worth specific attention, as follows.

Overview of Project and Goals

The RFQ should state why your utility seeks a contract service provider. The message should be clear and demonstrate the full support of the utility managers and elected officials, as appropriate. The RFQ should *not* be used as an exercise to solicit ideas about a potential undefined project in the future. If specific information about the project and/or the facility is available, interested contractors should be instructed on how to get it.

Services Sought

The RFQ should refer to your forthcoming RFP, which will outline the detailed scope of work required from the contractor. Information about the type of firms sought is important. For example, if engineering and construction services are required, this should be generally described in the RFQ. The anticipated scope of services should be described in a general way to make it clear what type of firm is being sought.

Outline of Procurement Process

For planning purposes, contractors desire an outline of the overall procurement process and schedules. Providing the process description and schedule demonstrates the comprehensive thinking you have invested in the procurement. The schedule may include:

- RFQ issuance date
- Conference for interested vendors
- Statement of qualifications due date
- Date for identification of a short list
- Issuance date for RFP
- Preproposal conference
- Dates available for plant tours
- Proposal submission date
- Date for completion of proposal evaluations
- Dates for oral presentations
- Dates for conducting negotiations with preferred firm(s)
- Date for completing negotiations and executing a contract
- Date for commencement of operations

Because the later dates are subject to change, the RFQ should state your utility's right to alter the dates as necessary.

In this section of the RFQ it is also advantageous to cite the applicable provisions of state law that enable the utility to conduct the procurement process. Any special features of the applicable state law should be mentioned.

Preliminary Statement of Contract Principles

The detailed contract requirements will eventually be articulated in a contract document (service agreement) between your utility and the contract operator. If you have some preliminary information about the contract principles at this early stage, include it in the RFQ. A risk matrix, as described earlier in this chapter, can be included.

Submission Requirements for Technical and Financial Qualifications

An RFQ typically requests information in two categories: technical qualifications and experience and financial qualifications. The qualifications sought must match the services desired from the service provider. Information typically includes:

Technical Qualifications

- Reference facilities of similar size and type

- Staff qualifications and résumés

- Applicable project-specific information

- Specialized technical expertise consistent with the services sought

Financial Qualifications

- Recently audited financial information

- 10-K statements filed with the Securities and Exchange Commission

- Proposed financial backer (i.e., project guarantor)

- A list of fixed and contingent liabilities, including outstanding litigation

Evaluation and Selection Criteria

Contractors want to know your evaluation criteria, and your evaluators need the same objective criteria for conducting an analysis of the submissions. There are several approaches for developing evaluation and selection criteria. One approach is to set minimum threshold limits on technical experience and financial strengths, while another simply states a generalized basis for shortlisting without providing specific threshold criteria.

For example, threshold criteria could require the respondent to have a specific net worth and to have operated at least three plants of comparable size and of similar technology for three years. Such criteria enable unqualified respondents to be eliminated.

A generalized approach to financial criteria simply states that the contractor must demonstrate sufficient financial capacity to undertake and back the obligations of the contract operators. Generalized technical criteria might state that the contractor must demonstrate sufficient applicable experience in operating facilities similar to the subject of your solicitation.

Specific threshold criteria are useful when many contractors are already providing similar services elsewhere. More generalized criteria may be advantageous in potential service arrangements where direct experience is not available, but where certain tangible experience can demonstrate the contractor's ability to undertake the project.

There is a continuing discussion in the industry about point based ranking systems. Some state procurement laws require a point based ranking system, whereas other states allow more flexibility. While points may be useful, you must step back and consider whether or not a submittal has weaknesses not reflected in the point based system. For example, a firm could have outstanding technical qualifications, but marginal net worth for backing the contractual obligations. A possible approach for developing a point based system is assignment of 60 percent of the points to the technical criteria (i.e., experience, regulatory compliance history, reference facilities, and proposed personnel) and 40 percent to the financial criteria (i.e., equity, net worth, contingent liabilities, and leverage of the project financial backer).

Evaluation of Statement of Qualifications

Responses to the RFQ must be evaluated on established criteria. This evaluation should focus on the respondents' technical and financial qualifications, including reference facilities, experience, proposed project team, reference checks, net worth, liquidity, equity, and profitability. The evaluation team can be developed in various formats, depending on the policy and governance structure of your utility. Sometimes an evaluation and selection committee includes two groups: one to do the technical review and the other to make the selection. One group acts as the official governmental organization that selects a short list of firms to receive the RFP while the other group, comprising technical, financial, and legal experts (who may be staff members or outside experts), provides summary analyses to those making the final decisions.

Preparing matrices to compare the evaluation criteria with the information submitted by each respondent is useful because it will enable you to make a side-by-side comparison of each proposer's response relative to the criteria. The results of the RFQ evaluation process should be formalized by a resolution of the governing body with jurisdiction over the process.

A reasonable outcome of the RFQ process is a short list of between three and five firms that will be invited to submit proposals in response to the RFP. A larger number of firms creates difficulties in managing a thorough process while a limited short list conveys an important message to the market that your utility can make tough decisions.

SOLICITATION DOCUMENTS—REQUEST FOR PROPOSALS (RFP)

The RFP is the most important document in the procurement process. It incorporates your utility's policy decisions, the project approach and scope of services determined in earlier planning sessions, and reflects the results of the RFQ process. Appendix B includes an outline of a typical RFP document.

There are many approaches to requesting and evaluating proposals and to arriving at an executed service agreement. An RFP can be several pages long or it may contain hundreds of pages. The brief approach simply states in general terms the services required and provides maximum flexibility to the service provider on how to provide the services. The most extreme approach provides details of all the services required, leaving little room for creativity in the proposal responses. In order to maximize the benefits of public/private partnerships there must be a proper balance between flexibility and creativity on the one hand, and specificity about the desired results on the other.

Earlier RFPs sought proposals without including the terms and conditions of the engagement, particularly the risk allocation between the project sponsor and the service provider. As a consequence, the service agreement or contract had to be developed in detail during a negotiation period after selection of the preferred proposal, resulting at times in incomplete contractual provisions. In addition, the process was complicated since each term and condition had to be developed during negotiations.

Now, however, many modern procurement processes include the service agreement terms and conditions in the RFP, especially for longer-term contracts that may include capital improvements. This facilitates the negotiation and creation of a contract.

It is useful to separate the RFP into four parts:

Part 1: Front End Document

Part 2: Proposed Form of Service Agreement or Contract Principles

Part 3: Performance Specifications and Guarantees

Part 4: Appendices to the RFP

Part 1: Front End

The front end of the RFP document is usually an instruction to proposers that sets forth the rules of the procurement process. It should include:

Services sought. A description of the services you are seeking should be provided. If the terms and conditions (i.e., service agreement and/or contract principles) and the performance specifications are annexed to the RFP, this description can be brief.

Evaluation criteria. As with the RFQ document, proposers want to know your rules for selection. In addition, evaluators need objective or comparative criteria for conducting an evaluation of the proposals. The criteria normally include both technically based criteria and business/financial criteria, as follows:

Technical Criteria

- Approach to O&M services
- Quality of proposed project staffing
- Approach to reducing environmental impacts
- Demonstrated ability to implement services as proposed
- Quality and reliability of proposed designs or capital improvements (if applicable)

Business/Financial Criteria

- Capital and/or annual costs
- Cost effectiveness (i.e., net present worth or life cycle cost analysis)
- Transition plan for existing employees (if applicable)
- Compliance with minority business or other special requirements

The qualifications and experience submission requirements of the RFQ process could be reintroduced, together with the listed criteria for the proposal stage, into the RFP process to support a comparative evaluation of qualifications. This may be necessary since proposers can typically provide more detailed information about the proposed staff at this stage that is closer to the anticipated date of operations.

Proposal submission requirements. The RFP should give instruction about the proposal submission requirements. In order to facilitate fair, side-by-side comparisons of proposals received, a required format (i.e., table of contents) of the proposals should be provided. It may be useful to request technical information in a section separate from the business, financial, and cost information so that evaluations can be performed by separate parties. Some states require cost proposals to be submitted separately from technical proposals to avoid influencing the evaluation of the technical proposal with costs.

A page limit can be placed on proposals to minimize the receipt of excess information. Proposals for O&M contracts without capital improvements could be limited to about 50 pages, excluding proposal forms, résumés, drawings, and other background information. The limit could be extended to about 100 pages or more for projects with capital improvements, depending on the extent of the programs.

Project contact. The RFP should provide the name of the individuals in your organization to whom questions and comments should be addressed. Require these inquiries to be in writing. The RFP should also indicate the location where proposals must be sent, the number of copies, and the date and time for submission.

Available background information and facility/site access. Proposers should be provided access to copies of previous reports and studies about the facility and system. A list of available documents, ranging from engineering drawings and specifications to capital improvement programs and operating data, should be made available to enable thoughtful and comprehensive proposals. It may be worthwhile to establish a central location for inspecting and requesting copies of information.

If the contractor is required to employ some or all of your existing employees, detailed information about salaries, titles, benefits, years of service, and other related information must be provided. The available reference documents should include any collective bargaining agreements between your utility and unions that have jurisdiction over the existing services.

With respect to operating data, the information to be made available should include monitoring reports, sampling results, existing permits, regulatory enforcement documents (i.e., consent decrees and consent orders), chemical usage records, existing budgets, and other available documents that depict existing performance. In addition, any existing contracts with outside service providers should be included in the list of available documents.

To supplement the review of existing documents, proposers should be given access to the facilities and systems. The RFP should state how to arrange for these visits and the number of days allowed. Since your utility wants thorough documents, proposers should be allowed liberal access rights, especially for long-term contracts where the proposer must undertake extensive services.

Part 2: Proposal Forms and Contract Principles

Proposal forms. One of the easiest ways to receive focused information is to include a series of proposal forms within the RFP document. Proposal forms should be provided for items such as:

- Noncollusion statement

- Acknowledgment for receipt of addenda

- Financial commitment of project backer

- Proposed O&M fee

- Proposed capital costs (if any)

- Pass through costs (e.g., utilities and insurances)

Proposal forms can also be used to solicit certain creative features of the proposals or to direct specific requests on project staffing. Such forms may include:

- Suggested reallocation of risks and proposed cost savings

- Alternative price proposals (must be in addition to base proposals)

- Detailed experience of proposal personnel

Service Agreement or Contract Principles

Most RFP documents include the proposed terms and conditions of the agreement. These documents are included as an attachment or appendix to the front end RFP document, which refers to this attachment for the terms and conditions.

Experience has shown that inclusion of a service agreement (contract) document or a contract principles document aids the negotiation process after the preferred service provider is selected. With the inclusion of these documents in the RFP, a contract can be executed within weeks after selection for a simple O&M contract, or within several months for a more complicated contract that includes capital improvements. For complex projects where contract principles or service agreements were not included, utilities have at times needed six months to a year for negotiating contracts.

An outline of the typical contents of a service agreement can be found in appendix C.

Part 3: Performance Specifications and Guarantees

Technical performance requirements and guarantees are attached to a service agreement and should be included with the RFP. While the details of the service agreement are found elsewhere in this chapter, certain key issues are addressed now.

Limits of performance in terms of quality and quantity are central to the performance guarantees or specifications. Quantity based specifications are output or process based, such as minimum gallons per day for water production or treatment, along with daily, weekly, monthly, and annual average capacities or treatment requirements. The performance specifications should tabulate such capacity, treatment, production, or delivery requirements.

Quality based performance specifications and guarantees vary for different water systems and include output based performance parameters for items such as turbidity, metals, organic and microbiological contaminants, and taste and odor requirements.

One of the challenges in developing performance specifications and guarantees is to make the distinction between regulatory requirements and more stringent standards that may be the desire or policy of your utility. For example, a utility may operate a water filtration plant that consistently achieves turbidities much lower than existing regulatory standards. Presumably, the utility would like to provide its customers with similar high quality water when the facility is turned over to a contractor.

Performance specifications should also address nonnumerical performance requirements such as:

- Sampling, monitoring, and reporting procedures

- Building and site maintenance

- Repair and replacement

- Housekeeping standards

- Sludge handling and disposal

Part 4: Appendices to the RFP

In addition to the service agreement or contract principles attached to the RFP, appendices may be included to provide useful information for preparing proposals. These appendices could include background information about the facility, existing permits, historical sampling information and flow rates, utility energy usage rates, existing budgets, and organizational charts. The availability of this information will enable proposers to begin preparing their proposals before reviewing other documents or visiting the facility.

Evaluation of Proposals

The RFP must include both specific evaluation criteria and the exact format of the proposals to enable a side-by-side comparison of the responses. The evaluation process should include the following steps:

- **Completeness check**: Determine the completeness of the proposal relative to the requirements of the RFP. Substantially incomplete proposals may be eliminated at this point.

- **Technical evaluations**: Proposals should be evaluated on project approach, technical reliability, technical viability, proposed staff, and overall quality of the proposed services.

- **Financial evaluations**: Evaluation of the proposed project costs can be conducted concurrently with the technical evaluations. The financial evaluations should look at proposed capital (if applicable) and O&M costs. Costs should be tabulated using a life cycle present worth analysis.

- **Business and legal evaluations**: Proposers should also be required to submit their business plans and comments on the proposed form of the service agreement (or contract).

Due Diligence

As part of the evaluation process it is beneficial to conduct due diligence or reference investigations of the proposal team and their respective reference facilities. A due diligence process typically includes two steps. The first focuses on telephone interviews of references about the proposed firm, team, and staff. A list of standard questions should be prepared, in order to gain comparable information from the references.

The second step is to visit reference facilities. This would include a tour of the facilities and personal interviews of qualified individuals. Proposers should be given the opportunity to "inspect" the project closely to determine the actual maintenance levels and status of all equipment and facilities. Contacts should be made with both the proposal firm's representatives at the facility and the host utility's representatives. In addition, it may be advantageous to contact the regulatory agencies that have jurisdiction over the facility. As with the telephone interviews, a checklist of standard information should be prepared prior to conducting the tours.

The due diligence process will provide substantial benefits in the evaluation, selection, and eventual negotiation process. It either validates the information contained in proposals or

293

uncovers discrepancies between proposals and actual performance elsewhere.

Request for Clarifications and/or Best and Final Offers

Complex projects and services, especially those that include capital improvements or that are for long-term contracts, may result in proposals that raise unanswered questions. Utilities should reserve the right to request clarifications from proposers. Clarifications, however, are not an opportunity for proposers to modify their proposals, and they are not an opportunity for bidding proposals against one another. Clarification questions may be posed to resolve discrepancies in the proposal and to request back-up information to support unclear statements or questionable assertions.

It is sometimes advantageous to request best and final offers from proposers. These are essentially requests for revised proposals, based on the information learned during the solicitation and proposal process. For example, some of the base conditions may have changed and therefore the original proposals may not be valid. Creative information from one proposer, however, should not be divulged to other proposers. The step of requesting best and final offers is advantageous for refining the proposals and reducing overall project costs. This process affords the greatest opportunity to tap into the expertise of the private sector for cost saving initiatives and technical enhancements. It is often unnecessary to request best and final offers in most contract operations solicitations.

If you decide to request best and final offers be careful to avoid information leaks that may be unfair and may even create the risk of protests or litigation.

Finalist Interviews

For many projects, especially complex ones with multiple facets of the proposed service, it may be advantageous to conduct interviews of the finalist contractors before selection. This will enable your utility to fully understand the proposals and what is being offered. A list of questions should be developed prior to the interview. An interview of two to four hours is appropriate for a contract operations engagement. Allow 30 minutes for a formal presentation, followed by an interactive question and answer period that focuses on:

- Confirmation of information contained in proposals

- Clear understanding of the proposed staff

- Level of commitment to the project by the proposer

- Assurance that a long-term partnership is possible

It is important, however, not to be swayed by extremely polished presentations given by the proposer's sales representatives. Therefore, when inviting firms for interviews you should clearly state a desire to have project personnel conduct the interview.

NEGOTIATION OF A COMPREHENSIVE SERVICE AGREEMENT

At the conclusion of the evaluation process, one or more proposer(s) may be selected for contract negotiations. Some states allow for simultaneous negotiations whereas other states require selection of a single preferred proposer before initiating negotiations.

Contract negotiations are usually conducted by a small group of negotiators that include technical and legal experts. Since contractors will likely include these experts on their team, you should have similar resources.

The proposed form of the service agreement (contract) in the RFP should be used as the basis for negotiations. This sets forth the risk posture and services sought by the utility, and therefore serves as the logical framework for negotiations. Appendix C includes an outline of the typical service agreement, and the balance of this section addresses the information typically contained in one.

The elements of a service agreement are illustrated in Figure 15-2.

Some of the provisions contained in the service agreement shown are only utilized for more complex agreements and may not be applicable in all cases, especially those of shorter duration. The option to purchase under the Terms of Agreement box is one

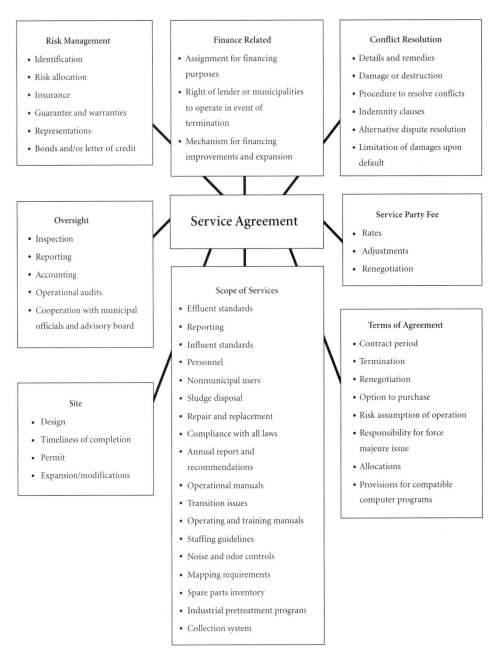

Figure 15-2   Elements of a service agreement

example of that, as is the right of lender or municipalities to operate in event of termination, which is listed in the finance-related box.

The following discussion highlights elements listed in the various boxes of Figure 15-2.

Scope of Services

This section provides the heart of the service agreement and is the building block for all other areas of the service agreement. The scope of services is typically included as performance specifications and standards and is attached to the legal requirements as an exhibit or appendix to the agreement. Performance guarantees will be based upon specification of raw water quality and the plant's ability to properly treat and meet current regulatory and utility-specific requirements. This section also establishes the objective standards and scope of work for which the contractor will be held responsible by the additional sections of the service agreement. Projects that include a design build component will introduce additional details to the scope of services to address design requirements, repair and replacement activities, construction features, and some type of post-construction testing. These matters are further addressed in chapter 17.

Site and Project Limits

The service agreement should address the limits of the contractor's responsibilities, which may also be depicted on a site plan. Where the project includes capital improvements and/or building a new site as part of a design build operate project, the details of the site become more important, as further described in chapter 17. The service agreement must permit you unlimited access to the site and facilities during the term of the contract.

Oversight

Oversight activities are appropriate for all forms of service agreements. Requests for weekly, monthly, and annual reporting are common. Access to all plant records, operations, and related information are typically addressed in this section. The ability to conduct third party reviews of the contractor's program and operation are contained in this section, as are provision for formal planning of facility expansions and upgrades. Further suggestions about oversight and performance monitoring are found at the end of this chapter.

Risk Management

The RFP process will have identified important elements in the area of risk management. While permits are held by the owner of the facility, the utility should obtain indemnification from the operator. Performance guarantees, as previously discussed, make the contractor responsible for meeting current regulatory requirements, provided the raw water quality is within specified criteria. Future regulatory requirements are often accommodated via work change mechanisms where responsibility for new facility capital (if any) is specified along with any change in operating costs.

In general, it is not reasonable to expect the contractor to assume open-ended, unlimited liability. The costs are likely to be prohibitive and the issue of accountability (and therefore real liability) difficult to assess. Be aware that where unlimited liability and risk assumption has been obtained from the contractor, he or she may have utilized a wholly owned subsidiary that has limited assets and limited financial capability as an alternative way to cap legal and financial exposures on the project.

Finance Related

Generally, the basic contract operation and maintenance agreement will look to the public sector owner to provide new capital. The failure to do so can result in a breach of contract and relieve the vendor of additional obligations. Where the facility is owned by the contractor or the capital is provided by the contractor, the service agreement should address how the capital (type and amount) and compensation are authorized. Typically the utility is able to request (at its expense) a third party review of the proposed capital additions or modifications to address the design, capital, and operating costs.

Conflict Resolution

As in all relationships, conflicts may occur during the term of the service agreement. The service agreement should provide an appropriate pathway for resolution. For example, failure to provide a weekly report in a timely fashion is usually not cause for contract termination but may be addressed under a fines and penalties section of the conflict resolution section. It is common to specify increasingly stringent requirements for conflict resolution based upon the severity of the issue and the frequency of occurrence. For example, chronic inability to provide specified reports and information, and failure to correct the situation after proper written notice, may escalate into an issue of contract termination for failure to perform. Many service agreements specify the use of arbitration rather than the court system to facilitate quicker and a less expensive resolution of issues. While insurance coverage can be used to assign responsibility and costs for many items (e.g., general liability, automobile, errors and omission, and fire and earthquake), there are matters that are not insurable risks. These are generally covered under a force majeure provision.

Service Fee

The service fee includes the basic program costs, generally expressed as a fixed price for the first year of the program. It covers costs for the treatment of a specified influent quality and quantity and for the delivery of a specified effluent quality and quantity. Cost reimbursement compensation methodologies may also be developed but are less favored since they do not readily transfer the financial risk of performance to the private sector. In using the first year, fixed price methodology, the service agreement must specify the financial basis for adjusting compensation during the term of the agreement. Such fixed costs are typically subject to annual escalation by indices such as the Consumer Price Index.

Develop provisions to address:

- Changes in raw water quality and quantity that increase or decrease the cost of plant operations

- Changes in the quantity of water requested by the utility

- Adjustments for inflation so that the first year price and the related scope of work are accomplished for the same constant dollar amount throughout the term of the agreement

- Adjustments for cost reductions through incentive saving programs

- Cost adjustments for a change in scope of services required by additional facilities, new regulatory or legal requirements, or negotiated revisions requested by the utility

- Adjustments for the provision of capital replacement items beyond the scope of the routine, preventive, and predictive maintenance requirements of the agreement

The format is fairly well established in existing agreements for these compensation adjustments.

Some utilities request an up front or concession fee payment by the contractor. Similar to the provision of private sector capital, this payment would be amortized based upon the length of the contract as part of the compensation methodology proposed by the contractor. While a concession fee has the advantage of supplying up-front cash, it may be a less desirable way to raise money than other options available to a utility.

For longer-term contracts, pass-through costs (i.e., payments made to the contractor to reimburse for actual costs incurred from third parties) are typically used. Such costs may include electrical charges, insurance costs, and other costs from third parties that may not be predictable over a long-term. Pass-through costs usually do not include any mark up by the contractor.

Terms of Agreement

This section of the service agreement contains many provisions that constitute the financial and legal basis of the agreement. Specified conditions for contract termination are detailed in this section. For facilities owned by the contractor, the utility should require the option to purchase the facility. Since tax exempt debt may not be available without a bond issue, the utility should require language that addresses this time delay and possible obtains a bridge loan commitment from the private sector owner for the time period needed for the issuance of the debt. This approach will allow the legal transfer of the facility much earlier in the process.

The service agreement must also recognize that conditions will change during the term of the agreement (e.g., due to new laws and regulations) and should provide for a renegotiation mechanism. Often addressed through a change in scope of services provision, the objective is to provide an explicit pathway and process for modifying the service agreement to accommodate changes without the need to develop an entirely new agreement.

PERFORMANCE MONITORING

Once a contract is executed, monitor the performance of the contractor to determine that the requirements of the agreement are being met. Although performance monitoring requirements and options should have been articulated in the agreement, they should not interfere with the contractor's rights to operate and maintain the facilities and systems and to design their own procedures, provided that such procedures are consistent with the service agreement provisions. You should be monitoring for performance based results and standards, although the methods are left to the contractor's expertise. With this in mind, some of the goals of performance monitoring are:

- To determine the degree that the original objectives for engaging a contractor are being realized.

- To provide an independent, third party assessment of the contractor's performance, and to identify factors contributing to both problems and successes.

- To provide recommendations leading to action plans for improvements when necessary.

Relevant Issues

Monitoring performance provides information that enables you to assess how well the contractor is performing. Contractors will normally seek to operate as efficiently as possible from a cost standpoint while satisfying their responsibilities. Therefore an effort needs to be expended during the term of the contract by both the contractor and the utility to execute and monitor those responsibilities. There are two general types of issues that typically arise for the utility:

1. Issues that seemed clear when the agreement was negotiated but are sufficiently vague to allow alternative interpretation.

2. Conflicting goals between your utility and the contractor that can result in the contractor operating in a manner not in your best interests, but that still appears to satisfy the contract.

Contract oversight is needed to protect your interests and to ensure that the original objectives of the service agreement are realized. This requires a combination of technical expertise in the area being reviewed, as well as experience in managing the performance of a contractor.

Auditing contract operations and maintenance will provide a detailed understanding of plant and system conditions, and a perspective on whether or not the plants and systems are operating properly. An audit of an O&M contract will provide answers to questions regarding:

*   **Performance:** Is the operator meeting its contractual performance guarantees?

*   **Costs:** Are the facilities being operated so as to minimize all costs?

*   **Preservation of assets:** Are facilities being maintained properly to maximize their useful life, and to reduce the need for capital improvements/retrofits?

- **Environmental compliance:** Are the facilities operating in compliance with all environmental permit requirements? Are compliance action plan schedules being met?

Audit Approach

A successful service audit or assessment reduces uncertainty, provides good operational practice incentives, and encourages innovation and improved efficiency over the life of the contract. Typically for contract operations, performance monitoring includes the following key elements: contract review, facility inspection, performance audit, invoice review, and environmental compliance review.

Contract Review

If the audit is being conducted by someone unfamiliar with the contractual arrangements, initial efforts will involve a detailed review of the service agreement to familiarize him or her with the overall delineation of responsibilities and performance requirements. This review includes performance guarantees, incentives, pass-through cost allowances, operation and maintenance responsibilities, reporting requirements and other appropriate contractual provisions. In particular, an understanding of the allocation of risks and assignment of control is essential.

Facility Inspection

Site visits to the facility(ies) should be conducted by technical experts experienced in the type of operation being reviewed. The focus is to observe the general condition of the facility(ies), observe operating practices, and to review operating and maintenance records. Observations are documented via field memoranda and checklists, and could be supplemented by photographs or video tapes.

Performance Audit

Using available documentation, facility performance is reviewed and compared to contractual responsibilities to assess compliance with all contractual guarantees. This is generally accomplished by a detailed review of documentation submitted in compliance with the service agreement's monitoring and reporting requirements.

Invoice Review

The invoices submitted by the contractor can yield valuable information about how the facility is being operated. Invoices should be reviewed for pass-through costs and supporting documentation, as well as for other relevant information. A comparative evaluation of all costs (including chemical and energy usage) among the different facilities and with similar facilities provides a means of determining how efficiently operations are conducted.

Environmental Compliance Review

The results of continuous and periodic sampling, as required by the service agreement, should be checked against the performance standards of the service agreement. In particular, compliance action plan schedules and milestones required by any regulatory consent orders must be achieved to avoid exposure to significant legal and financial penalties and adverse public reaction. Contractor performance is critical in this regard, and must be carefully reviewed in both a technical and contractual context.

Evaluation of System Operations

Based on the outlined information and analyses, an overall evaluation of the operation can be made. Findings should be summarized in an annual report, including an assessment of the degree to which objectives are being achieved, identification of problem areas, and identification of factors contributing to the program's successes or shortcomings. Actions that can resolve any areas of concern should be developed.

SUMMARY

Contracting for services should be done with the greatest of care and attention to details. We call this *thoughtful contracting*, a process that is fair to the contractors yet protective of the utility's interests. The best contractual arrangements in the long run are those that are entered into by a knowledgeable utility and a reputable contractor who is willing to commit experienced staff to your project.

Our point is that good relationships can and *should* be developed between you and the contractor—when the decision is right in the first place; when the project, its work scope, terms, and conditions are well drawn; when the contracting parties are of equal competence; and when trust, fair play, and a commitment to quality public service characterize the relationship.

Valuable lessons can be learned from the experience of agencies that have entered into contract operations. Some reflect complete satisfaction, others less. Among the items that appear to be the most significant, positively or negatively, are:

- The use of a thorough, thoughtful process to define the scope of work.

- The development of mutual respect and trust among the parties.

- The ability of the on-site manager of the contracting party to fairly and effectively execute the contractual responsibilities.

- The care that was taken in developing a clear and comprehensive service agreement, especially in the areas of compensation, risk sharing, maintenance, and renewal and replacement of mechanical equipment.

If there is a single factor, outside the human elements, that produces the most satisfactory experience, it is the terms and conditions of the service agreement. It also appears that selection of a contractor on the basis of cost alone is least likely to result in success. The time and energy you invest in developing and executing thoughtful contracting processes will be rewarded during the evaluation and selection process, and during the service period of the agreement.

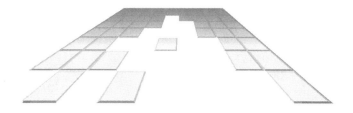

CHAPTER SIXTEEN

Managed Competition

A managed competition is a structured, formal process in which the employees of a public utility and contract service providers submit competitive proposals for a contract to perform one or more utility functions. In general, the public sector employee group does not have access to its own independent sources of capital, design, or construction expertise. Therefore, the managed competition procurement process has been used for the operation and maintenance of existing facilities and systems under a contract of modest length (three to five years). As a result, these procurements do not involve the provision of new facilities or capital, or up-front concession fee payments.

In general, managed competitions follow the same basic procurement approaches described in chapter 15 for thoughtful contracting processes with the addition of: (1) establishment of a public sector team that must function totally independently of the procurement management team; (2) funding for the activity and assembling the skills needed for the public sector team and its proposal preparation; (3) development of evaluation factors and criteria that create equality for the assessment of the public and contractor proposals; (4) recognition of the disparate abilities of the proposers to provide guarantees, assume risks, and offer liability protections to the utility; and (5) agreement that the public sector team cannot sign the standard contract used for an outside contractor but must operate under a memorandum of understanding as its contract equivalent.

Where managed competitions have been conducted, the results demonstrate that cost savings are achieved with the utility team being selected in some cases, and in others with the selection of an outside contractor.

Studies and analyses that compare public and private costs without the actual, formal solicitation of proposals are not considered managed competitions. Characteristics of an effective managed competition include the use of actual formal written proposals by public and contract service providers; a "level playing field" assessment of these proposals by the utility; and the advance publication of evaluation factors and criteria.

Since many of the processes associated with a managed competition are similar to those described in chapter 15, the focus of this chapter is to discuss the objectives, potential benefits, and risks of managed competition; to describe the process itself; and then to highlight how it differs from the processes described in chapter 15.

OBJECTIVES OF MANAGED COMPETITION

The primary objective of a managed competition is to reduce costs and/or to improve performance by expanding the proposal submittals beyond outside service contractors to allow public sector employees the opportunity to compete for the work. To achieve these objectives, the procurement process must be structured explicitly to address the inherent differences between proposals from the contractors and those from the public employees. While the costs for the service provision outlined by the work scope will be an important consideration in the assessment, comparisons based on the technical, financial, managerial, and organizational capabilities of each proposal team must also be evaluated. We have included a number of case studies on managed competitions in appendix A that illustrate the techniques being used. Interestingly, the City of San Diego has an explicit policy that while recognizing the importance of service quality and reliability, the pricing for private sector proposals may be 10 percent less than that for the public sector.

Assessment factors should also address the contractual terms and guarantees offered by the proposers and the risk allocation matrix achievable with each proposal. Additional factors may include a commitment of capital by the proposer for facilities and systems, minority business commitments, and other site-specific features.

A secondary objective of a managed competition is to bring a competitive element into public sector organizations. The process of developing a proposal to compete against a contractor alternative usually leads to the realization of efficiencies and cost savings that can have long-term and widespread effects throughout a utility. For example, a managed competition in Tulsa, Okla., resulted in the city team rethinking the necessity of having 24-hour staffing of a satellite wastewater facility. In other cities managed competitions are providing an impetus for incentive programs (gainsharing) for public sector workforces, flatter organizational structures with reduced levels of supervision, revised job classifications and increased cross-training, enterprise activity-type funding, and activity based cost accounting. These benefits accrue to the entire utility and potentially to other city departments and functions as well.

Finally, the use of a managed competition may overcome the political difficulties often associated with procurement processes that exclude the affected employees. Expansion of the competition to include utility employee participation may also provide a springboard for productivity improvements and cost reductions in other functions of the utility.

Although only limited empirical data are available on the long-term success of managed competition for water and wastewater services, results in cities such as Tulsa and Charlotte, N.C., show that the public sector team can indeed compete and win. Solid waste collection services provide additional examples of successful competition by a city/employee team. Properly done, a managed competition procurement adds another motivated participant to the competition. In general, when utilized for contract O&M programs where access to capital and new technology are not significant opportunities, over time this increased competition tends to decrease costs and increase efficiencies.

Managed Competition at Charlotte-Mecklenburg Utility Department

The Charlotte-Mecklenburg Utility Department (CMUD), Case Study 4.4, is a municipal agency that provides water and wastewater services to approximately 500,000 residents of Charlotte and Mecklenburg County in the southern Piedmont region of North Carolina. Spurred by inquiries from the private sector, unsolicited proposals to purchase system components, and political interest, the department decided to allow private firms to compete with city staff for the operation and maintenance of the department's eight treatment plants. In response, eight proposals were received from private firms and from the department's own bid team, referred to as "Charlotte Mecklenburg-Contract Operations," or "CM-ConOp." CM-ConOp submitted the lowest cost proposal, which will result in savings of

BENEFITS OF MANAGED COMPETITION

In Houston, Indianapolis, Charlotte, Tulsa, and other cities, managed competition has provided dramatic cost savings. However, achieving the lowest possible cost is not an automatic, or even the only outcome of managed competitions. Improvements in service quality and service performance guarantees are also obtainable.

Of course, several steps are required to realize these potential benefits. Long-term competitive pressure must be maintained while transaction costs are kept realistic. The impact of qualifications, risk assumption capability and capacity must be assessed, along with guarantee provisions on the quoted price. Above all the process created must be perceived as fair and honest.

Although each managed competition provides opportunities and incentive for cost savings, maintaining competitive pressure over a longer time period is undoubtedly the best way to keep costs as low as possible over time. Said another way, monopolies (whether public or private) tend to breed inefficiencies over time.

Careful consideration of maintaining the benefits of competition should be part of the contracting process. While often site-specific in final form, such decisions as contract term, number of contracts, utilization of capital and technology, incentive savings programs, and the issue of whether or not the utility should retain a core capacity to operate its system all affect the maintenance of a competitive situation.

As previously mentioned, the experience of having to compete can lead to fundamental changes in operational assumptions, methods, and organization culture. In Tulsa, facilities were converted to remote operations and 24-hour staffing was discontinued. In Charlotte, the internal competition team made process changes, cut staff by almost 40 percent, and interviewed existing staff to select those who would be part of the new downsized team. In Houston and Indianapolis, private sector proposals revamped service delivery without decreasing quality to achieve significant savings. In addition, having to compete as the municipal team can result in identifying inefficiencies, barriers, and institutional disincentives for achieving low cost operations. Among these barriers are procurement and purchasing policies, employee job classification systems, and the continuation of

310

facilities and processes that are costly and should be candidates for replacement.

THE CREDIBILITY RISK ASSOCIATED WITH MANAGED COMPETITION

A risk often overlooked in conducting a managed competition is that of losing utility credibility if the procurement process is not perceived by all participants (elected officials and the public) as explicit and fair. Because of the perceived conflict between the utility's role in submitting a proposal, evaluating proposals, and making or recommending an award, there is a greater risk of conflict, lawsuits, and controversy in a managed competition procurement. Furthermore, if the process is not seen as fair, fewer (or less qualified) contractors will participate, thus diluting the benefits of competition. Factors in creating this perception are the initial decisions regarding risk assumption, performance guarantees, liability coverages, budget guarantees, evaluation criteria, and the procurement process itself. The need to provide expertise for public sector proposal development, including access to ongoing technical, operational, and administrative expertise, should also be explicitly addressed as part of the public sector's proposal and price.

Specifications, performance guarantees, and contractual provisions must ensure that public health concerns and risks are adequately weighed in decisions about operations and processes. The proposal evaluation process should be structured to allow consideration of these issues and not just the lowest cost obtained from a proposer.

The contractor teams must be allowed sufficient access to the project and available operational information and data to enable them to prepare creative and responsive proposals.

ENSURING COMPATIBILITY AND FAIRNESS

In a managed competition, ensuring comparability and fairness between public and contractor proposals involves a more costly and complex process than other procurements. The scope of work and specifications must be carefully developed so that neither the public sector nor contract proposer has an inherent

$4.2 million over the five year term of contract. In addition, the Charlotte model for competition showed that the public sector can compete successfully with private firms and that the result can be a complete change in the way the agency does business.

Private Contractor Beats City Bid in Indianapolis

The City of Indianapolis, Ind. (Case Study 3.2) uses private operation and maintenance contract services at its two wastewater treatment plants.

In order to compare actual costs to projected costs submitted by bidders, the city established that in 1993 it spent $30.1 million to operate the two plants. This figure includes operating costs but does not include capital expenditures and some city and departmental overhead costs. The shortlisted proposals were as follows for the operating cost component: White River Environmental

311

Partnership at about $18.1 million, PSG at about $18.1 million, and city management employees at about $27.7 million.

The lowest bidder and winner of the contract was the White River Environmental Partnership, at 44 percent below existing costs for five years of operation. The city management team bid was approximately 10 percent below existing costs for five years of operation.

advantage. From a financial perspective, allocation of overhead in the public sector evaluation should be clearly defined in the RFP, so that the private sector understands how the public sector's costs will be distributed. This is particularly important if the public sector is only contracting one of a group of its facilities, and management and administrative costs can be distributed over the entire facilities base. A third party competition manager is often used to resolve technical issues that arise during the competition (such as evaluation of proposers' experience and qualifications). Finally, resources are often allocated to assist the public team in preparing a competitive proposal, and to audit the cost allocation and other aspects of their proposal. Regardless of how much these added functions require outside consultants, the time, effort, and resources expended will be significant, and greater than those required for a procurement restricted to contractor proposers.

Perhaps the most significant risk issue not adequately considered in advance with a managed competition is the risk associated with potential labor relations issues of a "win" by a contractor team and resultant "loss" for the public sector team. Since the contractor will most likely decrease the total number of employees and simultaneously augment the municipal workforce, the utility must now successfully integrate two organizations that were direct competitors the day before. Anticipating this potential outcome and advance planning for how to proceed under these circumstances should be part of the thoughtful contracting process. Even with a proactive transition plan the risk of significant labor and organizational turmoil exists.

THE MANAGED COMPETITION PROCESS

The managed competition process follows a similar path to that described in chapter 15:

- Building qualified project teams

- Developing a strategic approach

- Determining site-specific contracting alternatives and approaches

- Assessing market availability

- Conducting a solicitation process

- Negotiating a comprehensive service agreement

- Conducting performance monitoring

The solicitation process typically consists of a Request for Qualifications, shortlisting, and a Request for Proposals.

For further information refer to the discussions in chapter 15 on each of these subjects.

These added elements are essential to a successful managed competition process: integrity, clarity, and support. A high credibility (high integrity) process is important to a managed competition because it results in the greatest number of qualified proposals, and minimizes the risk that the municipality will lose credibility during the process.

Establishing Utility Teams

In order to preserve the credibility of the process, from both an internal and external point of view, it is essential to establish completely separate teams to manage the procurement and to submit the employee proposal. The procurement management team is responsible for overseeing the preparation of procurement documents, processes, and communication with all proposers, while the public proposal team assists your staff in preparing a competitive, responsive proposal. There should be no overlap between members of the two teams, and a protocol should be developed to assure their independence. The procurement management team is essentially the evaluation team described in chapter 15.

Sufficient funding must be available for the public proposal team to enable city employees to gain access to the necessary expertise for proposal development, technology, and proposal presentation to permit them to compete with the contractor teams.

Communication between Teams

Most utilities put procedures in place for communication with contractors during procurement of contract services. However, in a managed competition it is essential to the credibility of the process that the procurement management team (and other staff involved in conducting the competition) treat the public proposal team exactly the way a contractor would be treated. This includes attendance at meetings, access to information regarding the competition, and submittal deadlines and procedures.

Communication with the public proposal team about the progress and substance of the managed competition should follow these guidelines:

- Would this discussion or meeting be happening if the public proposal team member were a contractor?

- Will the content of this discussion be communicated to the private vendors?

If the answer to either question is no, then that particular communication or meeting with the public proposal team is inappropriate.

Providing Expertise

During the development of procurement documents it is essential to have input from the current operators (public or private) of the facility or services to assist in developing the scope of and thinking through contract terms and conditions. This involvement can be handled by assigning one or more persons from the public operation to provide this input. It is often helpful at this point to have independent technical advice on the feasability of specifications and contract principles, so that the desired scope of services is realistic, fair to all parties, and in the best interests of the municipality. Whatever decision is made regarding the need for outside expertise, once the competition is underway (at RFP or RFQ issuance), communication between teams must be consistent with these guidelines.

Team Composition

Because communication between teams will be limited during the competition, team composition requires some deliberate thought and planning. In general, the public proposal team should include all of the expertise and resources that it needs to compete and advocate for itself successfully. This often means that technical, management, and financial people from the utility will be assigned to the public proposal team. These resources should be costed and priced as part of the public proposal team's proposal.

The procurement management team may include representatives of the following functions:

- Utility manager's office

- Purchasing

- Legal

- Finance

- Related technical department head (e.g., Public works) who is not part of the public sector team

- Department director (depending on needs of public proposal team and credibility issues); providing the director is not part of the public sector team

- City council or other governing body

- Procurement management

Providing the public proposal team with the resources its needs to compete often leaves the procurement management team short on technical and procurement management resources. The gap can be filled by hiring experienced consultants to assist the procurement management team with tasks such as evaluation of the technical qualifications of proposers, resolution of technical issues, and development of procurement processes and evaluation criteria. The procurement management team should be chaired by someone who understands the goals and intent of the managed competition, is committed to an open process and timely completion, and is perceived as being able to make a clear, fair, and defensible recommendation. The San Diego and Charlotte (case studies in appendix A) managed competitions illustrate the use of outside consultants to assist the procurement management teams, whereas the example of Tulsa illustrates the use of city staff from other departments.

GETTING RESULTS

A number of decisions must be made about the competition ground rules early in the procurement planning process. These decisions relate to the steps to be followed, the documents used, the work scope and evaluation criteria, the quantification of the risk assumption capacity of the proposers, and the basis for selection. In San Diego for instance, procurement consultants were used to guide the City through difficult strategic decisions regarding procurement (e.g., selection criteria), financial/business (e.g., how risks would be allocated), and technical (e.g., geographical boundaries for facility operation) issues. This necessitated a strategy session at the beginning of the project. The decisions made at this session provided overall guidance for the RFQ and RFP development, and were refined throughout the procurement phase.

A number of these basic process parameters are now discussed.

Prequalification

Many states allow bidders or proposers to be prequalified. This prequalification process essentially makes a "first cut" of interested competitors based on an assessment of their qualifications of financial strength, experience, project team, understanding of the project, references, and compliance with legal criteria. Some state and local laws and ordinances limit the use of prequalification. Massachusetts law, for example, does not permit use of a separate prequalification RFQ, but instead requires that a single qualifications and proposal document be utilized.

The obvious advantage of the prequalification process is that potential proposers can be screened so that only those proposers demonstrating at least a minimum level of qualifications may proceed to actually develop and submit proposals. This also saves time and effort later for the proposal evaluation team.

In many cases the public proposal team is automatically qualified as meeting the minimum qualifications for providing the services in question, by virtue of the fact that they have been performing the services for some time (if that is the case). If your utility takes this position, that decision should be clearly and prominently disclosed in the RFQ, notifications, and other announcements. Properly done, this approach should eliminate challenges to the RFQ process since contractors know in advance that they will be competing with an automatically qualified public proposer.

Bids versus Proposals

The decision about whether or not to solicit bids or proposals in a managed competition is often guided by state and local ordinances. Generally speaking, an Invitation to Bid (ITB) is a possibility when the scope of services can be clearly defined, alternatives to the specified work scope are not being considered, and when prequalification can be used to create a short list of qualified bidders who are "equal" except for the price they will charge for their services.

For the most part work scope specifications developed for water plant operations and maintenance are not a good match with Information to Bid criteria, since water plant operations are unique in many aspects regarding risk issues when compared to other public services. For example, when a waste-to-energy plant cannot meet its emission permit, the solid waste contractor bears the financial risk of alternative disposal (e.g., landfilling) until the facility can be brought into compliance. In this instance there may be little risk to the public. On the other hand, if water plants do not operate correctly, the public may be at risk if the products are consumed before operations can be interrupted. Therefore, from a public health perspective, consideration of qualifications is critical to provide the desirable level of service. Proposer capabilities will differ in experience, on-site staff capabilities, systems, financial capability, ability and willingness to assume risks and liabilities, and to provide guarantees. Furthermore it is in the utility's interest to seek the best combination of service cost, reliability, quality, and risk allocation and to assess these factors in making its evaluation and selection.

Fortunately the RFP process is usually more flexible in terms of allowing the use and comparison of additional criteria beyond costs alone. Because this process can accommodate guarantees, liability coverage, and risk assumption, it is generally preferable

for procurement of O&M services involving the complex requirements of drinking water facilities.

Some procurements use an evaluated bid, in which technical
and financial criteria are considered in addition to cost. Evaluation factors and criteria are developed to reflect the risk allocation
capability inherent in each proposal. These criteria are outlined in
the RFP so that the proposers are cognizant of how the proposals
will be judged.

San Diego is proceeding in this direction and has assembled
an evaluation team for assessing both the RFQ and RFP. This
evaluation team is comprised of City staff representing the competition program and financial services, as well as individuals
from outside the City, including a representative of a water district that would receive water from the plant under emergency
conditions, other municipalities who have conducted managed
competitions, and southwestern water agencies who understand
the technical issues associated with supplying and treating the
sources available to the contract operator. Technical and legal
consultants assisted in preparing a summary comparison of
responses to the RFQ and RFP, but were not voting members of
the committee. The list of committee members was distributed
during a preproposal meeting with qualified bidders, making it
clear that contact with these individuals was prohibited during
the procurement process. Any contact by the proposers with the
evaluation team or members of the city council (who must
approve the recommendation), regarding the procurement process, would result in immediate disqualification.

Competitive Negotiations

Some states allow competitive, simultaneous negotiations with a
small number of competitors after receipt of proposals or bids.
Depending on the specificity of the RFP, ITB and proposals, this
may be a way to clarify and ensure the comparability of proposals
and to achieve the best possible option for the utility.

Competitive negotiations, however, pose special challenges
for managed competition. Take care to ensure that the negotiating team is independent, and that negotiations focus on clarification of competitors' proposals for the scope of services in the

RFP/ITB, not on the generation of lower-cost offers or scope alternatives that could lower cost.

The RFQ and RFP documents used by the City of San Diego illustrate the thoughtful consideration of these issues and explicitly state the basis and the factors upon which the procurement will be based. As illustrated by San Diego's policy decision, the pricing for the private sector proposal may need to be at least 10 percent less than the public sector price to be competitive. While other communities may not impose a similar hurdle, the San Diego RFP provides an excellent example of clarity and up-front statement of the procurement ground rules.

SUPPORT

The establishment of a high integrity process, clearly and explicitly documented and made available to all participants, is essential in a managed competition. Without this condition, it will not be possible to obtain the organizational, political, union, marketplace, and local citizen support of the process.

Additional steps should be taken to gain the support for the procurement process. For example, costs for the utility staff proposal and the basis for cost allocations from related utility departments must be clearly established prior to the actual proposal process. The methodologies should be openly shared with all interested parties and must attempt to fully allocate the costs for services received and costs avoided.

Equally important, the utility or other authorizing body must embrace the managed competition process. This entails approving the procurement objectives, scope of services, evaluation factors and criteria, as well as the key contract terms and conditions for the procurement. Failure to obtain these details before the procurement is begun may lead to a procurement process and selection that is subsequently not supported by decision makers.

Support of utility staff members who are proposers must be even-handed and perfectly clear if the staff is to embrace the process as fair and reasonable. Failure to achieve this support will create a cynical attitude toward the process, especially if the utility team does not win. In short, they will feel taken advantage of and that they were sucked into a highly manipulative process in which they could not be successful. The workforce integration issues under such a scenario increase dramatically.

Discussions with local community and business leaders regarding the objectives and process of a managed competition might also be part of the overall buy-in. Similarly, this same information needs to be shared in a positive manner with the local citizens and rate payers. Newspaper articles and TV news coverage are two vehicles typically used to communicate these issues with the public.

While it is unlikely to expect universal support for the managed competition procurement process, following these steps will help you gain support and approval from the utility's stakeholders.

SUMMARY

Managed competitions provide another procurement vehicle for obtaining contract services when they are properly structured and conducted. The results can be mutually beneficial for the utility and the community. This chapter has provided guidance in determining the appropriateness of a managed competition, and in structuring and managing it properly. A final word of advice: don't decide on a managed competition until you are certain you understand fully what it entails and you have commitments from all involved to do every detail right. Be sure to talk to those who have done it!

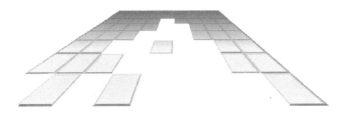

CHAPTER SEVENTEEN

Approaches to Facility Development

The need to construct new water facilities opens the door to considering the use of the private sector to perform a range of integrated services such as permitting, design, construction, long-term operation, and even private ownership and financing, all under a service contract. These ventures are called Design Build Operate (DBO) and Build Own Operate Transfer (BOOT), depending on whether ownership and financing remain with the public agency or are transferred to a contracted entity. These two approaches are used extensively worldwide but have been used on only a few occasions in the United States.

The BOOT scheme is appropriate for projects requiring expenditures beyond the financing or borrowing capacity of the public utility where there is a desire to minimize consumption of scarce public credit by using private capital. The DBO approach is applicable where private financing is not required or desired. Both approaches offer potential advantages of greater efficiency by integrating project elements and risk allocation in a single contractor.

This chapter describes how and why the DBO and BOOT models differ from the traditional approach to facility development. It outlines the processes of selecting a preferred approach and building a project philosophy for the solicitation process. It

Seattle DBO Results In High Savings

The City of Seattle (Case Study 5.4) used a Design Build Operate (DBO) process for its new 120 mgd Tolt Filtration Facilities.

The total value of the contract is about $101 million (net present value, 1998 basis). This includes $65 million for design and construction, and $36 million for 25 years of facility operations. The term of the contract is 25 years (15 years with options for two five-year renewals with the same contract terms, at the City's discretion). The obligation for design, construction, and operations lies with the CDM PHILIP consortium and is guaranteed by a single guarantor— Philip Services Corporation—for the term of the agreement. This innovative procurement process produced a savings of $70 million when compared to the estimated cost of building a comparable facility using the conventional process of consultant design, low-bid construction, and city operations.

also covers the steps associated with the solicitation process including contract negotiations.

Private sector DBO for new facilities is essentially a single source procurement for a work scope of traditionally separate contracts. The DBO agreement requires a "front end" component that deals with the design and build aspects of the project, and an operational section similar to a long-term O&M contract. Should private sector ownership of the facility be added to the DBO work scope, then an additional section would be necessary. The legal, financial, and contractual issues increase dramatically in this form of service agreement. In some states, design and build formats are not permitted. In all cases compliance with state and local procurement regulations for construction wages must also be addressed.

The use of DBO or BOOT methods for project implementation by public utilities is likely to be limited in the United States, with most applications by mid-sized and larger utilities undertaking new projects. Since the processes for DBO and BOOT implementation are more complex and costly than contracting for O&M services alone, most smaller utilities will be excluded.

The Seattle project (refer to Case Study 5.4) demonstrated that the use of DBO for a new 120 mgd water filtration plant could result in significant cost savings in both capital and long-term operating costs (savings of $70 million compared to the estimated cost of building and operating a comparable facility using a conventional design-bid-construct approach with city operation).

COMPARISON OF APPROACHES

The traditional model for developing and operating water and wastewater treatment facilities in the United States has been the design bid build operate model. Basically this model involves three parties: the designer, the constructor, and the agency or contractor responsible for operating the plant.

In the first step of the traditional model a private design firm is hired under contract to the utility. Its responsibilities include determining plant requirements and developing bid and contract documents. In the second step bids are tendered by contractors to meet the requirements of the contract and specifications, and the lowest responsive, responsible bidder is awarded the job of constructing the plant. In the third step the utility is responsible for operating the plant in accordance with performance standards, either with public staff or under contract with a service provider.

While the traditional model is well understood and has a number of advantages, there are some potential shortcomings including the following:

- Segmenting the design and construction process can lead to higher design costs and normally does not permit the designer and builder to collaborate in ways that would lead to reductions in construction costs.

- Risks associated with the failure of the plant to operate in accordance with the utility's intent are often ultimately borne by the utility.

- The low bid, fixed price method of selecting constructors heightens the risk of performance failures; this situation worsens when it is not clear who is responsible for a failure.

- Some municipalities are not equipped to efficiently operate complex plants, especially when they lack economies of scale in management and chemical procurement, and are subject to highly restrictive labor agreements.

RETHINKING ROLES AND RESPONSIBILITIES OF THE PARTICIPANTS

In today's dynamic environment, many of the basic roles and responsibilities implicit in the traditional approach are worth rethinking. A changing point of view about municipal versus contractor capabilities, the increasing attractiveness of integrating design, construction, and operating processes, and the emergence of capable contract operators have all contributed to a broader range of opportunities for utilities considering an alternative approach to the traditional model.

In considering an alternative approach to the development of a new project like a water filtration plant, utilities might consider:

Distinguishing Features	Traditional	Design Build (DB)	Design Build Operate (DBO)	Build Own Operate Transfer (BOOT)
Level of integration	segmented	partial	complete	complete
Operational responsibility	utility	utility	service contractor	contractor owner
Ownership and financing	utility	utility	varies	contractor owner

Figure 17-1 Distinguishing features of alternative approaches

- The advantages and disadvantages of integrating design, construction, and operations

- Options for operational responsibility: utility versus contractor

- Plant ownership and financing: utility versus contractor versus mixed model

- The utility's desired position on risk and control: primarily the utility's or primarily the single source contractor's

- How to specify deliverables: a performance based approach versus a prescriptive approach

Several basic approaches and their distinguishing features emerge from these options. They are presented in Figure 17-1, and display progressive departures from utility ownership and operations.

SELECTING AND FURTHER REFINING THE DBO OR BOOT APPROACH

A DBO approach has appeal to utilities that want to take advantage of an integrated approach that keeps design, construction, and operation under a singular responsibility, but still wish to retain ownership and financing control. These circumstances were consistent with Seattle's needs and thus formed the basis for Seattle's decision to use a DBO model for developing its Tolt Filtration Plant (refer to Case Study No. 5.4 in appendix A).

However, developing a project philosophy and an overall project approach involves more than simply choosing to pursue a DBO or BOOT model. Major issues to be addressed in the context of a DBO or BOOT approach include:

- Does the utility want to require results or prescribe a system or method (e.g., a filter plant built to the following specifications) or instead present desired results, in terms of *required outcomes,* with the process for achieving it left open to the vendor to propose?

- Should innovation in proposed technical approaches be encouraged in favor of potential efficiencies and costs savings, or discouraged in favor of more predictable traditional approaches?

- Does the municipality want its role in plant operation to be an arm's-length relationship, or a close partnership with participation in plant operation and in smaller scale decision making?

These choices are interrelated and reflect tradeoffs between perceptions of control and costs, and the consideration of short- and long-term impacts. The project philosophy drives these decisions in one direction or another, i.e., in the direction of more prescriptive solutions, based on the belief that this control will lower risk and costs, or in the direction of more performance oriented solutions, based on the belief that this path will reduce risk and costs.

As an example, consider the following three DBO/BOOT projects undertaken in the United States. (Two of these case studies are described in detail in appendix A.)

1. *City of Chandler, Ariz. (Case Study 5.2):* A 5 mgd wastewater reclamation plant in central Arizona that contracted for private financing, ownership, construction, and operation.

2. *Puerto Rico Aqueduct and Sewerage Authority (PRASA) North Coast Superaqueduct Project:* A project that consisted of a storage reservoir and intake facilities, a raw

water pumping station, a 100 mgd water treatment plant, a major transmission pipeline, and storage facilities.

3. *Tolt Treatment Plant, Seattle, Wash. (Case Study 5.4):* A 120 mgd water filtration plant that the city financed but contracted for permitting, design, construction, and long term operation.

In Chandler, final design and specifications were complete and, although the proposers were given the opportunity to modify the design, the facility was built approximately as originally designed. The city had firmly established its requirements and the provisions to be included, particularly to facilitate the significant expansion that would be required in the relatively near future. Chandler exercised a high level of control over what it received.

In the second instance, PRASA merely outlined the overall project and its elements in the RFP, giving the proposers wide latitude to configure the facilities. For example, the water treatment plant component was described in the RFP as follows:

"Water must be treated using one water treatment plant prior to delivery to achieve acceptable chemical, microbiological, and aesthetic standards as defined in, and required by, all applicable federal and local codes and regulations, including but not limited to the Safe Drinking Water Act. The consortium must submit a plan to achieve this goal using the best available technology to comply with quality, quantity, location, type of process, . . ."

It further required "The design of treatment processes and devices shall depend on evaluation of the nature and quality of the particular water to be treated and the desired quality of the final product."

This approach placed nearly all of the responsibility for facility planning and design on the proposers, with PRASA retaining minimum control over the end product except for final performance.

In the third case, Seattle already had a preliminary facility design and several project-related studies. However, proposers were encouraged to use creativity to design a facility to meet the specified performance criteria, and only basic specifications were included in the RFP to assure a desirable, long-life facility.

In an effort to advise proposers of its desires, Seattle inserted the following statements in its RFQ:

> "In developing this project using the DBO process, the city hopes to benefit from the knowledge and experience of respondents in minimizing cost and maximizing performance."

and

> "Seattle Water Department desires to maintain creativity and cost-competitiveness in the DBO process and therefore provide flexibility to the responders with respect to the selection and configuration of treatment processes. Nevertheless, it is critical that the treatment systems proposed be comprised of unit processes that have been proven to provide treatment to meet the regulatory requirements outlined in Section 2.2, which will be subject to approval of WDOH."

Each case had its own circumstances and any of the spectrum of options might have worked. The market of service providers clearly favors the approach that maximizes their opportunity to use their experience and creativity to provide the most effective and efficient solutions to the advantage of the utility and its customers. This approach makes evaluating and selecting a contractor more complex and challenging. However, a utility willing to release a measure of control over the facilities and to undertake the difficult task of developing carefully crafted performance requirements and a proposer selection process can potentially reap significant rewards.

STEPS IN THE PROCUREMENT PROCESS

The procurement process, from project definition through contract negotiation, can be defined in five major steps similar to the process described in chapter 15. Each step in turn has a number of elements, which are outlined here.

Step One: Defining the Project

- Establishing a project philosophy
- Determining the utility's needs and the project objectives
- Establishing an overall project approach and process
- Determining the potential impact on the process by state legislation

Step Two: Soliciting Vendor Qualifications and Developing the Shortlist

- Broadly defining the performance requirements and desired risk allocation
- Determining the evaluation criteria
- Constructing the request for qualifications (RFQ)
- Evaluating statements of qualifications and developing the shortlist
- Determining the impact and role of state and local regulatory and permitting agencies

Step Three: Developing Performance/Risk Specifications and the RFP

- Defining the performance specifications and minimum plant requirements
- Determining other risk allocation parameters
- Determining proposal evaluation criteria
- Developing a draft service agreement
- Finalization and release of the Request for Proposals (RFP)

Step Four: Proposal Development, Evaluation, and Contractor Selection

- Working with proposers to further clarify utility's needs and expectations

- Conducting preliminary evaluations of proposals and undertaking due diligence review of the proposers' references and previous work

- Soliciting Best and Final Offer (BAFO) (optional, but likely to be used)

- Conducting final evaluations of proposals

Step Five: Contract Negotiation

- Negotiation leading to finalization of performance specifications, minimum requirements, and risk allocations

- Negotiation leading to a final version of the service agreement

- Final agreement to proceed

SPECIAL DBO CONSIDERATIONS

While the five major steps in DBO procurement are similar to the thoughtful contracting process described in chapter 15, there are important elements that require consideration since significant design and construction work is included and the contract is long-term.

In the following sections we discuss these elements:

- Overall project approach and process

- Solicitation and evaluation teams

- RFP considerations

- Due diligence

- Best and final offer

- Evaluation ranking and contractor selection

OVERALL APPROACH AND PROCESS

A utility's needs and objectives can be used to construct a philosophy around the following concepts:

Seattle's DBO Rationale

After Seattle made a preliminary choice to pursue a DBO approach to developing the 120 mgd Tolt Filtration Plant, it was clear to the project team and elected officials that there was not a sufficient base of experience from which to firmly project cost savings or likely market interest in the project. The City was interested in proceeding with the DBO alternative only if sufficient savings could be obtained from capable firms with proposals that could meet Seattle's stringent water quality requirements. At the same time the vendor community was concerned about the potential for an arbitrary and prolonged decision process after they had made a substantial investment in the development of proposals.

This uncertainty led the City to develop a process by which final adoption of the DBO approach would be subject to a threshold determination of its viability. This process was intended to protect both the interests of the City and those of prospective proposers.

Under this threshold determination process, the City agreed to proceed with the DBO approach if one or more proposers could guarantee:

• *Cost savings of at least 15 percent compared to a realistic life cycle cost estimate of a benchmark plant Seattle would have built under a conventionally sequenced design and construction with City operation.*
• *Water quality, quantity, and cost performance through the application of proven technology.*
• *Agreement with the City's published risk allocation requirements.*
• *Conformance with the City's standards of operation and plant maintenance.*

This threshold assessment would be made in the context of proposal evaluation.

Having tentatively chosen the DBO

- Needs should be defined in terms of performance requirements and standards, which allows contractors to propose how to achieve those requirements.

- The definition of these standards in terms of performance, reliability, and quality, should be developed through a team process within the utility involving managers and technical staff, supported by consultants (if necessary) with legal, financial, and privatization procurement expertise.

- Technological innovation should be encouraged within the range of proven technology, and competition be used to achieve both technical innovation and lower cost.

- Risk should be allocated between the utility and the contractor so as to minimize overall project costs (i.e., assign the risk to the party best able to manage it).

The use of an alternative approach to project development, like the DBO or BOOT models, places additional importance on having a well-designed approach to the project. In these cases utility staff and management, elected officials, and the public at large are usually not entirely sure of all of the steps so a good level of communication is necessary.

Important features of an overall project approach might include:

- Establishing clear points of involvement for elected officials in project development, objective setting, decision making, and project oversight.

- Assembling a knowledgeable support services contractor team to assist the utility in each significant phase of the project.

- Creating a multidisciplinary project team consisting of: utility staff, including legal, financial, and legislative departments; members of the support services contractor team; and for some elements, representatives of any wholesale customers.

- Having the project team, utility management, elected officials, and other stakeholders work together to define the required standards of performance, quality, and reliability.

- Using the two step selection process of RFQ and RFP that benefits both the contractor's community and the proposal evaluation team.

- Considering providing an honorarium payment to unsuccessful contractors. (Seattle included an honorarium payment of $100,000 to each vendor that, upon invitation, submitted a final proposal.)

option subject to the threshold determination, the City proceeded to create the process that would lead to the successful development of the Tolt Filtration Plant.

SOLICITATION AND EVALUATION TEAMS

Because of the complexities of a DBO or BOOT project, an important element in project success is the careful designation of both a solicitation team and an evaluation team. The solicitation team conducts all aspects of the procurement process, while the evaluation team is responsible for developing the shortlist from the statements of qualifications and for establishing the ranking order of the shortlisted firms.

Both the evaluation team and the solicitation team may have some common team members, whereas others on the solicitation team may be used as advisers to the evaluation team.

The solicitation team consists of utility management and staff along with experts in the technical, legal, and financial elements of the project, and the proposal and selection process. Depending on a given utility's resources, the needs of some solicitations might be satisfied with utility and municipal staff and advisers who are already retained. However, it is likely that because of the specialized nature of the DBO and BOOT process, that outside independent experts will be required to comprise a solicitation team.

The evaluation team consists of unbiased individuals who are committed to an open and fair evaluation of proposers in accordance with the criteria established in the RFQ (to develop a shortlist) and the RFP (to rank proposers for negotiations and ultimately recommend a single proposer for contract). The

The Seattle Evalua-tion Team consisted of 12 voting members as follows:

• *From Seattle Pub-lic Utilities staff*
• *City Council staff*
• *Purveyor represen-tative*
• *City office of Man-agement and Planning*
• *Outside attorney*
• *Consultant representatives*

evaluation team might consist of as few as five individuals or, as in the case of Seattle, 12.

RFP CONSIDERATIONS

While the contracting process outlined in chapter 15 is generally applicable to the DBO and BOOT processes, some enhancements to the process are necessary to address the additional services required. The procurement documents may include the following additional features:

Facility Specifications

In addition to the performance standards and guarantees, the RFP and resulting service agreement should include generalized facility specifications that establish a minimum level of quality. For example, the specifications for materials of construction, including a required useful life, might be provided. While exact equipment specifications may not be necessary, some minimum quality standards should be addressed (i.e., hydraulic ranges, redundancy, valve materials, and tank materials). Where interface is required with the balance of the utility's system, especially in instrumentation and process control, more exact specifications may be required.

Development Period

Unlike an O&M contract, a service provider is required to under-take numerous development activities such as preparing environ-mental impact statements, permit applications, preliminary designs, and related initial activities. Typically such preliminary activities are conducted as part of a specified development period, which covers the time between when the service agreement is exe-cuted and when the contractor receives the necessary approvals to commence construction. This period may take one or two years, depending on the level of regulatory approvals required. The util-ity's and the contractor's obligations during this period, including payment limits and incentives to accelerate the development period, should be addressed in the solicitation documents and in the eventual contract.

Design and Construction Monitoring

While the design and construction activities are delegated to the contractor, the utility (especially for a DBO project) will need to have opportunities to review designs and construction activities for compliance with the service agreement provisions. While the level of oversight is less than a conventional public works project, periodic reviews of design packages and fairly routine construction monitoring may be appropriate. This is especially necessary to review activities completed relative to payment requests. The design and construction review procedures should be specified in the solicitation documents and service agreement.

Acceptance Testing

The transition point between construction activities and facility operations should be a rigid acceptance test. This test is typically used to determine whether or not the facility has been designed and constructed to meet the quality and performance requirements of the agreement. Final construction payments, including any retainage, should be withheld until the contractor passes this rigid test, and the O&M fee during the operating period should not be paid until the acceptance of the facility is successfully achieved. The acceptance test procedure should be addressed in the solicitation documents and the service agreement.

Project Financing

The utility typically provides financing for DBO projects. For BOOT projects, however, financing is provided by the service provider. The terms and conditions of this financing should be addressed in both the solicitation documents and the eventual service agreement.

There are advantages and disadvantages to issuing a draft service agreement with the RFP. The primary disadvantages are that developing a service agreement takes time and may inhibit some of the proposer's creativity. The major advantages are that a well-developed service agreement makes it easier to compare proposals, (especially the cost elements) provides a foundation for the negotiation process, and usually shortens the time from receipt of proposals to contract execution. While these advantages are clear when contracting operational services are discussed in chapter 15, they are even more helpful here because of the increased complexity of a DBO or BOOT project.

Issuance of the service agreement with the RFP permits the utility to require a base proposal on acceptance of all service terms. Proposers are permitted to offer alternatives and enhancements to their base proposal along with a description of resulting advantages. The following language worked well for Seattle in their RFP:

"If a Proposer believes that significant cost savings or other benefits to SWD are possible by a different allocation of risks, and wants to propose alternative contract language, the Proposer shall submit Form 11. Substantive changes to the technical performance requirements generally will not be considered, although Proposers may use Form 11 to suggest beneficial changes or enhancements that could be of value to SWD. Submittal of Form 11, without submission of completed Proposals A and B, will not be considered. In Form 11, Proposers shall identify the suggested reallocation of risks and the related cost savings, and/or the suggested beneficial change and the related additional cost."

DUE DILIGENCE (CHECKING REFERENCES)

Due diligence or reference verification and checking of an individual's proposed and project experience is extremely important in the selection process of a DBO or BOOT project because of the contract's inherent higher value and the long-term contractual relationships. The foundation for due diligence is the information submitted in the proposer's Statement of Qualifications and Proposal in the form of references, project summaries, or individual résumés. Due diligence can be accomplished by telephone interview and/or site visits. In order to ensure fairness and receipt of similar information on all proposers, the evaluation team should adopt a series of queries for use by those conducting the due diligence investigation. Practical inquiries should be made by a limited number of individuals, ideally members of the evaluation team. Written reports should be distributed to all members of the evaluation team.

An additional and important extension of due diligence is the interview process. This permits clarification of proposals and facilitates a better understanding of the proposed project teams and corporate relationships and responsibilities. Typically each proposer team is scheduled for an extended interview of two to eight hours. The selection team conducts the interviews in accordance with a framework made available to the proposers before the interview.

A format for a four-hour interview might be:

- Introduction of selection team by chairperson

- Thiry minutes to introduce the proposer to the team staff and to present highlights of their proposal

- Three hours of questions from the interviewer panel. This could consist of about two hours for a set of 10 questions that are asked of all proposers, and about one hour of questions specific to the individual proposer

- A 15 minute summary and closing by the proposer

- Brief closing remarks by chairperson

The effort expended during due diligence provides significant insight into the proposed staff and the degree of previous client satisfaction with both individuals and firms. The information obtained is useful in:

- Developing a request for a best and final offer

- Developing questions and discussions for the interviews

- Evaluating proposers based on the criteria described in the RFQ and RFP

BEST AND FINAL OFFER

A Best and Final Offer (BAFO) element of the evaluation process can be effectively utilized after receipt and initial evaluation of the proposals. Seattle found it beneficial to complete some of the due diligence elements before issuing the request for BAFO. Even the most carefully crafted proposals can contain some errors and

Seattle Due Diligence

For the Seattle project the following due diligence was performed for the four proposals:

- *40 phone interviews of individuals identified as key staff*
- *33 phone interviews about referenced projects*
- *Four site visits (One per proposer)*

ambiguities that need to be corrected and/or clarified. In addition, the utility might decide after reviewing all the proposals that certain enhancements to the submitted proposals might be in their best interest. The purpose of requesting a BAFO is to obtain certain clarifications and corrections of the proposals, obtain any specified proposal enhancements, and to give proposers an opportunity to submit a best and final offer.

EVALUATION RANKING AND CONTRACTOR SELECTION

The solicitation and evaluation teams must establish a fair process and method for evaluating all proposals in accordance with the criteria specified in the RFQ and RFP, so as to identify which proposer(s) should participate in negotiations. The discussions in chapter 15 also apply here.

The utility must keep in mind that the DBO typically has three major components, often individual entities—design, construction, and operation; while the BOOT has four: ownership or financing in addition to design, construction, and operation. The multiple entities must function under an effective overall leader and have well-conceived leadership structure to enable the benefits of the DBO or BOOT methods to be realized through appropriate integration of functions and project execution.

SUMMARY

While the use of DBO and BOOT methods of project implementation by public utilities is likely to be limited in the United States, the experience in Seattle demonstrates that significant savings in both capital and long-term operating costs can be achieved. The Seattle experience also provides needed additional information and knowledge about development of long-term operating contracts and the inclusion of capital works in service contracts.

The processes for implementing DBO and BOOT projects are similar to those for any thoughtful contracting of O&M services, although special attention must be paid to the increased complexity of procurement and contracting.

PART 4

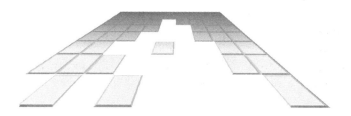

Implementation Issues

Foreword to Part 4:
Implementation Issues

Several issues of critical importance to municipalities and other public agencies surround the operational and management alternatives confronting water utilities today. Four of these issues are addressed in this part of the book.

Political issues have always existed. One might expect that we would be better at handling them than we are. In many localities water utility managers are situated a few steps away from city council yet have been limited in their freedom to interact with council members. Other public utility managers report directly to their governing body, demonstrating varying degrees of skill and various philosophies of board-staff relationships. In any case, the sort of organizational changes discussed in this book cannot possibly be implemented successfully without the full engagement and support of the governing bodies involved. Chapter 18 delves into this topic with practical, realistic advice.

Missteps in legal and financial matters have been the downfall of many public officials. Management and operational alternatives have obvious and not-so-obvious legal and financial strings attached to them and these are discussed in chapter 19. Another challenge is the matter of conflicts of interest. The ethical considerations associated with contracting, managed competitions, DBO, and BOOT are serious and, under the light of public scrutiny, not very subtle. Public officials, utility managers, and private sector contractors need to behave in ways that are above suspicion and reproach, using the sort of guidance found in chapter 20.

Changes in utility operations and management cannot be done without reference to compliance with the network of regulated obligations that surround us. These requirements must be provided for in any outsourcing of utility O&M. The community's obligation to protect public health, the environment, and public and worker safety, can never be relinquished. In delegating them, the utility must assure regulatory agencies of their ongoing compliance to protect the public interest. Chapter 21 identifies the principal concerns that must be attended to as the utility restructures its operations.

These chapters cannot replace the need for site-specific, financial, legal, and regulatory advice and counsel. But the authors trust they will serve as reminders and initial guides as to what should be done locally to protect utility owners, managers, staffs, and the public they serve.

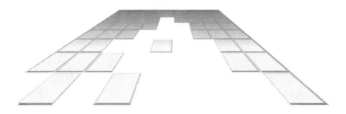

CHAPTER EIGHTEEN

Political Issues

Many utilities today are challenged when representatives of large private sector companies go to their elected or appointed boards and encourage the board to privatize some or all of the utility, thereby promising large financial returns to the community. The tensions created by the appointed or elected officials asking for higher levels of efficiency from utilities, or actually being willing to engage in privatization efforts, have left utilities scrambling to prove that they are efficient and well managed. The purpose of this chapter is to explore the issues surrounding political bodies and ways that utility leaders can cope with the public pressure for efficiency. In this case the "political" being discussed is in the form of stakeholders and the public environment in which large (or medium and small) utilities operate.

ORGANIZATIONAL STRUCTURES: HOW UTILITIES ARE DIFFERENT

Throughout the United States utilities have developed in different ways with different political governance structures. Some are fully public with an elected body such as a city council acting as the board of directors. Such utilities are often the equivalent of city departments but usually have a separate fund structure to accommodate the independence required to account for rate income. Some have either independently elected boards or boards appointed by a public official such as a mayor, and are treated as relatively independent public utility districts (PUDs). With this type of independence also comes some freedom from the give and take of an elected political body. In addition, appointed board members may be oustanding citizens who have expertise in business or some aspect of utility management. Elected political bodies are less likely to include individuals with extensive business experience.

The nature of the political environment within an organization will differ depending on its structure and the history of the organization. However, as government has become more complex with more demands placed on the public sector to meet more needs, all government bodies have become more political and must become more sensitive to the needs of a variety of interests. Even private utilities have had to become more responsive to political stakeholders and to customers. The following information about political issues is important to today's organizations, but readers must pick and choose among the points included here to find those that are relevant for their political environment.

Case Study 1.2: Los Angeles, Calif.

The Los Angeles City Council, concerned about the Department of Water and Power's (DWP) electrical power system in light of deregulation, called for an audit of the system to determine its optimal size. The audit was widened to include the water side of the DWP and the findings determined that the water system could benefit from benchmarking. The audit provided a specific directive for DWP to move forward on benchmarking and to focus on measuring performance.

ROLE OF A BOARD OR ELECTED OFFICIALS

A utility's major stakeholder is its board of directors, whether that be elected or appointed. Organizations can enjoy a period of harmony with a board when most issues—particularly technical ones—are delegated to the technical staff. However, we have all heard stories of the day the leadership on the board changes, or one new "difficult" member is appointed or elected, or when some outside interest lobbies the board and comes between the board and the leadership of the organization. Because such stories are numerous and have often ended the career of a well-respected utility director, all managers should learn the skills of handling board relationships. In two case studies, both Los Angeles Department of Water and Power (1.2) and Austin's Water and Wastewater Department (2.13) initiated their efficiency studies because the board or a political body began to respond to public interest in efficiency.

Board members want to feel fully informed and warned in advance of problems or issues that may become politically charged. A basic rule in dealing with elected officials is "no surprises." They always want to know about good or bad news so they do not get caught by surprise by the press or an important constituent. Although individual board members or elected officials often have special interests or projects that they follow or care about, they all want to be respected and treated in a manner that makes it possible for them to be responsive to public opinion. One important guideline for utility managers who are working to be successful with their elected or appointed board is to get to know the board members and their interests. If the utility manager understands the interests of the board, he or she will be more likely to anticipate issues and problems about which the board members will want information and briefings.

In particular, managers should be sensitive to the fact that any political body today is under pressure to respond to the public demand for efficiency and effectiveness in spite of diminishing resources. The current national criticism of government makes it difficult to be an elected official. To be responsive to customer demands, an elected official has to demonstrate that he or she is making changes in the public arena that constituents desire. Expect boards to ask for increased efficiencies and to ask for proof that an organization is acting like a business. They will ask for attention to the bottom line and for reductions in costs to hold rates down.

Anticipating these questions and preparing for them as a preventive management strategy is the most effective way to manage a board. Initiating an investigation into how to increase efficiencies or ways to deliver a more customer friendly service is always better than waiting to be asked to come up with strategies and then having to defend your organization. The business community's assumption that there is always a more efficient way to deliver a product has become a basic tenet of management and operations. In both Los Angeles and Austin, utility leaders began internal assessments in response to board interest in saving money or concerns about competition from the private sector.

OTHER STAKEHOLDERS

The political environment is becoming increasingly complex and managers must pay attention to a wide range of interests and make efforts to accommodate them. Utility managers are likely to

Case Study 2.13: Austin, Texas

Independent of the work done by the Austin Quality Forum to conduct peer reviews based on the Baldridge criteria, a 1996 study conducted by Price Waterhouse of the city-owned electrical utility caused the Austin City Council to be concerned about how competition in the water industry could impact the Water and Wastewater Department (WWD). While WWD had a de facto monopoly with respect to water supply and distribution, a local river authority had been expressing interest in providing competitive water supply services. The city council instructed WWD to conduct metric benchmarking.

341

have a long-term horizon in working to solve water problems because planning for and constructing water facilities takes a long time. Elected officials and special interest groups are likely to have short-term or more narrowly focused interests. The utility leadership team should make an effort to evaluate its political environment, to understand the motivations of each active interest group and to have a strategy for managing the interests that will be active within it. Some of the more active stakeholders today are:

Executive Officials

In some utilities the water department manager reports to a city manager, a director of public works, or a utilities manager. These officials, who have other obligations, are likely to focus their energies where there are issues of public or media concern. Thus a utility that has a good service record and low rates is not likely to get much attention, while a utility that has higher costs or experiences quality or reliability problems will receive greater scrutiny on the part of the managing executive and the governing body. This chapter urges cooperative team building at all levels. It may be difficult to get the attention of elected officials, the executives, or elements of the manager's utility staff because of competing agendas. However, the water manager's challenge is to generate the kind of attention and consideration that will, in the long run, create an efficient and well-run utility.

Water managers should remember that they literally hold the keys to the water service system and that executives who are usually generalists must depend upon them for technical as well as organizational advice. This offers an opportunity for water managers to structure good strategic plans and to recommend appropriate priorities to improve performance. Remember that management means blending the best of the team's efforts, whether those efforts are a combination of departments of which the water department is one, or a combination of divisions within a water department. Blending inevitably means compromise and it is the responsibility of the water department manager to seek those compromises that will best provide for good long-term service. In order to have the type of relationship where a higher executive official relies on the water manager's advice requires constant attention to relationship building with the higher officials.

Utility Staff

It may not seem obvious at the outset that a manager's staff is an "interest group" to be analyzed and managed as carefully as an outside interest. This is most likely because the manager thinks about the staff's interests almost automatically. However, at times the staff's interests may not be in alignment with the customers' or some other significant outside interest and then the manager must manage the staff much as one would an outside interest. In issues relating to change, inside staffs are likely to seek maintenance of the status quo because it is comfortable and understandable. Most utility staffs are also proud of their work and see efforts at change management as some type of criticism of their work. Introducing a change agenda within the staff—especially one that is being initiated because of a political or outside interest—requires special care and attention to the manager's staff. Chapter 14, People and Change, addresses many of the issues the utility leader will have to address in introducing change to a staff.

Labor

Utilities range from those in right to work states with no unionized activity to others with large numbers of unions to deal with. Unions are becoming more strident in their opposition to outsourcing and privatization. However, if the political board is asking the utility leadership to demonstrate creative ways of contracting out or to initiate managed competition, the leadership must work to gain union support. The successful Charlotte-Mecklenburg (Case Study 4.4) effort at managed competition entailed developing a careful working relationship with unions and with affected employees. Utilities need to have some form of labor management committee that creates a forum for ongoing communication and mutual understanding.

A labor management committee can decide how best to involve the union in the strategic planning teams and work process teams proposed elsewhere in this book. Such involvement in the utility's planning may eliminate the need to submit job changes to the bargaining process. If a utility is going to become efficient and bottom line oriented, it will likely have to look at personnel classification and procedures. Changes in these matters will require union understanding and support. Union leadership will sometimes cooperate with management to support change if it believes that the change will be good for employees and will save existing jobs or make existing working conditions better. Utility management must work closely with union leaders to explain what pressures are on the utility and why change is necessary. Such explanations should address all the issues related to employees and how management intends to take care of them. If management understands and pays attention to union issues and concerns, union officials are more likely to understand the situation and try to work toward a mutually acceptable outcome. The imperative for change discussed in chapters 1 and 2 will affect the utility's survival and thus the union local's survival. Chapter 14 includes more information about the impact of labor unions on the utility.

Some utilities have experienced challenging labor situations, including a strong role of the unions or employee groups in the election of policy officials. These employees' representatives have a real incentive to ensure the election of officials who are sympathetic to their desires relating to compensation and work rules. Then the elected official feels an obligation that may affect that official's attitude toward management and toward issues of improved performance and cost reduction. Some utilities operating in the context of a strong union environment (e.g., the Department of Water and Power of the City of Los Angeles, Case Study 1.2,) have had strong policy level support for thoughtful downsizing. In an employment environment where labor representatives and activists want to leave things as they are, except for benefits, it is difficult to be an agent of change. Nevertheless, the utility manager must remember that in the end it is the customer that must be served by the utility, not the employees. The majority of employees want to do an excellent job, and to meet the needs of the community they serve. The manager's job is twofold. First, to convince the committed union leadership that change is in their members' long-term interests; and second, to make sure that the majority of the employees are allowed an opportunity to participate in the communication process and feel free to express their views.

Environmentalists

Utility operations have become more complex as communities have become more conscious of environmental issues and the need to balance resource conservation with economic development. This consciousness is represented by: state and national regulatory agencies that are issuing more stringent requirements to protect environmental resources, e.g., watersheds (fish and endangered species); water becoming a precious resource that is shared among more interests (fish, agriculturists, and business and recreational interests); the greater importance placed on water quality by the health conscious; and the 1996 reauthorization of the Safe Drinking Water Act, which has public involvement as its linchpin. All of these issues require the utility manager to have an active program to work with environmental interests in the community and to provide them a part in advising on policies adopted by the organization. If engaged in a dialogue with utility leaders, environmental interests are more likely to see the utility as a reasonable ally than an enemy, and they are more likely to support requests for resources that might have an impact on the rate structure. It is important to realize that environmentalists are a special interest group and get support from their membership by strongly advocating for specific issues. Utility managers are rarely in a position to completely satisfy environmental interests because management is constantly balancing interests among many constituents. However, by interacting with environmental groups, management may be able to understand their positions and to move in a direction that environmental groups will see as positive.

Business Interests

Business interests will have a range of requests for utilities from added services and higher quality water to lower rates. Some of the large private water companies may have an interest in being involved in your territory. Just as the business interests themselves are varied in any jurisdiction, the response to those interests must also be varied. Often a good approach is to get involved in business organizations (chambers of commerce, downtown associations) and to make the business community aware of services available. Some utilities have developed customer service centers which handle "key accounts" or special teams to deal with commercial sector requests. One good approach is to share your strategic plan that details the organization's goals and objectives and shows that the utility is thinking in the same strategic way that any business might. It is always smart to communicate with your business customers and to find out what they are planning and what their interests are so that you can develop a responsive strategy and new services that will be recognized as valuable and responsive.

The Press

In today's world, where the press gets many of its headlines from controversy, it is important to think of the press as an interest group. Utilities must keep the press informed of its activities especially where there are issues of public concern, or if your improvement plans call for an ambitious capital program that will require rate increases. Media training can be helpful to utility managers in meeting the challenges of the press.

Water is a buried infrastructure, a silent service that gets little attention unless there is a problem. Having a proactive press strategy that involves publicizing new customer services, conservation, resource protection efforts, and other issues can keep the organization identified with positive activities in the public's mind. Citizen surveys that ask how the public feels about aspects of the utility identify issues that may become public assets or liabilities for the utility. When strategies are developed to respond to those issues, the press can become an ally in communicating your plans and changes to the public.

The General Public

Citizens are interested in quality services and a quality product for a reasonable price. The public seems to have a high level of understanding of environmental issues and why it may be important to spend more money to get quality water or to preserve the environment. The hardest problem with dealing with the public is to find them. When everything is going well and people turn on the tap and water flows, they do not think much about what is involved to get it there. When rates go up, their service has a leak, the utility has a construction job underway on their street, or some sort of controversy erupts, they get involved and express their opinions. It is smart business for a utility director to begin preventive information gathering campaigns with the public to find out the level of satisfaction with services by doing large-scale citizen surveys or focus groups on targeted issues and planned programs. Citizen advisory committees can be useful if they are given a role that that will make the members feel valued and that will generate substantive advice for you.

Regulators

Regulators are a type of special interest group because it is important for utility management to keep them informed about programs and improvements they might be interested in. It is also important for management to stay up to date on current information and technical issues that regulators will be watching. At the federal level, both AWWA and AMWA have extensive programs to keep management abreast of all current regulatory issues and are able to put utility management in touch with other utilities facing similar problems. Again, the best approach for dealing with regulators is to keep them informed of your programs and to stay in touch on a regular basis to know what they view as the current hot issues.

Consultants

Consultants are a type of interest group because they tend to keep in touch with you as a "business development" activity. Different utilities have different strategies on how close their relationship with consultants is. Some utilities like to maintain a keen competitive environment in an effort to keep costs down and therefore will work with a number of consultants and keep them competing for available projects. Other utilities like to develop and maintain an environment of trust with one or two consultants who will then know their business and their staff. If one selects the latter relationship, it is important to have strategies for ensuring that costs are competitive and that one does not get too dependent on a particular firm. The number of consultants a utility can work with is also dependent on the size of the community and the number of firms that are active in the area. Smaller, less dense areas will have fewer firms and less choice.

End of the Era of Technocrats

One of the results of the complexity of the new era in which politics and stakeholders are so important is that the technical expert can no longer hold as much power and influence. Public bodies will no longer turn to an engineer, a computer expert, or a department head and say "fix it" or even to ask for an answer. As problems become complicated, responses also become complicated. It is important in any organization to have diverse people and points of view when engaged in solving a problem. Teams should solve problems and be constituted with diverse points of view and requisite types of expertise. Decision making that reflects all stakeholder points of view will produce recommended solutions that meet the objectives of the public at large.

Role of a Board of Elected Officials

As efforts at cooperative management through work groups and teams progresses there will be increasing communication demands on all stakeholders, particularly those at the upper end of the management structure, including the governing body. It is therefore important to make the best use of limited time available for meetings, workshops, and other discussion activities. Because most discussions with elected public officials must take place in open, public meetings, careful thought should be given to the structure of an agenda, priorities for discussion with elected officials, effective presentation of important information, and preventing unnecessary interruptions and conflicts. An essential ingredient for achieving an effective discussion in public with elected officials is to have brief individual discussions with stakeholders before the meeting so that each party will participate meaningfully, and not just attend. If local open meetings laws permit retreats, they can be effective mechanisms for creating an environment in which complicated issues can be discussed in more informal settings. Even if the retreat includes only local officials and interest groups without speakers or experts from outside, the setting offers an opportunity for people to be on a more common ground and gives them the feeling of investing in a group endeavor.

ANALYZING THE POLITICAL ENVIRONMENT

Doing an explicit analysis of your political environment will help to develop solutions to problems that your organization will face. Such an analysis contains identification of the issues that will be coming forward, and interests served and interests hurt by the potential decisions to be made. The analysis then forms the basis of developing a strategy to reach those interests. Basic citizen concerns about good service and low rates should be included in your assumptions of issues that will arise.

Once issues have been identified, a plan should be developed for how to reach those interests. The plan should include both an external communications strategy and a method for involving key interests in the decision making processes. It is also important to know if any of the vested interests are supporters of members of the board of directors. It can be assumed that those interests will talk regularly with the board member that they support. Involving said interests in decision making can be accomplished through citizen advisory committees or by an outreach strategy that includes either workshops on the issues (at which interest advocates can talk about their concerns) or through focus groups (at which you learn additional information about how to find solutions that will meet their needs). Because of the pace of today's busy lifestyle, general, ongoing citizen advisory committees are not as helpful or useful as ad hoc groups organized around specific issues of interest to the local community such as environmental impacts or watershed and creek basins.

BUILDING RELATIONSHIPS WITH ELECTED OFFICIALS

As privatization has become an issue in the water industry, more and more stories are told of a large private company going to elected officials and selling them on the concept of privatizing or outsourcing large pieces of a water system.

Case Study 1.3: Bergen County Utilities Authority (BCUA)

The BCUA formed a privatization committee to assess potential improvements to more cost effectively operate its wastewater treatment system and to continue to improve the quality of service provided to its customers. The decision to conduct such an analysis was a response to inquiries from various vendors about owning and operating the plant and an earlier County study that recommended looking into privatization. A detailed analysis was conducted to assess alternative management and ownership structures that could provide operational and economic benefits to the BCUA. Based on the results of the preliminary analysis, BCUA is in the process of proceeding with an intensive concurrent approach consisting of optimization of its operations and a comprehensive contractor procurement process.

The selling is often done with promises of greater income or greater efficiency than can be achieved by public utilities. Given the national rhetoric about finding efficiency in government, some elected bodies respond to these overtures positively and direct the utility manager to engage in an operations contract or managed competition process. Problems are exacerbated in situations where there is a communication gap between the elected body and utility leadership. Not surprisingly, utility leadership will often take such requests personally and feel that they are being criticized for their work and productivity. However, the best strategy is a preemptive one to demonstrate to elected officials that your agency is (or can be) more efficient by undertaking the types of assessments described in earlier chapters of this book. The following are strategies to avoid or manage difficult situations if they arise:

Communicate Frequently with Elected Bodies

The best strategy is to avoid the problem by having excellent communication with the elected body in the first place. Most elected officials raise issues they have heard from the electorate. Respecting this situation and helping them to meet what they perceive are expectations of government is a useful strategy. Your views of how to run a water system may be different, but usually there is a reasonable point of compromise where both interests can be met. Community liaison programs involving the utility workforce can be helpful in facilitating communication.

Education of elected officials in water issues and utility management principles is necessary. AWWA offers special sessions for public officials at its annual conference. Take your board or commissioners to the annual AWWA meeting or to other events where they can meet their counterparts and refine their understanding of their roles. Another strategy is to bring experts to your community by setting up a meeting or retreat where both outside experts and elected officials are invited. In this environment the elected officials can ask questions and be educated by experts concerning what is going on around the country on a particular issue. If one selects this strategy it is important to bring in experts who have good skills at teaching people outside the industry in simple and understandable terms.

Develop an Efficiency Campaign

By developing an efficiency campaign and committing to look at all aspects of operations to assess performance and find ways to save money and improve services, directors can avoid having the elected body ask them to do it. Most utilities find that there are better ways to carry out operations and there are effective modern techniques for facilitating change. Much of this book is devoted to explaining how to do this, and even how to involve your board members in the process. Chapters 3 and 4 provide a good overview for managers. Chapters 11 through 14 describe in detail how to assess your performance and undertake a comprehensive improvement program. All of part 2 of the book, chapters 6 through 10, address principal functional areas of utility operations and how they can be made more competitively efficient. When elected officials see you undertaking a program like this, when they recognize your goals of better service and lower costs, and when you make them a part of the improvement effort, the relationship between you will be strong and mutually supportive. They will also have more confidence in your management. (See Case Study 1.3 on Bergen County in which the utility initiates a study in response to interests expressed by elected officials).

Take a Leadership Role in Carrying Out Proposals of Elected Officials

Follow the advice of the old adage, "If you can't beat them, join them." If the elected body insists on some type of outsourcing effort, then take a leadership role in making it happen. One of the most common mistakes elected officials make is to think that any type of operations contracting or managed competition can be done quickly. Utility leadership should start by talking to other utilities that have gone through such processes (see appendix A for case studies). They should request consultant assistance because a consultant is often in a better position to explain to an elected body the length of time it will take to do the job right. The consultant can also identify successful strategies and procedures.

Having the consultant assist in developing a reasonable plan and schedule protects the utility leadership from seeming defensive.

SUMMARY

It is important to craft the best possible relationship with an elected or appointed board that feels it is responsive to the public at large. Such officials will be receiving diverse points of view and managing the range of issues that the public has a right to be involved in. Today's elected officials are not likely to rely solely on engineers and technocrats to find the right solution. Today's right solution is one that balances a wide range of interests. Utility leadership must be involved in crafting strategies that meet the interests of other groups such as labor, business, environment, and the press. Understanding their points of view protects management against unexpected attacks from special interests. It is also useful for utility leadership to take the initiative in defining efficiency as a major goal of the utility and to communicate with all stakeholders about improvement plans and results.

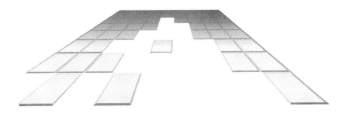

CHAPTER NINETEEN

Legal and Financial Issues

The ability of a utility to enter into contract agreements with the private sector or another service provider for the provision of services is governed by both financial and legal restrictions and requirements.

While experienced legal and financial expertise should be a prerequisite for any proposed selection process and subsequent service agreement, a number of basic guidelines can be offered as helpful background. Successful structuring of contract services, particularly the procurement of capital related items as a part of the service agreement, must be carefully considered to satisfy the procurement laws of each state, as well as local laws and ordinances.

The information in this chapter is not intended as site-specific guidance that replaces the need for qualified and experienced legal and financial expertise. It will, however, provide an overall framework against which the site-specific procurement and service agreement can be developed.

Federal, state, and local procurement laws often preclude certain forms of public-private contractual arrangements. Union agreements and changes that could affect public employees provide another set of factors that must be examined and satisfied for successful implementation of a procurement process and service agreement. The early identification of these constraints is necessary. The service agreement, discussed in chapters 15, 16, and 17, provides the basic necessities for legal and financial issues covered in this chapter.

ISSUES RELATED TO THE PROCUREMENT PROCESS

There are a number of federal regulations that must be considered for all operation and maintenance contracts:

1. Safe Drinking Water Act (SDWA)—establishes *minimum* regulatory requirements for treated water. State safe drinking water regulations must comply with these minimum standards and, in some cases, are more restrictive and expansive.

2. Ownership—since permits are issued to the owner of the works, the facility ownership specified in the service agreement will impact permitting requirements and related liabilities. Where the contractor is not named as the permittee, the service agreement should provide guarantees for treated water quality and indemnification of the permit holder by the contractor.

3. Wages—when construction services are part of the work scope, the applicability of the Davis-Bacon Act regarding prevailing wages must be considered.

4. Tax exempt financing—financing of facilities with tax exempt debt involves a series of federal laws, regulations, and constraints. Generally, unless tax exempt Bond Cap Allocation can be obtained, tax exempt debt cannot be used to fund a facility owned by the private sector. Bond Cap Allocation is a specific amount of tax exempt financing set aside for each state that can be made available for projects in which the assets are owned by the private sector. A designated amount of such financing, based on a minimum amount adjusted for population, is available for each state. Historically these allocations have been used primarily for highly visible projects such as public housing but may be available for water utility projects. Additionally, where tax exempt debt is outstanding on a facility, the terms and conditions for contracting with the

private sector for contract operations are subject to compliance with federal requirements (specifically Rev. Proc. 97-13).

5. State laws—in some states, Massachusetts for example, special legislation is required for each contract that seeks to include provision of capital items as part of the contract scope of work. Massachusetts also provides specific guidance on the procurement process under state law titled MGL 30-b. The state law specifies that a single document must be used for qualifications and proposal submittal and requires a separate and independent evaluation of price. Under MGL 30-b, the price for each proposer's service is then evaluated separately, having been submitted in a separate, sealed envelope. In other states such as New Jersey, a well-defined process for contracting with service providers is established by state law. In both states, local ordinances and union agreements may be more restrictive than the state or federal requirements.

6. Lowest bidder requirements—further complicating the decision path and the legal and financial requirements are state and local laws that specify that the contract must be awarded to the lowest priced bidder qualified to submit a proposal. In other jurisdictions a contract may be awarded based upon an proposal process in which price is just one of the criteria used in evaluating proposals from qualified bidders.

7. Record keeping—documentation of the RFQ and RFP process, as outlined in chapter 15, will ensure the development of explicit and objective work scopes and evaluation criteria. This documentation is essential to effectively deal with potential legal challenges to the procurement process.

8. Legal challenges—legal challenges often arise when procurements have not been carefully managed by the rules and when the flow of information to potential proposers has not demonstrated equal and consistent treatment of all

proposers Additionally, the lack of explicit evaluation criteria may leave the procurement open to legal challenge.

The development of a well-specified work scope, evaluation factors and criteria, and a specified evaluation and selection process will reduce the likelihood of legal challenges and provide a documentable defense should such a challenge occur. This same approach of full documentation and disclosure should be used when dealing with questions and information dissemination to proposers, i.e., accept only written questions or requests and respond to them only in writing.

9. Contractor access—during the procurement you should also control the access of proposers to city staff and elected city officials. Consistent with this managed equality approach, proposers should be explicitly instructed regarding the contact person with whom they must interface during the procurement.

While no procurement process can prevent third party legal challenges, adherence to these guidelines will prepare the proposal team and legal staff with documented procedures that can be used to address a legal challenge should one occur.

EXECUTIVE ORDER 12803 AND REVENUE PROCEDURE 97-13

Recent administrative actions by branches of the federal government have expanded the structures and approaches available for partnering with the private sector. This chapter focuses on the capabilities created by two administrative actions:

1. Executive Order 12803 (adopted 1992) provides for the sale and long-term lease of public sector, publicly financed facilities to the private sector with sale proceeds prioritized to the local water utility and not to the federal government.

2. IRS Rev. Proc. 97-13 (adopted 1997) rewrites the IRS interpretation of federal tax laws for the use of publicly financed facilities by the private sector. It also makes it

possible for the private sector to enter into 10-, 15-, and 20-year contract service agreements with municipalities for facilities financed with tax exempt debt.

Since 1986, public-private partnerships utilizing municipal assets having outstanding tax exempt debt have been limited to relatively short-term (less than five-year) contracts for the operation and maintenance of water and wastewater treatment facilities and of distribution and collection systems.

Investor owned water utilities are subject to regulation and rate setting at the state level through Public Utility Commissions. These utilities must use taxable debt and shareholder equity to finance their facilities. They operate outside of the control of the local municipality and are not considered a form of public-private partnership for the discussions in this chapter.

Where obtainable, tax exempt debt may be utilized by the private sector for the ownership of municipal water and wastewater projects, provided the project qualifies and the tax exempt debt is available within the limitations established for each state (i.e., the bond cap allocation amount). Such structures have been used and continue to constitute a viable alternative, where the bond cap allocation is available, for private sector ownership that utilizes tax exempt financing.

The option of selling municipally owned assets to the private sector and entering into long-term service agreements has always existed for the municipality owning the facilities. However, unless tax exempt bond cap allocation was obtainable, such long-term asset leases and transfers involved the elimination of all tax exempt debt associated with the facility. Elimination of the debt involves repayment of principal and interest (bond defeasance) and may constitute a considerable expenditure at the time of the transaction. Leases of assets to the private sector have triggered similar restrictions.

The most commonly used approach for utilizing tax exempt, municipal financing with a private sector partner has involved the design, construction, and operation of a new facility. Referred to as design, build, and operate (DBO), these partnerships have combined the advantages of the lower financing costs obtainable with municipal tax exempt financing and the benefit of the private sector partner's provision of single source responsibility and risk assumption (budget, schedule, and performance) obtainable with a DBO contract. Each of these projects involves the qualification of the project under the bond cap allocation requirements of the specific state in which the project is located.

FEDERAL EXECUTIVE ORDER 12803

Executive Order 12803 provides a new approach for the sale of publicly financed facilities to the private sector. Assuming the availability of bond cap allocation, tax exempt debt may continue to be the financing vehicle or taxable debt may need to be used to replace the tax exempt debt. In either case the ownership of the facility has changed hands, from the public to the service provider, and the public sector may now enter into a long-term service agreement for the provision of services by a private sector owner. In addition, the sale of the facility may generate up-front proceeds to the municipality. Restrictions apply on the amount of these proceeds and their use (rate implications and fair market value for the facilities are two examples); these are discussed later. Long-term leases where the lessee assumes substantially all the risks and benefits of ownership also fall under Executive Order 12803 and the restrictions regarding the use of tax exempt debt with long-term agreements.

The Executive Order has seen limited use since its adoption in late 1992. Numerous reasons exist to explain this inactivity. These include:

Perceived Loss of Control

Sale of municipal facilities involves significant complexities regarding future provision of service and the perceived conflict with public policy. While the Franklin, Ohio, asset sale and the more recent Wilmington, Del., and Cranston, R.I., lease transactions illustrate the ability to contractually resolve this issue, the complexity remains significant.

Need for Regulatory Approvals

Once the transaction is successfully developed at the local level and the parties have reached agreement, the following additional sign-offs are necessary:

- Municipal bond council—retention of tax exempt debt or bond defeasance

- Local and state regulatory agencies—all permits must be transferred to the new owner

- Federal Environment Protection Agency (EPA) and Office of Management and Budget (OMB)—each agency must review and approve the transaction

Unfortunately, the Executive Order could not, and the EPA and OMB have not, codified the Executive Order. As a result, each proposed partnership arrangement, beyond those qualifying under a restricted safe harbor provision, must be reviewed and approved by EPA and OMB. Significant time, expertise, and expense are therefore necessary without assurance of gaining approvals.

Cost of the Financing

While municipalities may receive significant up-front cash payments, rate stability, regulatory compliance guarantees, and facility expansion and upgrade commitments, the existing transactions have each involved defeasance of outstanding tax exempt debt and the substitution of taxable, private sector debt financing. This interest rate differential remains a major hurdle for the successful sale of assets to the private sector.

In summary, the Executive Order provides a pathway for the sale of municipally owned and financed assets to the private sector. Repayment of federal government grants is now limited to only the unamortized portion of the grant utilizing a 20-year, accelerated depreciation schedule; sale proceeds beyond those needed for grant repayments and bond defeasance are made available for local use with few restrictions beyond the funding of additional infrastructure and its operation. Often the enterprise fund character of the local water utility will impose additional restrictions on the use of sale proceeds. These restrictions need to be thoroughly evaluated and understood but are beyond the scope of this chapter.

REV. PROC. 97-13

Responding to the request for additional flexibility for contract term the IRS presented this procedure in early 1997. With 10-, 15-, and 20-year service agreements now possible for facilities with tax exempt debt outstanding, the public sector has a new alternative to the pathway outlined by Executive Order 12803 for a long-term partnership with the private sector.

As a result of these restrictions, public-private partnerships have focused on the relatively short (less than five-year) contracts allowable under the IRS interpretation of the 1986 tax law. Rev. Proc. 97-13 attempts to expand and liberalize both contract length and contract terms, allowing for the continued use of public sector, tax exempt debt alongside private sector debt.

Rev. Proc. 97-13 allows the benefits of long-term service agreements without the need to sell the facility to the private sector and/or eliminate any outstanding tax exempt debt associated with the facility. Limitations are once again prescribed regarding the form of the service agreement, including the types of compensation methodologies that are and are not acceptable.

In considering the suitability of these long-term contract alternatives, part of the site-specific assessment should include a careful consideration of the factors that differentiate the long-term contract arrangements from the shorter term three- to five-year contract O&M agreement. These factors are summarized in Table 19-1.

The most important advantage of Rev. Proc. 97-13 centers on the contract term presently available. The private sector partner now has a significantly longer term for the amortization of capital and systems provided by the private sector. Payback on these investments is now possible with the longer term agreement. Additionally, up-front payments in the form of a concession type fee can now be amortized by the private sector over a considerably longer contract term.

Table 19-1   Features of long-term contracts

Item	Benefit	Issue
10–20 year term	Longer cost saving and capital amortization time periods	Compensation requirements
Incentive savings	Longer term for participation	Conflicts with compensation methodology
Tax exempt debt	Remains in place; no defeasance or bond cap requirements	Must meet 97-13 requirements
Retention of Municipal Ownership	Retention of ownership control, avoidance of public perception issues associated with a sale	Provision of future expansion and facilities for new regulatory requirements
Up-front Concession Fees	Ability to obtain significant up-front dollars with longer contract term	Properly structured, the concession fee is not part of the fixed/variable compensation requirement

However, Rev. Proc. 97-13 also provides new requirements regarding the compensation methodologies that may be utilized. Specifically, the Rev. Proc. outlines the need for compensation to be classified as "fixed" and "variable" with limitations on the percentage of the compensation that may be variable.

Compensation based upon per capita charges or gallons-produced methodologies are not allowed by the Rev. Proc. 97-13. While annual adjustments for inflation are not classified as variable compensation, significant additional items appear to be so designated. For example, incentives and compensation may be based on revenues or costs, but not both. Increases in volume of water produced or wastewater processed that require additional compensation appear to be categorized as a variable cost once again. Similar issues exist regarding the financial treatment of new or expanded facility costs.

The restrictions on contract term and facility ownership that exist wherever tax exempt debt is outstanding on the facility represent basic legal and financial requirements. Hence, if properly followed, Rev. Proc. 97-13 offers the potential of long-term contracts without the need to retire (defease) the outstanding tax exempt debt. The requirements of Rev. Proc. 97-13 regarding the financial structure of the service agreement, limitations on incentive payments, and the need to fit compensation adjustments into a fixed/variable formula partially offset these advantages. Facilities without tax exempt debt or tax exempt debt issued to the private sector under revenue bond cap allocation are obviously not governed by these restrictions and may enter into long term service agreements and ownership structures outside of these restrictions.

The IRS recently indicated that it would not allow an "Expectations Test" as the basis for satisfying the fixed/variable test. Should this interpretation continue, communities utilizing Rev. Proc. 97-13 will need to comply throughout the term of the agreement and/or develop termination methodologies should events develop that contradict the requirements of Rev. Proc. 97-13.

At this time, only the Evansville, Ind., contract has advanced forward for review and approval under Rev. Proc. 97-13. However, San Diego is proceeding with an RFP with an envisioned 7–10-year term and other communities are developing longer term agreements under Rev. Proc. 97-13 that will provide examples of methodologies that comply with the requirements.

RELATED CASE STUDIES

The following case studies included in appendix A may be helpful in providing experience of others that have or are in the process of obtaining contract services.

Seattle, Wash.

Procurement of design, build, and operate services for drinking water under a 25-year service agreement. Rev. Proc. 97-13 was incorporated into the service agreement.

Franklin, Ohio

Procurement of design, build, own, and operate services for a drinking water plant under a 20-year service agreement. Single source guarantee for project with private sector financing.

Miami Conservancy District

Asset sale of wastewater facility to private sector under a 20-year own and operate service agreement. Surplus funds paid to communities and public sector bond defeasance.

New Haven, Conn.

Contract operations with capital provision by the contract operator.

Cranston, R.I.

Long-term service provision under a 20-year lease agreement with up-front concession fees using Executive Order 12803.

New Bedford, Mass.

Long-term contract operations (15 years) with up-front concession fee payments by the contract operator. Rev. Proc. 97-13 was incorporated into the procurement.

SUMMARY

The two federal rules discussed in this chapter provide new opportunities for public-private partnership arrangements. This is a clear advantage for the industry. However, the legal and financial boundaries of these provisions require expert interpretation and careful management. As industry experience with these opportunities accumulates, those interested will have more precedents to examine and more examples to learn from.

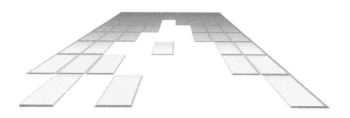

CHAPTER TWENTY

Conflict Of Interest

Our country's history is full of scandals featuring charges of conflict of interest at the interface of public-private relationships. The interaction of private interests with the responsibilities of a person in a position of public trust, such as a political leader or utility manager, provides potential for conflict of interest allegations. But facts and perception are often indistinguishable, and whether a conflict of interest is real or perceived, at a minimum it can undermine trust and can be an issue or a barrier to the success of efforts to improve the effectiveness and efficiency of water utilities. Consequently, rules on behavior to avoid conflict of interest are common for public servants.

With the private sector actively seeking expanded roles in public water utility ownership and/or operations, the potential for conflict of interest is increased. Many of the currently proposed forms of privatization or public-private partnership present new and less well-trodden paths with less historical guidance to how such relationships may affect behavior or how such relationships may be perceived. It is appropriate, therefore, to review conflict of interest perceptions in light of the changing face of public-private relationships. Behavior must not only be right, it must be *perceived* to be right.

The central issue is that the public's business must be carried out with the public's interests foremost, and no one involved should use these relationships improperly for personal benefit.

The following paragraph is from an editorial in *Science* Magazine by C.K. Gunsalus about ethical conduct in research to apply to our current concern about conflict of interest:

> *Somehow it has come to seem unfashionable, almost priggish, to talk about concepts of honor, duty, and obligation. At the same time, increasing funding pressures have created perceived incentives to behave unethically or unprofessionally. What are the boundaries of ethical conduct, and whose job is it to set them and make them stick? Self-policing is a difficult task that few professions seem to have mastered. But the fact that it is difficult doesn't mean that our community shouldn't try. Each of us has the obligation to confront ethical issues and their implications for our personal conduct. Each of us also has the duty to address the ethical aspects of our work with colleagues and students. Institutions have a responsibility to articulate standards for ethical conduct and to see that they are put into practice. These standards should be higher than the merely legal; they should define professional standards of behavior. The goal should be to provide guidance to the well intentioned—those who may want to do the right thing but who genuinely do not know what is right in a complex situation or how to determine it.*[1]

It is with this in mind that we attempt to address the conflict of interest issue in this chapter, the contents of which are organized around the following issues:

- Conflict of interest definition

- The various stakeholders' interests

- Conflict of interest considerations

- Approaches to avoiding potential conflicts of interests issues

- Practices being used to minimize conflicts of interests issues

---

[1] C.K. Gunsalus, "Ethics: Sending Out the Message." *Science* Magazine, April 18, 1997, 335.

CONFLICT OF INTEREST:
WHAT IS IT?

A discussion of conflict of interest or ethical behavior inevitably seems to get bogged down in issues of definition.

"Ethics" and "conflict-of-interest" are often used interchangeably. This is not surprising because the common usage of both ideas is broad and seems to overlap. Early Greek philosophers discussed ethics as the theory of the nature of "The Good" and how it can be achieved. While that definition is still sometimes used, current usage more generally means "the study of the standards of conduct and moral judgment," as well as "the system or code of morals of a particular person, profession, etc." On the other hand, when discussing conflicts of interest we are usually describing "the conflict between one's obligation to the public good and one's self-interest." Note that having a conflict of interest does not inherently imply that one does something wrong. Rather, it only implies that there exists *the opportunity* to serve more than one interest when those interests might not be identical. The ethical question arises when judgment must determine whose interest is best served.

Every conflict of interest issue is an ethical issue—which is why conflict of interest and ethics seem to overlap—but ethics are far broader and can involve much more than just conflict of interest. Suffice to note that neither ethics nor conflict of interest in themselves need rethinking. It is only that opportunities for conflict of interest grow as public-private relationships become more extensive and more complex.

CONFLICT OF INTEREST:
A GROWING ISSUE

We recognize that there is a growing imperative to make changes in the way public water utilities provide services. The imperative is an outcome of increasing public demand for its utilities to provide an abundant supply of higher quality water, higher levels of service, and at less cost to the rate payer. In essence, the demand is for better effectiveness and efficiency in providing water service.

In combination, significantly improved effectiveness and efficiency often require utilities to seek new opportunities to improve and/or change traditional technical and management practices.

When one of these new opportunities is a form of privatization, it is important that both the public agency and private service provider understand the entrepreneurial spirit that drives privatization. Under such circumstances, the zeal of both public and private sectors to be entrepreneurial enhances the risk of creating new conflict of interest situations.

The water industry has long recognized the value of the public's confidence in the quality, reliability, and cost of water supplied under public ownership and operation. Often this confidence is reflected in the confidence the customers have in the public utility itself and in its leaders. Unfortunately, there exists a view that, "In most recent years, public cynicism about government has been widespread. Negative attitudes include distrust in the ability of government to lead or contribute to solving complex societal issues, as well as concern about the integrity of public officials."[2] It is against this backdrop of distrust that the public will view actions that are being taken by public utilities to become more efficient and effective. Consequently the playing field for a privatization program can become a virtual minefield of conflict of interest possibilities, and surprises can be devastating to the public trust.

We recognize that "there is not a set of ethical standards for privatization activities and contracts, although several state agencies appear to be moving in that direction."[3] It is our intent that discussion of conflict issues will further sensitize elected officials, public agencies, and private service providers. Further, we hope it will aid them in maintaining a high level of public confidence in the quality of drinking water and those responsible to provide it.

---

[2] Evan M. Berman, "Restoring the Bridges of Trust: Attitudes of Community Leaders Toward Local Government." *Public Integrity Annual,* The Council of State Governments, 1996.

[3] Totty, Michael. "Privatization–Push Creates an Ethics Obstacle Course." *The Wall Street Journal,* January 27, 1997.

THE STAKEHOLDERS

The interaction of public and private interests occurs in three primary areas:

- The decision to privatize or not privatize a historically public service

- The procurement process for private services

- Structuring the public-private service agreement

There are a large number of interested parties, or stakeholders, involved in the expanding role of private sector involvement in public water utilities. These include:

- The general public

- Rate paying customers

- Regulators

- Bondholders/shareholders

- Political leaders

- Utility managers

- Utility staff

- Business service providers

- Independent advisers and consultants

Unfortunately, each stakeholder has both similar and conflicting interests with other stakeholders. Each has unique characteristics:

The General Public

The public consists of the many individuals who consume water, discharge water, or who are otherwise economically impacted in both the near and long-term by the actions of the public utility. It also includes individuals who are interested in how well government officials execute their responsibilities; environmentalists who may be interested in protection of the watershed; other communities that may be served by the watershed; business, land, and agricultural interests; and so on. The public expects that their officials never use their authority or public funds for their personal benefit or to favor an individual.

Rate Paying Customers

The customers of the public utility largely consist of a diverse group of rate payers including homeowners, commercial, and industrial establishments. This group is largely interested in high water quality (safe water), adequate quantity, good service, reliable fire protection, reasonable rates, and customer-responsive policies, procedures, and operations.

Regulators

Usually state or federal governmental organizations ensure compliance with law, rules and regulations, health codes, financial procedures, bond sales, rate setting, and other matters protecting the public interest.

Political Leaders

The political leaders who may become involved in decisions affecting the operation of a public water utility include the entire span of elected officials from the governor's office to the local mayor and local governmental councils, supervisors, utility board members, and others. Political appointees may also be grouped into this category and include: city managers and public works directors. Characteristically, most of the political influences in a water utility are local.

Utility Managers

Water utility managers may be professional or politically appointed leaders who plan, direct, manage, and oversee the utility operations.

Utility Staff

Utility staffs consists of managers, supervisors, and employees who carry out the day-to-day operations of the organization.

Business Service Providers

Business service providers can be a single firm that offers selected business services such as equipment servicing, billing, meter readings, groundskeeping, and cleaning or a singular or multidisciplined entity that provides complete operations and maintenance, financing, construction, and other privatization services.

Independent Advisers and Consultants

There are numerous firms and individuals that provide expert services to the public utility that include: planning; engineering and design; evaluation and recommendations on methods to improve effectiveness and efficiency; legal, financial, and technical support in developing and implementing solicitations; service provider evaluation; and contractual processes involved in public-private partnerships, etc.

CONFLICT OF INTEREST CONSIDERATIONS

Each of the principal stakeholders described is expected to operate within the principles of a code of conduct. Conflict of interest avoidance is an important part of these codes. Each profession also creates its own rules of practice to assure that the practitioner's behavior is ethically appropriate. Typical of these are the American Society of Public Administration (ASPA) Code of Ethics and the American Public Works Association (APWA) Code of Ethics.

To properly execute a code of conduct, every public servant needs to understand what is expected by the public in order to maintain its trust. This is usually far more than strict conformance with laws. Rather, public servants must recognize that to avoid the appearance of impropriety their relationships with stakeholders must be above reproach. Rules that protect the integrity of the process must go beyond normal legal expectation. Public servants must be concerned about the perception of their own behavior as well as the behavior of service providers, advisers, and consultants.

A measure of a state's public integrity laws is how clearly and comprehensively they cover the variety of ethics decisions facing state officials in state government. An article in *The Wall Street Journal*, "Privatization Push Creates an Ethics Obstacle Course,"[4] raises the specter of the need for government officials to closely adhere to the intent of existing ethics legislation and suggests that privatization contractors also should have a written ethics policy that "covers conflicts of interest and contains policies on offering and accepting gifts and favors; offers of employment; and the conduct of personal business, as well as professional and investment activities, that 'may conflict with the contractor's duties to the state.' The company would also have to set up ethics-training programs for its employees."

It may be prudent for a utility to request that proposers on contract services furnish their written policy on ethics or conflict of interest or, lacking such a policy, a procedure that they will strictly adhere to while in pursuit of a project.

Independent legal and financial advisers are of course expected to put their client's interests ahead of their own interests, and their ethics codes deal with conflicts of interest. For independent consulting engineers, architects, and public accountants the issues are still more difficult. Such professionals have even higher standards: while they, too, are required to put the client's interest ahead of their own, they are also expected to put the *public interest* ahead of the client's interest.

How do these observations on conflict of interest relate more directly to the newly developing businesses centered around improving a water utility's efficiency and effectiveness? It is the multiple hat problem.[5] It's the problem of separating the professional activity from the more conventional business activity.

Let's look at construction. Why are consulting engineers employed by owners during construction? Resident engineers

---

[4] *The Wall Street Journal, Texas Journal,* January 27, 1997.

[5] Busch, Paul L. "Consulting Engineer: Time for Another Look at Conflict of Interest." *Environmental Engineer,* July 1996.

observe construction to ensure that the project is built following the plans and specifications in accordance with the designers' intent. Together with office engineers, they also correct errors of commission or omission that inevitably occur in the planning process. Consulting engineers who represent the client should be well experienced with contractors who offer alternate designs, substitute materials, and differing equipment under the "or equal" clauses usually included in construction contracts. The independent resident or office engineer must examine the details of the proposed changes to determine whether or not they are in fact equal to the original specification. In making that determination, the engineer considers the proposed change in terms of its full life cycle expectations compared to the specification. Performance, longevity, operating/maintenance costs, safety, aesthetics, and all things that are in the best interests of the public and client must be considered. While the contractor is expected to deliver a completed project in accordance with the plans and specifications, an *independent assurance by a disinterested professional* is the norm in most construction projects.

The contractor's behavior is expected to be based on both the terms of the contract and the desire to maximize profit (remember, contracting is a risky business). The engineer, on the other hand, is expected to represent the client, assure fulfillment of the contract intent (that may sometimes exceed the contract terms) and protect the public health and safety. While we would like the goals of the contractor and the project to converge, it is the responsibility of the independent engineer who represents the client to see that the "right" project is finally delivered and that appropriate change orders are issued to adjust for actual project delivery.

When design-build is the method of project delivery, however, who represents the client and the public interests? This is a tough question because while there are many commonalities, there are also differences between the public and client interest of the independent design professional and the economic interest of the design builder. Therefore, the utility has a special responsibility to institute measures that guard against even the appearance of conflict of interest.

Under the U.S. economic system we accept that in business we are permitted to put our own interests first, provided we behave lawfully and with appreciation of the public good. For example, it is certainly acceptable in business to minimize capital costs even though this may not create the best life cycle cost. (Think of examples such as automobiles, electric appliances, shoes, and luggage) Although such a decision might not be in the best long-term interest of the client or the public, this is a *business* issue, not an ethical one. Of course, one way to manage design build programs in the public and the client interest is to prepare a well thought out request for proposals and a detailed contract. However, inasmuch as contracts can never convey everything intended, one cannot expect a design builder to deliver the same product or protection to the client and to the public that was included under the conventional design bid build approach (which is one reason why selecting a design builder based on low bid rather than experience and trust doesn't make sense). After all, it should be expected that delivery systems that include players with different responsibilities will yield differing results. Therefore, new safeguards must be introduced.

Of course, wearing multiple hats—professional and conventional businesses—raises another issue. When an "independent" engineer or architect (a professional) recommends design build, and plans to be part of the design build team (a business), what guarantees his independence in the recommendation itself? Conflict of interest arises when one is in a position to serve both seemingly conflicting interests.

APPROACHES TO AVOIDING POTENTIAL CONFLICT OF INTEREST ISSUES

It is clear that old and familiar relationships take on additional complexity when considered in the light of the multiple hat issue. Political leaders and utility leadership will be expected to reexamine their relationships with their independent legal advisers, financial consultants, consulting engineers, and architects if that adviser or consultant is also on a team that design builds construction projects, acquires utilities, or privatizes utility operations. What does an individual support when he or she has roles with significant differing interests? In such cases there is a difficulty in separating the independent judgment of the professional from the self-interest of the business person. Further, the public may be expected to perceive that the multiple roles of independent adviser or consultant versus business service provider create a conflict of interest even though there has been no unethical conduct. The potential for abuse may trigger suspicion of abuse. This perception can undermine the public trust in political leaders, utility managers, and utility staff, which is *an unacceptable result*. Consequently it is imperative that disclosure of potential conflicts be made part of agreements with independent advisers and consultants and each utility's procedures for dealing with potential conflicts of interest be clearly identified and periodically emphasized as discussed in the following paragraph.

The basis for a good public-private partnership is trust, fairness, open communication, and public support. Each of these can be undermined by real or perceived conflict of interest issues. Consequently political leaders and utility leadership and staff should prepare formal conflict of interest avoidance procedures for their internal use that include protocols for disclosure of potential conflicts. Independent consultants and advisers to the utility should also be required to adhere to clear and comprehensive conflict of interest avoidance procedures prepared by the community or its public utility.

PRACTICES TO MINIMIZE CONFLICT OF INTEREST ISSUES

A number of approaches can be considered to maintain high ethical standards and to minimize the potential for conflicts of interest, especially those that might be associated with public-private partnership. These include:

- Increased awareness by both political leaders and utility managers of the potential conflict of interest issues, adoption of disclosure procedures to be followed, and declaring the unacceptability of certain relationships.

- Development and strict adherence to a communication protocol used during the procurement process. The protocol should be made available to and be respected by all parties participating in the project.

- Use of truly independent consultants and advisers who are not directly or indirectly involved in the business of offering or undertaking contract services of the nature being recommended or procured.

- Special efforts to communicate with the media and public to increase their understanding of the process and the need for confidentiality of certain information until final selection is made.

- Establishment of a fair and open procurement process (see chapter 15).

EXAMPLES OF CODES AND CONFIDENTIALITY PROTOCOLS

In chapter 15 we discuss the potential need to obtain the services of experts as part of the project team during planning, selection, and contracting phases of a thoughtful contracting process. Most public agencies recognize the importance of the consultants and experts being unbiased and independent. Many include requirements to guarantee this in their selection of the consultants.

Springfield (Mass.) Water and Sewer Commission, reflecting a concern to avoid conflict of interests, required an "unbiased and independent perspective with respect to privatized operations" in its Organizational Evaluation and Assessment RFP for a consultant to assist with its solicitation and selection process for contract operations and maintenance of its water and wastewater functions. The commission included the following requirement in the RFP:

> "Proposals from firms which have contract operations or full privatization business as any part of their corporate structure will not be entertained."

The Commission further defined "contract operations or privatization as a public-private partnership which involves the shifting of some or all of the operational or ownership responsibilities for water or wastewater service from the public to the private sector and usually includes "risk sharing activities." The Commission presented hypothetical examples to clarify its intent.

Seattle (Wash.) Public Utilities has a General Consultant Contract Protocol to ensure fair and equitable process to consultants competing for projects. It provides as follows:

"Since the competition for contracts often begins long before the formal issuance of a contract solicitation, the Department must apply and maintain the same principles for information contacts during the pre-solicitation period as it applies and maintains during the formal solicitation period.

"While the Department is always interested in the qualifications and ideas of firms who want to do business with the City, it should be understood that it cannot offer specific insights or responses, *outside the formal RFQ/RFP/Interview process*, on the appropriateness of a firm's qualifications, its proposed team or its proposed approach to a project (proposal strategies)."

The principles guiding project-specific information dissemination are:

- All information necessary for firms to prepare their statements of qualifications/proposals will be included in the RFQ/RFP or made available in materials cited in the RFQ/RFP.

- At the time of solicitation for consultant services, the Department will identify and make available specific documents and reports for interested parties to review and rely on during the development of their contract proposals. Also at the time of solicitation, the Department will maintain a list of those firms who request the RFQ/RFP, so that if new and relevant information relating to the project or process arises, the Department can issue information updates to those firms.

    NOTE: *The list is also made available to other firms who may be interested in forming or joining a team.*

- The Department, at its sole discretion, may also offer pre-proposal conferences for interested consultants, where questions may be asked about the project and where clarifications of the process will be given. Notice of these conferences may be published in the solicitation advertisement and/or the solicitation instrument (RFQ/RFP). Clarifications made or new project-specific information disseminated at any such conference will be documented and available to those who are unable to attend the conference and request the information.

- During solicitation, any individual contacts with the Department must go through the person (usually the project manager) identified in the solicitation instrument or advertisement as the Department Contact Person. Such contact, whether by phone or in person, will be only for the purpose of clarifications and general inquiries about the solicitation process or schedule. No project-specific information should be requested, and none will be given out, except that already available in the solicitation documents.

- Private, one-on-one meetings or telephone conversations, whether with the project manager or other Department staff, are not permitted, once the intent of the Department to solicit a consultant contract is known.

- If a meeting or private conversation takes place between a firm and a Department staff person during the solicitation or pre-solicitation period, the occurrence of the meeting/conversation, the subjects discussed and any information divulged by the Department (except that already available in the RFQ/RFP or in cited Department reference materials) will be documented and disseminated to other firms participating in the competition.

- After a finalist has been selected and a contract executed, interested parties may contact the Department to discuss their performance and rating. Staff workload and schedule will determine how much time can be made available for these debriefing sessions, but every effort will be made to give inquiring parties constructive and useful information. At a minimum, a report on the relative scores and ranks of the finalists will be made available on request."

Seattle Public Utilities inserted the following restrictions in the Request for Proposals for solicitation support services for the Tolt Design Build Operate project:

"Consultant understands that it is being retained by the City to assist in development and conducting of a competitive process related to the Tolt River Filtration Plant Project, and that it is essential that there be no questions about the integrity of that competitive process. Consultant agrees, therefore, that it will not participate in any way as a proposer or consultant to any proposer on the Request for Proposals for the Design/Build/ Operate Contract for the Tolt Filtration Plant Project, nor will it provide any information to any persons about this Project without the express approval of the City. Additionally, Consultant will require that all of its officers, employees, agents, and subcontractors associated with this Project to agree to these restrictions in a manner satisfactory to the City."

Similar language, modified to fit the situation, is applicable for all sizes of projects.

SUMMARY

As the chief executive involved, the utility manager has a special responsibility to guarantee that the issue of conflicts of interest is handled with scrupulous care and precautions. Public trust in the utility and in the officials involved can be quickly devastated by charges of improper conduct. And in this litigious era, serious damage—financial and otherwise—can easily be wreaked upon individuals and organizations. The stakes are so high and the ethical principles so compelling that all parties involved in public-private partnerships of all kinds must take pains to ensure that the public interest is unquestionably protected and well served.

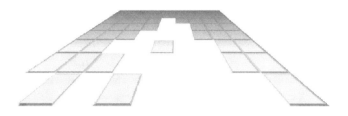

CHAPTER TWENTY-ONE

Regulatory Compliance Issues

Water quality goals, standards, and regulations have a significant impact on the cost of service and the utility's ability to deliver adequate quantities of water. These requirements that protect public health and provide an aesthetically pleasing water supply are part of a larger body of federal, state, and local laws that impact the provision of water and wastewater service. In addition to the Safe Drinking Water Act, important requirements potentially affecting the utility's cost of operations are contained in the Endangered Species Act, CERCLA (Superfund), and the Clean Water Act.

Many states have environmental and drinking water quality statutes that enforce more restrictive requirements. In the recent past these requirements have been considered nearly absolute. Once they are adopted, practical considerations like cost or new technology are given lower priority, except perhaps in the schedule for achieving compliance. As each utility seeks to optimize its performance, regulatory requirements should be included in the process. Representatives of regulating agencies should be considered stakeholders in developing consensus, so that the goals for more efficient performance can reflect anticipated regulatory requirements. The objective should be optimized compliance, even if this requires exceptions from normal state or federal standards. The ultimate result can be updated regulations based on the most cost effective and reliable service.

WINNING THE CONFIDENCE OF REGULATORS

This book describes a range of activities to improve performance. Performance improvement requires flexibility or at least an open mind toward potential changes in current operating practices. We have often assumed that regulations are fixed and only the manner in which they are achieved through design, construction, and operation is flexible. Some of our case studies show that regulators are increasingly willing to work cooperatively with utilities to achieve the best results. This may mean exceptions from long-established performance and quality standards. A utility should consider the importance of each regulatory requirement as it affects the cost and achievement of service goals. When a particular regulation, interpretation, or statute creates a significant impact on performance optimization, the utility should raise the issue so that a more practical and potentially better technique or process can be used to achieve compliance and reduce cost.

Concepts of participatory management should apply in the regulatory environment as well. Working groups of utility and regulatory representatives can be formed to review a new facility's specifications, a program for long-term compliance, or an analysis of alternatives for treatment or distribution system improvement. Regulators can be included with representative community groups that are developing new project or performance goals so that they are exposed to the same set of attitudes and positions that the utility itself faces. This will pose an additional burden on the regulators, which can be decreased with assistance from the utility. However, if this effort is properly scheduled it will result in a more successful project that is delivered at a reduced cost. Some of the activities that can involve regulators are demonstrated by the Tolt project experience in Seattle.

The following sections outline some of the key areas of regulatory activity that can affect performance. The regulators in each state, frequently overburdened with heavy workloads, should be given an opportunity to consider the issues from the utility's perspective without sacrificing regulatory independence. By enhancing the factors of safety that assure long-term reliable compliance, perhaps they can reduce the risk of standard violations that many changes bring. Utilities should seek ways to create a higher level of regulatory confidence through the use of comparable examples, pilot testing, demonstration projects, or other techniques to demonstrate that there is good reason for the government administraters to be confident in your system.

COMPLIANCE ASPECTS OF PERFORMANCE IMPROVEMENTS

The principal regulatory framework for water utility operations is the Safe Drinking Water Act. Most states have received the statutory delegation to administer the Safe Drinking Water Act; consequently the federal regulations have been supplemented by a range of state regulations. However, EPA reserved the right to intervene in a particular regulatory activity, should a state fail to perform. Most states have adopted requirements for the design and construction of water works (relying in part on the standards of the American Water Works Association), which provide a baseline for utility specifications and operations. These requirements sometimes limit new and creative technology that might achieve a more desirable result, both from a regulatory and from a cost standpoint.

Following are some potential regulatory impacts of performance improvement efforts.

Reorganization

Organizational improvements may shift the responsibility for performance to a less experienced team to run complex and sophisticated treatment and delivery systems. This approach may be desirable from an administrative standpoint in terms of overall management efficiency, but it may have undesirable safety implications. This potential weakness can be overcome by using a matrix type of organization with clearly established lines of authority and a good communication system. A question always to be asked in reorganizing treatment plant staffs is: Can we offer the same assurances about water quality with this approach?

regulatory compliance by individual water quality constituents.

- *Technology that departed from health regulatory design standards led to obtaining agency options prior to contract negotiations, to reduce contractor risk.*
- *Informal confidential discussions were held on the relative merits of competing technologies.*

License Requirements

In granting utility permits, states usually specify the kinds of licenses that must be held by operators of plants of various sizes. These requirements may conflict with the use of SCADA and other automated systems to reduce staffing needs. Civil service rules and union contracts may also limit your freedom to change personnel requirements, train staff for new and additional tasks, or restructure schedules. These barriers to change can often be overcome by involving all interested parties in working out solutions in the best interests of all. Where there is a will to achieve the utility's goal of competitive performance, there will be a way to overcome the hurdles.

Unattended Operations

Automation may create regulatory concerns regarding reliability. In attempting to optimize treatment plant performance, the City of Seattle in the Tolt facility RFP, allowed unattended operation at the option of the proposers, providing they accepted the compliance risk. Some proposers suggested unattended operation, but the final contract included 24-hour operations. However, instrumentation and control for unattended operation was included in case the plant's actual performance would justify its use. Automation, if properly designed and maintained, can enhance reliability with redundant systems and multiple levels of oversight. For instance, Yorkshire Water in the United Kingdom reports that centralized oversight for its more than 100 treatment facilities results in an average attended time of 37 hours per week per plant. In addition, operator action or inaction has been the cause of some of the more serious contamination incidents. Skillfully designed water quality monitors and limited fail safe and routine controls (see chapter 9) can potentially save money and reduce the risk of violation.

Reliable Communication

New communication links through microwave systems or fiber-optic cables can provide vital information during catastrophes. It is possible that additional investment in communication and automation equipment could reduce staffing cost, and at the same time give regulators a higher level of confidence during routine and emergency operations. The human systems and staff procedures surrounding the use of these technologies must be well thought out to cover all possible contingencies.

Improving Standards

Prescriptive construction requirements can severely limit the opportunities for optimizing the design and construction of new facilities. The traditional sequential method of meeting regulatory requirements is to provide a facility design to meet those requirements, then to secure competitive bids for the project construction, and finally to hire and train operators to operate the facility. Alternatively, as discussed elsewhere in this book, a single contractor selected in a competitive process can be used for multiple responsibilities including: operations, design build, or design build operate. Irrespective of the management system used, inclusion of regulators in the design of performance specifications and in the review of proposals from competing firms could potentially minimize the cost of compliance and improve performance.

ALLOCATING COMPLIANCE RISK

All water utilities must protect public health and the local system owner bears the responsibility for compliance. If a serious violation of standards occurs either due to misoperation of the facility on a day-to-day basis or due to an emergency or unforeseen event, the agency, its staff, chain of command, and ultimately its governing body are collectively responsible. If a service affecting compliance is delegated, the contractual standards for performance should include all potential regulatory impacts of failed performance during emergencies or routine operations, excepting only circumstances clearly beyond the control of the contractor.

There are generally two types of responsibility that can be affected by outsourcing treatment facility responsibilities: design and construction, and operations. The advantages inherent in combining responsibilities for design, construction, and operations in a single contract are discussed in chapter 17. There are also regulatory advantages. These include having a single entity to accept risks of compliance for the total program, not just for a single phase. However, the principal risk occurs in the operations phase when operator failures and/or emergencies can result in public health risk, public notification requirements, and fines. The utility's transfer of this responsibility is usually characterized by:

- The city's retention of responsibility for raw water conditions that exceed defined limits due to unpredictable events. The limits should include only extraordinary conditions that occur infrequently.

- The contractor's acceptance of performance within limits including all penalties associated with failure.

- Supplementary noncompliance penalties for failure to perform above regulatory requirements, accompanied by rewards for consistently superior performance.

This model should be modified for each local facility since most plants have unique conditions that should be considered in dividing the risk. An emerging issue is the increasing reluctance of local consumers to accept water that may meet current federal or state standards but contains detectable amounts of compounds that are toxic in larger concentrations or that result from industrial processes. This gap between regulatory requirements and public acceptance can affect treatment costs and the usability of the affected waters. The result is a de facto non-detect level regulation that becomes a moving target. The risk of noncompliance becomes greater, and a public agency's ability to shift the risk by outsourcing can be significantly reduced due to both technical and regulatory uncertainty.

There can be a perceived long-term conflict between optimizing performance and maximizing public health protection. This challenges managers to find ways to improve performance *and* to improve public health and environmental protection. This should be achieved where possible not by applying more workforce, but by taking advantage of new technologies. The principle of considering goals rather than specific historic standards or procedures applies not only to the direct effects of treatment facilities, but also to other effects including reservoir operations,

pumping facilities, distribution system operations and maintenance, customer service, and other related functions.

ENVIRONMENTAL REGULATIONS

There is a wide range of environmental regulations at state, federal, and local levels. Each regulation puts an additional burden
on the utility seeking to meet its performance and service objectives. These regulations invariably include some form of public
participation process. The resulting public discourse on environmental issues can be unbalanced or even one sided. However, this
process offers opportunities for the utility to involve interested
citizens in its activities with the objective of mutual understanding. Committed advocates of special interests may not have considered the full implications of their position, and the interests of
citizens normally not active in public affairs may not be fully considered. Environmental impact statements and reports and other
public regulatory events can be used to expose groups to the benefits of community-wide consensus on improved water service.

In the development of a treatment plant or other project that
must meet environmental requirements, it is better to complete
the environmental review process prior to contracting or outsourcing. The risks associated with the process are not ones easily
assumed by a private company without significant charges and
increased potential for delay. However this is highly dependent on
the relationship between the utility and the regulators, particularly local agencies or departments. In some cases a private contractor may have more success in obtaining the necessary
approvals. Both internal and external operations improvements
should be accompanied by an effort to anticipate environmental
regulatory demands as part of the change process.

A new treatment plant may offend a local neighborhood.
However, if the design and performance of the plant are optimized,
there may be sufficient savings to fund local impact-mitigation
measures. Similarly, pipeline replacement and construction work
can be done in ways that reduce local impact during construction;
it may also offer an opportunity to enhance local site conditions.
The latter may require the active participation of the local public
works departments that regulates the construction in developed

areas. Challenges to local rules may be counterproductive, but finding ways to comply efficiently may lead to modified requirements and lower costs.

Environmental regulation can also affect maintenance activities. Routine cleaning of reservoirs may produce materials that become hazardous wastes. These have included old paint residuals and caulking materials. The planning of maintenance activities should anticipate these regulatory and disposal obligations as a part of the optimization process.

SELF-IMPOSED REVENUE RESTRICTIONS

Most utilities operate under constant pressure to reduce rates, frequently in the face of rising costs and performance requirements. Although not a formal requirement as such, this puts an effective restriction on utility optimization. If a utility's investment in new technology is budget limited or if treatment enhancement is achieved piecemeal when a general improvement would be optimal from a long-range standpoint, then the control imposed by revenue limitations is counterproductive. Utilities are here to stay. The organizational structure and equipment they use may change, but the service is permanent. Therefore, the optimization of service requires a long-term view. If rate restrictions are arbitrary, the outlook becomes short-term and may not result in the best performance at the least cost. If long-term performance is to be optimized, public rate setting should be based on plans, budgets, reports, strategies, and achievements that include the long-term economic impact of changes in operations and schedules for capital improvements.

ZONING AND BUILDING RULES

Functional improvement of utility services will require varying levels of local government oversight and approvals. This applies to facilities of all types, including treatment pipelines, pumping plants, and administrative facilities. These rules may increase cost of projects and operations. Local regulations should be approached in a fashion similar to that identified for environmental and public health rules. The application of these rules to water utilities varies among states, localities, and even administraters. Participation of utility planners in early phases of water project development can reduce future conflict. Experience has shown that if significant efforts are applied in anticipation of a community concern about a particular change, the potential for future crisis can be reduced. *The objective is to align the utility's program with the goals of the regulating agency rather than with the letter of each and every rule or standard. This may mean a shift from standards based on statistics to standards based on performance.* The general concern in society for improved performance and service will apply increasing pressure on regulators to be flexible without altering their objectives.

OCCUPATIONAL SAFETY AND HEALTH ADMINISTRATION (OSHA) AND LOCAL SAFETY REQUIREMENTS

OSHA and local workplace requirements, while increasing safety performance, can increase costs and reduce performance of work tasks. Effective compliance with the safety imperatives requires continuing utility vigilance. State and local rules may, in some situations, be more stringent than OSHA requirements. This subject is also part of training and performance monitoring discussions in this book. While safety requirements are generally overriding, the objectives of the rules may be achievable in alternate ways and these should be discussed with the agency representatives with the goal of improving utility performance and worker protection. When tasks are outsourced to a contract service provider with more experienced personnel and ongoing experience with inspection and compliance, OSHA and local safety rules may be more effectively and efficiently achieved.

SUMMARY

Death, taxes, and regulations are today's facts of life. The utilities that have the best, most productive relationships with regulators are those who communicate most fully with them. Involve them early in major projects. Work with them to meet the goals of their programs, not merely the letter of the law. Above all, the utility's posture must always be in support of protecting the public health and welfare.

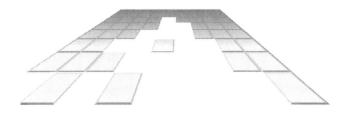

Case Studies

INTRODUCTION

This appendix summarizes case studies of water and wastewater utilities that have implemented actions to improve efficiency. The purpose of these case studies is to provide:

- Examples of how individual utilities have responded to the imperative to improve efficiency

- Results of improvement efforts

- Utility contacts for additional information and insights

In an effort to cover a broader range of experiences, drinking water case studies have been supplemented with several case studies of wastewater and water reuse/reclamation systems, which are categorized into five groups:

1. Utility management

2. Internal improvement using public works staff

3. Contract operations

4. Managed competition

5. Other contracted services

Each case study is presented in a similar format to provide the reader with basic information on the following:

- Name of agency and type of facility

- Key elements of the project

- Brief background

- Features of the approach

- Identification of parties involved

- Major lessons learned

- Results

- Contacts for further information

INDEX TO CASE STUDIES

Case Study 1.1

Work Management Improvement Processes and Benchmarking; Los Angeles Department of Water and Power

- Following the recommendations of a decennial audit, the Los Angeles Department of Water and Power (DWP) initiated a work management improvement process in its water operating division.

- Within five years its workforce was reduced by more than 20 percent, while productivity increased by 30 percent.

- The continuous process improvements were stalled by a buyout process that led to the exodus of key upper level managers, in particular the Water Operating Division Manager, who had initiated the improvement process.

- Separate from the decennial audit, a city council audit of the electrical power arm of the DWP was broadened to partially include the water utility. The findings from the audit suggested that the water utility part of the department could benefit from performance benchmarking.

BACKGROUND

The DWP provides water to roughly 3.6 million people within the political boundaries of Los Angeles, in addition to generating and distributing electrical power. They have no wholesale water customers.

Every 10 years the DWP hires a consulting firm to conduct an audit of the department. The audit concluded in 1990 determined that the department could benefit from implementing work management improvement processes.

Parties Involved

Los Angeles Department of Water and Power

APPROACH

The manager of the Water Operating Division initiated an improvement process, which led to: reduced crew sizes, the establishment of working supervisors, changes in the way material was ordered and packaged, and ongoing employee training. This was complemented by a group-based reward system. Data were consistently gathered in order to measure performance. Within five years of initiating the process, the workforce was reduced from 980 to 780 employees, while productivity increased by 30 percent.

The effort began to plateau in '94 and was hampered by an employee buyout process by the department, which led to the departure of key upper level managers, such as the Water Operating Division manager, who had supported and championed the work management process. Additionally, while the division set productivity goals that were met and exceeded, no goals to improve beyond what was originally established were set.

Preceding these events, the city council, concerned about the DWP's electrical power system in light of deregulation, called for an audit of the power system to determine its optimal size. The audit was widened slightly to include the water side of the DWP and the findings determined that the water system could benefit from benchmarking.

The audit provided a specific directive for DWP to move forward on benchmarking and to focus on measuring performance. It also served as a transition from focusing on specific work management improvement processes to benchmarking performance relative to other utilities. DWP sent out a survey to gather data on other utilities and received responses from 14 utilities. DWP started to extend the benchmarking to include process review, but the effort stalled due to the buyout process previously mentioned. DWP has since begun to work with the West Coast Benchmarking Group on process redesign.

LESSONS LEARNED

- DWP experienced difficulty in analyzing the responses to its survey because of differences in bookkeeping and operations between utilities.

- When the work management improvement and benchmarking processes stalled with the departure of lead personnel, DWP realized the importance of institutionalizing improvement processes.

- Focusing on operations is a good first step because, in the case of DWP, most of the work done in operations has data associated with it, and because money saved in operations has a greater impact than money saved elsewhere.

CONTACT

Marty Renert
Director of Administrative Services
Los Angeles Department of Water & Power
213-367-1131
213-367-0995 FAX
mrener@ladwp.com

RESULTS

Through the work management improvement process the workforce in the water operations division decreased by more than 20 percent, while productivity rose by 30 percent. A number of operational improvements were made. The focus shifted to conducting a benchmarking survey, which yielded data from 14 utilities. Though DWP has had trouble translating this data into a useful benchmarking process, it is building on the results of the survey through its involvement in the West Coast Benchmarking Group.

Case Study 1.2

Strategic Planning
Seattle Public Utilities, Wash.

- Seattle Public Utilities, working with a consultant, developed a strategic plan for the newly merged organization.

- The strategic plan and associated action plans are available on the department's internal web page and in hard copy.

BACKGROUND

The Seattle Public Utilities (SPU) is a municipally owned utility that provides water, sewer, drainage, solid waste, and engineering services to retail and wholesale customers. SPU was established on January 1, 1997, through the merger of the Seattle Water and Engineering Departments. To set the course for the new utility, SPU developed a strategic plan that outlines its vision, mission, goals, objectives, and strategic efforts (projects) to help it achieve those goals.

Parties Involved

Seattle Public Utilities Leadership Team
Social Marketing Services (consultant)

APPROACH

The SPU hired a consultant to facilitate the development of a new strategic plan. Working with the consultant, the executive team (branch directors) defined the audience for the plan and established a framework, i.e., the components of the plan, and a process and timeline for developing each of those components.

The first step in the process was to define the new vision statement of the organization. The vision was drafted by the executive team and then reviewed in focus groups of individuals at all levels of the organization. Those focus groups found the statement too wordy and suggested several alternatives that the executive team used in finalizing the statement.

With the vision and mission (which was developed for the merger proposal) in hand, the executive team established six areas of importance to the utility around which it drafted SPU's goals. The leadership team (division and branch directors) reviewed these goals and set out in small groups to establish objectives related to those goals. Working with a consultant, members of the leadership team refined these objectives for presentation to their direct reports. Throughout the process the leadership team was encouraged to share the progress on the plan with their staff for feedback.

At a half-day session small groups of these direct reports identified projects the department needed to complete over the next two years to accomplish these goals.

The consultant and staff from SPU reviewed the components of the plan, ensuring that it was consistent with other departmental documents and with direction from the Office of Management and Planning and elected officials.

LESSONS LEARNED

- Involve employees throughout the organization in the process. Although the greater the number of participants in the process, the more difficult the reconciliation of ideas, participation throughout the organization helps staff to take ownership of the product.

- Define the audience for the strategic plan at the outset of the process. The content of your strategic plan may differ depending on its audience, so knowing the audience will help focus content development.

- Define your goals and objectives with performance measures in mind. If your success as an organization is defined through achieving your goals, making sure that you can measure them is important.

CONTACT

Paul Reiter
Director of Strategic Policy
Seattle Public Utilities
710 Second Ave., 10th floor
Seattle, WA 98104
206-684-5852
206-684-4631 FAX

RESULTS

SPU printed its strategic plan and created an internal web site, which includes action plans and leads for each of the projects listed in the strategic plan. These action plans are updated and progress reports are developed regularly.

Case Study 1.3

Comprehensive Evaluation of Utilities Management
Bergen County Utilities Authority, N.J.

- The BCUA and consultants undertook a comprehensive evaluation of current costs and developed an analysis of external improvement options.

BACKGROUND

The Bergen County Utilities Authority (BCUA) provides wastewater treatment to 46 municipalities making up part of Bergen County, N.J. The BCUA owns and operates a 109 mgd wastewater treatment plant, including sludge dewatering and processing facilities, an interceptor network and associated collection system, and an on-site laboratory.

The BCUA formed a privatization committee to assess potential improvements to more cost effectively operate its wastewater treatment system and continue to improve the quality of service provided to its customers. The decision to conduct such an analysis was in response to inquiries from various vendors about owning and/or operating the plant, and an earlier County study that recommended looking into privatization. In addition to in-house staff, the BCUA was assisted by the consultants with additional support provided by legal, labor, and bond counsel.

Parties Involved

Public agency

Bergen County Utilities Authority

Consultants:

Malcolm Pirnie, Inc.
Raftelis Environmental Consulting Group
Value Added Management Services

APPROACH

A detailed analysis was conducted to assess alternative management and ownership structures that could provide operational and economic benefits to the BCUA. This analysis consisted of evaluating the potential improvements to current systems to optimize operations, assessing the feasibility of implementing privatized services at the wastewater treatment system, and investigating legal, regulatory, labor, and financing issues.

The consultant took the lead in conducting the analysis of BCUA's wastewater treatment system, including collection, treatment (headworks and tanks), analytical, sludge (treatment and disposal), and other administrative/nontreatment services. This required detailed review of personnel, utilities, maintenance and repairs, chemicals, debt service, permitting fees, contracted services, insurance, health and safety, and capital programs. Prior to proceeding with any of the alternatives being considered, an analysis of current system costs that were broken down into these areas, was necessary to establish a baseline.

Those functional areas where savings could be made through optimizing internal operations and outsourcing particular functions (lab services, sludge dewatering, vehicle maintenance, etc.) were identified and projected capital cost requirements and annual O&M savings were estimated.

The information resulting from Malcolm Pirnie's optimization analysis was then included in the overall assessment with the various privatization alternatives (e.g., asset sale, asset lease, and long- versus short-term contract operations). Raftelis Environmental Consulting Group compiled the results from the optimization study phase of the project, reviewed the potential privatization alternatives, and analyzed potential impact on operations costs. Value Added Management Services conducted a benchmark analysis, in which they compared BCUA costs with those of other peer utilities including Indianapolis, Essex/Union Counties, New Orleans, Ottawa-Carleton, Philadelphia, Wilmington, Westchester County (Yonkers), and others.

Based on the results of the preliminary analyses, the BCUA commissioners agreed to pursue a two-track approach. Under the first approach, the BCUA is implementing the recommendations to optimize BCUA operations. This includes implementation of telemetry and SCADA systems, process modification, and outsourcing of several functions. The BCUA is simultaneously soliciting proposals from private firms to operate and maintain the wastewater treatment system. This involves a procurement process consisting of Request for Qualifications, prequalification of vendors, Request for Proposals, evaluation of proposals and vendor(s) selection. The most cost effective proposal(s) will then be compared to the optimized BCUA operations to determine if privatization is worth pursuing.

LESSONS LEARNED

Due to the magnitude and complexity associated with this project, the BCUA determined that additional information was required to identify the most advantageous approach.

CONTACT

> Michael Oldham
> Executive Director
> Bergen County Utilities Authority
> Foot of Mehrhof Road
> Little Ferry, NJ 07643
> 201-641-5801
> 201-641-5356 FAX

RESULTS

BCUA is in the process of proceeding with an intensive concurrent approach consisting of optimization of its operations and a comprehensive contractor procurement process.

Case Study 1.4

Self-Assessment and Benchmarking
City of Austin Water and Wastewater Department

- A quality improvement effort initiated by the City of Austin led the Water and Wastewater Department (WWD) to undergo a self-assessment process that led to process improvements, reorganization, delayering, and workforce reductions.

- A study of the City's electrical utility conducted by Price Waterhouse included recommendations for how the electric utility should respond to meet the challenges of deregulation. Since the bonds of the electrical utility were combined with those of the WWD, the study also affected the WWD and raised the question of how competition may impact it.

- The self-assessment effort led WWD to initiate process benchmarking, while the audit of the City's electrical utility led the city council to instruct WWD to undergo metric benchmarking in order to gauge its performance against industry standards.

BACKGROUND

The City of Austin WWD provides water to 620,000 people on a retail and wholesale basis, in addition to wastewater treatment to a slightly smaller customer base. WWD's water system has a peak capacity of 215 mgd, but averages 110–120 mgd. The wastewater system's capacity is 120 mgd, with an average of 85 mgd treated.

Parties Involved

City of Austin Water and Wastewater Department
City of Austin City Council
Austin Quality Forum

Consultant

Malcolm Pirnie, Inc.

APPROACH

In 1990 Austin began a quality improvement effort, initiated by the City Manager, with each department responsible for establishing a quality improvement process. In 1992 the WWD participated in a self-assessment process developed by the Austin Quality Forum, a consortium of government agencies and private companies that leads organizations through self-assessment processes and conducts peer reviews based on the Baldridge criteria. One of the requirements of the Baldridge criteria is that an organization undergo process benchmarking. The first report from the self-assessment was issued in 1993 and recommended that the WWD do more benchmarking and develop performance measures. Influenced by the report, the WWD decided that it would conduct self-assessments every three years and initiate process benchmarking.

Independent of the work done with the Austin Quality Forum, a study conducted by Price Waterhouse in 1996 of the city-owned electrical utility caused the city council to be concerned about how competition in the water industry could impact WWD. While the WWD had a de facto monopoly with respect to water supply and distribution, a local river authority had been expressing interest in providing competitive water supply services. The city council instructed the WWD to conduct metric benchmarking.

The WWD initiated its metric benchmarking effort by distributing surveys to 50 utilities to gather information on industry standards. They received responses from 25 utilities and are analyzing the raw data collected and plan to report to the city council in the fall of 1997.

The metric benchmarking effort reinforced the need to focus on process benchmarking and has led the WWD to conceive of an effort to develop a consortium of other regional utilities to process benchmark selected processes.

LESSONS LEARNED

- Considering the varying approaches to accounting, work descriptions, and functional definitions, the WWD found it difficult to conduct its metric benchmarking survey and establish meaningful comparisons between utilities because of the inherent nature of metric benchmarking.

- They realized the need for projects such as QualServe in order to compare utilities within the context of a consistent, common framework.

- The WWD found that some utilities just did not want to embark on metric benchmarking and were concerned about making operational data readily accessible and associated with individual utilities.

- Deregulation of the electrical utility industry has awakened elected officials and some of the general public to the possibility of competition for public water utilities.

CONTACT

Randy Goss
General Manager
City of Austin Water and Wastewater Department
Tel: 512-322-2916
Fax: 512-322-2842
email: randy.goss@ci.austin.tx.us

RESULTS

The first self-assessment report generated in 1993 led to a reorganization of the WWD. The first phase of the reorganization has just been completed and has led to the elimination of vacant supervisory positions, delayering, and an increase in span of control. The WWD also began looking for opportunities to explore privatization, and was preparing to issue an RFP to operate one of their biosolids plants, but held off from doing so by directive of the City

Council. The WWD is in the midst of preparing additional analysis and anticipates that the trend of eliminating vacant supervisory positions will continue. The WWD's involvement in the self-assessment process has led to steady improvements since 1992.

Case Study 1.5

Strategic Planning/Internal Self-Improvement
City of Des Moines, Iowa

- The Des Moines Water Works of the City of Des Moines, Iowa, reassesses its strategic plan every three years.

- The Des Moines Water Works Utilities team drafts work plans and makes suggestions to identify and analyze current costs and staff capabilities. The goal is to improve efficiency, cut costs, retain staff, and identify areas of expansion and market opportunities for the utility.

- Findings are reviewed by the director team and general manager and judged against the strategic plan.

- The utility has been able to identify growth areas in neighboring communities where full service water management can be provided, as opposed to simply wholesaling water, which was the utility's former approach.

BACKGROUND

The Des Moines Water Works Public Utility (DMWW) is a municipally owned utility that provides water, engineering, and management services to more than 300,000 retail and wholesale customers in the metropolitan area, in addition to contract operations for three service areas which generally have less than 8,000 customers each. DMWW was formed in 1919 to provide water services for the City of Des Moines. It was not until the mid-1980s that a new course for the utility was developed. A strategic plan that outlines its vision, mission, goals, objectives, and strategic efforts (projects) to help it achieve success has been in effect since 1989.

Parties Involved

Des Moines Water Works - Utility Teams

APPROACH

The utility develops a new strategic plan every three years. A team of the CEO, directors, and board members define the components of the plan, and a process and timeline for developing each component. It is then used as the guiding document by staff team members in drafting work plans. The work plan process creates a comprehensive review of many aspects of operations.

The process is accomplished in a series of steps:

First, the strategic planning process begins with the CEO, directors, and board members reviewing and understanding the regulatory and market forces affecting the utility.

Second, the strategic planning team identifies up to six key result areas that define six major goals for the utility for the next three years. The strategic planning team then develops action plans that set objectives for each key result area.

Third, work plan teams, made up of utility staff with oversight of a director, draft work plans covering every aspect of the utilities work. The teams are encouraged to be entrepreneurial in their thinking, find new ways of doing things, and challenge existing ways. The work plans that get funded for the new budget year are those that most closely meet the directions set by the strategic plan.

Fourth, the team considers new areas of growth for developing and marketing a range of services beyond the metropolitan area. Such areas include, for example, meter reading services, contract operations, long-range planning studies for neighboring communities, etc.

Finally, the skills and flexibility of the utility's employees are evaluated, with the goal of maximizing the flexibility of staff. The goal is to retain employees and retrain them as necessary.

LESSONS LEARNED

- Existing procedures can be streamlined to improve efficiency, while new growth markets can be realized.

- Involve employees throughout the organization in the process.

- Define your goals and objectives with performance measures in mind.

CONTACT

L.D. McMullen, Ph.D.
General Manager
Des Moines Water Works
2201 Valley Drive
Des Moines, Iowa 50321-1190
515-283-8794
515-283-6160 FAX

RESULTS

Operations were streamlined and cost savings were realized while new growth areas for services in the market were developed.

Case Study 1.6

Total Quality Management and Internal Self-Improvement
Irvine Ranch Water District, Calif.

- The water district implemented a comprehensive, well-defined management system based on the principles of Total Quality Management, focusing on customer service and project management.

BACKGROUND

The Irvine Ranch Water District (IRWD) provides potable water, sewage collection and treatment, and produces tertiary-treated reclaimed water. Wastewater is treated at the Michelson Water Reclamation Plant, providing reclaimed water for landscape irrigation, agricultural, industrial, and commercial needs. Day-to-day operations are supervised by General Manager Ronald E. Young and his staff.

Located in the south-central portion of Orange County, IRWD encompasses approximately 76,000 acres of which about 18,000 acres are currently developed for urban use. The District serves the city of Irvine and the unincorporated areas of Foothill Ranch and Newport Coast. In addition, IRWD serves portions of Tustin, Santa Ana, Newport Beach, Orange, and Portola Hills. IRWD extends from the Pacific Coast to the foothills.

The IRWD board and senior staff decided that being a well-run water district was not enough. With the increase of privatization and consolidation, public agencies need to become accustomed to operating on standards similar to private companies.

The District recognized that most well-managed private companies utilize a comprehensive, well-defined management system based on principles of Total Quality Management. They decided that the process they would use to improve their management system would include these principles.

The most critical aspect of IRWD's approach to developing a Total Quality Management program was to focus on IRWD's overall management system. IRWD utilized the services of a consultant who worked with them to ensure that the organization was self-sufficient in implementing its TQM program within 24 months.

Parties Involved

Irvine Ranch Water District

Consultant

The Farrell Group

APPROACH

The major objectives of the program were to:

- Instill quality as the basic business principle at IRWD, and ensure that continuous improvement becomes the job of every employee.

- Ensure that all employees provide IRWD's internal and external customers with the products and services that fully satisfy their requirements.

- Establish, as a way of life, management and work processes that all employees use to pursue continuous improvement in meeting customer requirements.

In order to achieve these objectives, a number of areas were emphasized within the organization. These areas include:

- A supportive management practice that establishes clear, consistent objectives and creates an environment of openness, trust, respect, discipline, and patience.

- Consistent and expanded use of management tools, processes, and employee involvement.

- A process for identifying, agreeing to, and actively meeting customer requirements.

- A process for estimating and focusing on quality and operational measures to achieve continuous improvement.

- A commitment to do it right the first time.

The implementation of the program included:

- The creation of a quality council, consisting of IRWD's directors and senior managers, to oversee the implementation of the program.

- Training of all IRWD employees by Xerox Quality Services on common meeting management, problem solving, and process improvement tools.

- The development of an organizational measurement system that included the revision of the information presented to the board to encompass operational and quality measures, as well as financial measures.

- The development of a systematic, well-defined three year work plan listing all major milestones for the program.

- The creation of a significant number of cross-functional process teams to address critical District-wide issues.

Initially, only the quality council participated in the program. However, through the development of teams and the deployment of training, all District employees have been involved in and have contributed to the program.

The initial teams formed as part of TQM included a Customer Service Report (CSR) tracking team and an Engineering Project Management System (EPMS) team. Both teams developed software applications that streamlined and improved operations. The CSR process was improved with computerized tracking of all customer requests and problems. This allows a designated IRWD "customer advocate" to follow the resolution of the problem and it provides a single point of contact for the customer. The process includes provisions for measurement of customer satisfaction indices to assess the performance of the system. These indices show continuing improvement in the efficiency of problem resolution (problems handled without further referral) and overall satisfaction by customers. The EPMS team developed a project tracking system that allows individual project managers to track the progress of design and construction projects in relation to schedule and budget. This has streamlined and improved the capital project budgeting process as well as reduced the administrative overhead associated with managing these activities.

Since the formation of these initial teams, numerous additional teams and mini-teams have been formed to examine other operational areas of IRWD. These include an energy management team that discovered significant cost savings and a ride sharing team that implemented cost savings and dealt with sensitive employee benefit issues. These and other teams have saved IRWD hundreds of thousands of dollars in documented cost savings while improving the overall quality of service to the community. Monthly tactical and strategic measures allow management and the Board of Directors to continuously assess the progress of these improvements and to provide feedback to the line employees doing the work.

Some other examples of the success of teams include a customer satisfaction team that worked with an outside consulting firm to design and implement the first comprehensive survey of IRWD's customer base. The results confirmed IRWD's perceptions of overall customer satisfaction, but also pinpointed specific opportunities for IRWD to further satisfy its customers. An employee satisfaction team worked with an outside consulting firm to design and implement the first comprehensive survey of IRWD's employees.

LESSONS LEARNED

The primary lessons learned from the project are:

- The confirmation that clear, unwavering support from the general manager is critical to the success of the program.

- It can be difficult to prioritize the importance of issues facing the District in a noncompetitive environment. Hence, the program had a temporary setback when it tried to address too many problems at once.

- Tracking team participation is important to ensure that all employees are ultimately included in the process.

Contact

Ken Thompson
Irvine Ranch Water District
3512 Michelson Drive
Irvine, CA 92612
714-453-5852
714-476-1187 FAX

RESULTS

The primary result of IRWD's TQM program was the development of a team based decision making culture. For the first time staff members from all levels and departments of the organization have worked together on teams to address key issues facing the District.

Case Study 1.7

Cincinnati Water Works
Cincinnati, Ohio

- The Cincinnati Water Works views itself as a business much like any other business, with the exception that it happens to be owned by the local municipal government.

- As with any successful business, the Cincinnati Water Works developed a strategic business plan as a framework for guiding the organization toward superior capabilities and performance in the areas that are most important to its customers.

BACKGROUND

The Cincinnati Water Works is a municipally owned and operated water utility with approximately 240,000 accounts and 900,000 consumers. As a regional system covering 400 square miles, it serves all or portions of four counties in southwest Ohio. The utility operates two treatment plants with a total capacity of 260 mgd.

In October 1992, the Cincinnati Water Works began operating one of the world's largest granular activated carbon filtration plants for treatment of potable water. This was the culmination of more than a decade of research, design, and construction. Having completed a major milestone in advancing treatment technology, the utility needed to focus its resources on improvements in other parts of the organization. The challenge became one of trying to develop consensus throughout all parts of the operations on which priority projects should be addressed first.

At the same time discussions on the potentials and pitfalls of privatization began to appear throughout the water utility industry. Locally, discussions on privatization of other governmental operations started to take shape. The Cincinnati Water Works decided to employ a technique that has been the cornerstone of successful private operations—the development of a strategic business plan. The expectation was that, by going through a strategic planning process with the entire organization, a consensus

415

on priority initiatives could be reached and at the same time the utility could be prepared for any challenges of privatization.

PARTIES INVOLVED

Public agency

Cincinnati Water Works

Private party

Numerous customers representing various customer categories.

Consultants

Black and Veatch
O'Neal and Associates
Balke Engineers
Institute for Policy Research, University of Cincinnati
Organizational Horizons, Inc.

APPROACH

The strategic planning process began with issuing requests for proposals from consulting firms. With strong and differing opinions throughout management, union, and employees, an independent and neutral staff was needed to help develop a consensus. The selection process resulted in one prime consulting firm with subcontracting firms for customer survey and research, employee surveying, and benchmarking analysis of other utilities and companies.

The foundation for developing the strategic business plan began with formulating guiding principles for the organization. These principles took the form of a vision, mission, and values statement for the utility. Considerable time and effort were expended to make sure that the statements had a strong customer focus and shared understanding and acceptance throughout the organization. To accomplish this, approximately 125 employees representing all levels were directly involved in all-day, off-site sessions. These sessions occurred over a number of months. Once this group had developed final drafts of the statements, open sessions with all employees were conducted for a final review and, where necessary, redrafted.

With the vision, mission, and values statements developed, the same process was used to develop specific goals for the utility followed by objectives and then strategies. As guiding principles, the goals, objectives, and strategies had to support the vision and mission of the organization in order to be considered worthwhile expenditures of resources.

To assist the organization in the strategic planning process, a number of resources were developed to provide information during the process.

- A survey was conducted of the utility's customers to gain an understanding of their priorities, expectations, and perceptions. Many of the survey questions mirrored questions included on the American Water Works Association Research Foundation's national customer survey. In addition, half-day customer focus groups were held to provide a more in depth understanding of particular areas of concern.

- A survey was also conducted among employees within the utility to help ascertain their opinions about the strengths and weaknesses of the organization, perceived priorities of the customers, and desired direction of the utility.

- An analysis of strengths, weaknesses, opportunities, and threats was conducted by an outside consulting firm to determine the effectiveness of the organization's operations.

- A benchmarking analysis of key general indicators was conducted with other similar utilities in the United States. This analysis provided a comparative examination of the Cincinnati Water Works to other water utilities.

Once a consensus was developed throughout the organization on the goals, objectives, and strategies, the process was used one more time to prioritize the business plan's strategies. Action plan teams were then formed to begin work on the top 25 strategies. The teams were made up of employees throughout the organization, both union and nonunion. Training was provided to the teams to help them develop a structure within which to operate. A team sponsor from top management was assigned to assist the team in overcoming any roadblocks encountered. A steering committee made up of the utility's top management was formed to guide the teams as they worked on formulating and implementing solutions for their particular strategy.

LESSONS LEARNED

- The development of a strategic business plan is a time consuming process that cannot be done quickly. Consequently it is a process that should be used to prepare an organization for the challenges of privatization, but not as a last minute reaction to a privatization overture.

- It is essential for all levels of the organization to be involved in the development of the plan in order to gain the necessary shared acceptance and support for the strategies. Organizations that are driven by a clear purpose and shared values have a far greater ability to succeed.

- Active support and patience from top management is vital as the organization works through the consensus-building process. The divergence of backgrounds and opinions will result in some surprising and exciting alternative strategies.

- Top management must be committed to following through with implementation of the plan once developed. If top management is not completely committed, do not start the process. Otherwise the result will be a less motivated and more cynical organization than existed prior to start of the process.

CONTACT

Cincinnati Water Works
David E. Rager, Director
4747 Spring Grove Avenue
Cincinnati, Ohio 45232
513-591-7970
513-591-6519 FAX

RESULTS

The Cincinnati Water Works developed a strategic business plan that will guide the organization over the next five to seven years. The planning process succeeded in developing a consensus throughout the organization, both union and nonunion, on priority directions. Action plan teams are working on solutions to address a broad spectrum of strategies. The planning process also had a synergistic affect by effectively communicating the utility's desired vision for the future. Thus as day-to-day decisions are made throughout the organization, they are made with a better understanding of how all parts fit together to support the organization's overall vision, mission, and values.

Case Study 2.1

Internal Self-Improvement
Fort Wayne, Ind.

- The city council received an unsolicited private service provider proposal for contract operations and management.

- Utilities offered a counter-proposal that showed greater savings than the private service provider had proposed.

- Results have demonstrated that utilities can be competitive with the private sector proposal.

BACKGROUND

The Fort Wayne Division of Utilities provides water and wastewater services to a population of approximately 250,000 in Northeast Indiana. In late 1992, a private contract operator submitted an unsolicited proposal for O&M services to the city council. The City formed a citizens advisory committee to study the merits of privatization. The citizens group hired an independent consultant, who had not previously worked in the City, to further investigate privatization issues. The consultant found areas of inefficiency in the utilities, and recommended that the City consider privatization, particularly the water filtration plant.

At about the same time, the utilities had EMA perform a competitiveness assessment to determine the highest priority areas for improvements. Changing work practices was defined as a necessary component to help the utilities be more competitive. The utilities and EMA developed a plan to integrate any planned technology tools with reorganizing staff and reengineering each department's work practices and procedures.

Based on the assessment findings, the utilities prepared a counter-proposal for the Council that showed that they could accomplish even greater savings than the private contract operator had proposed. The competitive gap, that is the difference between Fort Wayne's cost of operation and that of firms employing the industry's best practices, was 23.5 percent or approximately $3.9 million in 1994 expenditures. Eliminating this gap will give Fort Wayne more flexibility to reduce costs, meet increasing service demands, improve customer satisfaction, and become competitive with private industry service providers.

PARTIES INVOLVED

Public agency

Ft. Wayne Utilities

Consultants

EMA Services, Inc.
Rust Environmental Consulting
Donahue & Associates
Malcolm Pirnie, Inc.

APPROACH

Primary recommendations in the assessment addressed the water treatment plant, Water Maintenance Services (WMS) Department, Water Pollution Control (WPC) Maintenance Department (sewer collection maintenance), and the wastewater treatment plant. Specific recommendations included how to reinvent practices in these areas as well as the need for developing an action plan and measuring progress. Workforce flexibility coupled with a skill-based compensation approach was recommended to reward employees based on their acquired skills and to provide the organization with a more broadly skilled staff.

Strategic use of technology is an integral part of supporting the new work practices. Though fewer staff are available, the utilities can reliably monitor, control, and save chemicals and energy at both the water and wastewater plants with advanced control systems. Using computerized management tools, the in-plant and field maintenance crews can plan and schedule work, track and optimally order supplies and inventory, and analyze and predict equipment repairs and costs. For example, the WPC Maintenance Department has started using geographic information system technology to support a planned maintenance program for monitoring and removing tree roots from sewer lines before they cause problems.

LESSONS LEARNED

- Use a basic plan to accomplish change: determine where you are, determine a vision of where you want to be, and systematically implement plans to close the gap.

- Benchmark activities against the best privately operated companies, not other public utilities.

- Empower employees and provide incentives for sharing in performance.

- Demonstrate early successes by beginning with the most visible, high impact goals.

CONTACTS

Greg Meszaros
Associate Director
Fort Wayne Utilities
City County Building
1 Main Street
Fort Wayne, IN 46802
219-427-1381
219-427-2540 FAX

Joe Springer, P.E.
EMA Services, Inc.
5404 Tunbridge Crossing
Fort Wayne, IN 46815
219-485-7746
219-486-2786 FAX

RESULTS

- Utilities management negotiated a workforce flexibility agreement in the 1996 round of union contract discussions with five utility unions, enabling cross-training and improved flexibility.

- The number of job descriptions at the wastewater treatment plant has been compressed from 17 to five.

- A process reengineering team analyzed work practices of the WMS Department and identified practice improvements. Examples include:

  - Remote meter installation productivity was improved by more than 300 percent.

  - Pressure testing and disinfecting new water mains has been streamlined by eliminating more than a dozen steps in the process.

  - Labor savings of more than $100,000 were achieved by eliminating two middle management positions following staff retirements.

  - Training programs resulted in five employees receiving state certified large distribution system operator licenses, a first for employees in WMS.

Case Study 2.2

Internal Self-Improvement
Colorado Springs, Colo.

- Through an internal optimization program, Colorado Springs Utilities has reduced its annual operating costs of $10 million by more than $2 million.

- The key to success was reengineering work practices and implementing new practices to utilize new technology.

BACKGROUND

Colorado Springs Utilities is a four-service utility providing electric, gas, water, and wastewater services to Colorado Springs, Colo. In 1995 the Colorado Springs Water Resources Department (WRD), Treatment Services Division began an optimization program aimed at improving productivity and being competitive with private companies. With five water treatment plants (combined capacity of 175 mgd) and two wastewater treatment facilities (combined capacity of 60 mgd), WRD believed that if privatization were to attract Colorado Springs, it would most likely be in the treatment area.

PARTIES INVOLVED

Public Agency

Colorado Springs Water Resources Department

Consultant

EMA Services, Inc.

APPROACH

Colorado Springs Utilities management understood that employees could contribute significantly to optimization efforts if they were given more participative, broader roles. At the same time, management was committed to protecting the jobs and financial security of employees. With these two guiding principles, a cultural change process began. A team was formed that represented all employee ranks from water and wastewater operations and management. The treatment optimization advisory team worked closely with consultant staff to learn new methods.

The team conducted a competitiveness assessment that compared current practices to those used by the most productive utilities. Additional training was provided on communication, conflict resolution, and other teamwork skills. Baseline metrics on which to monitor progress were defined. Each treatment facility also initiated a facility team to focus on specific plant issues. The treatment optimization advisory team presented its findings and recommendations to all employees; their findings indicated potential savings of $2.2 million annually.

Recommendations to capture the savings included reengineering work practices and implementing new practices in parallel with new technology. Object modeling was used to demonstrate visually what could be improved with current work practices. Scenario development and analysis were used to document the current situation and to move to a future desired situation. A program-driven maintenance initiative was undertaken to adopt managed maintenance concepts. These concepts include increasing preventive and predictive maintenance activities and reducing reactive maintenance. Maintenance activities are driven by treatment process criticality and agreed upon standards.

Automation strategies, which reduced human intervention, were adopted. The Treatment Services Division staff have implemented many strategies already, including converting chemical feeds from day tanks to bulk storage.

Workforce flexibility and skill based compensation are being implemented to support the new organizational infrastructure. The optimization advisory team developed broader structures for positions and determined skill levels, training, and certification requirements. Now skills and competencies, as well as the market, determine salaries. Employees have the incentive of sharing in savings through the pay for performance concept.

LESSONS LEARNED

Cultural change requires educating employees, obtaining commitment, and embracing new philosophies. By substantively involving employees and facilitating reengineering, new behaviors and routines are adopted. A skill based compensation program supports concepts such as workforce flexibility, which are needed for productivity improvements.

CONTACT

Leah Ash
Treatment Services Manager
825 E. Las Vegas Street
Colorado Springs, CO 80900
719-448-4414
719-448-4495 FAX

RESULTS

In approximately two years, more than $2 million in recurring annual costs have been saved, primarily a result of reducing staff through attrition. Thirty-nine positions were eliminated, and reductions will continue during the next two years.

In addition, more than six staff years of time have become available to support workforce flexibility and program driven maintenance efforts. More than $20,000 in chemical costs was saved in the first two months after making a treatment change.

Case Study 2.3

Internal Self-Improvement
City of Houston, Texas

- The Public Utility Group of the City of Houston's Department of Public Works and Engineering operates a large system of water and wastewater treatment facilities that serves more than 2 million people and employs 2,000 people.

- In order to improve efficiency and cut costs, and still retain staff, the Department is evaluating existing procedures, observing the practices of other successful operations, and analyzing current costs and staff capabilities.

- Preliminary findings indicate that retraining staff, changes in policies and procedures, and centralization/decentralization of certain functions can improve efficiency and lower cost.

BACKGROUND

The Public Utility Group of the City of Houston's Department of Public Works and Engineering is responsible for the water treatment and wastewater facilities serving about 2 million people in the greater Houston area. Facilities include 48 wastewater treatment plants, numerous surface water and groundwater treatment plants and supplies, 5,500 miles of water supply lines, and 6,000 miles of sewers. The Public Utility Group employs about 2,000 people. It is the largest water and wastewater system in the state of Texas.

Prompted by expressions of interest from private companies in owning and operating the Department's facilities, and in an effort to improve efficiency and productivity and to reduce costs, the Department undertook an analysis of its workforces in water and wastewater. Specifically, the Department suspected that it could increase the flexibility and efficiency of the workforce, better utilize equipment, and improve computer applications.

427

PARTIES INVOLVED

Public agency

> City of Houston
> Dept. of Public Works and Engineering
> Public Utility Group

APPROACH

The Department was also motivated by employee pride, and the desire to do the job as well as any private company would. Out of loyalty to staff, the Department was committed to retain existing staff to the extent possible.

The Department has undertaken a comprehensive review of many aspects of operations. First, existing policies and procedures were reviewed and evaluated in order to improve efficiency and eliminate wasteful practices where possible. Some policies and procedures originate from the Department itself, and will be relatively easy to change. On the other hand, the policies that originate externally (from the City of Houston), will be more difficult to change.

Second, the Department evaluated the true costs of various functions and processes to determine where savings could be created.

Third, the Department observed the technologies and methods used by private operators at other facilities to learn how such practices could be used to improve its own facilities. Learning from the private sector, which is the competition, proved useful.

Last, the skills and flexibility of the Department's employees were evaluated, with the goal of maximizing the flexibility of staff. The 2,000 employees on staff are diverse with respect to skills, salary, and responsibility, and changes have to be equally diverse to address all levels of the workforce. The goal to retain employees and retrain them as necessary, rather than replace them, was an important factor.

LESSONS LEARNED

- Existing procedures can be streamlined to improve efficiency.

- Retraining employees to widen their range of skills and flexibility improves efficiency while allowing staff to be retained.

CONTACT

Peter W. Dobrolski, P.E.
Chief Engineer
City of Houston
Dept. of Public Works and Engineering
1801 Main St., 13th Flr.
Houston, TX 77087
713-754-0807
713-754-0524 FAX

RESULTS

As of this writing, the study is partially complete and several findings have emerged. First, it is possible and advantageous to have a flexible workforce, which involves retraining employees to be multiskilled. Second, there are certain noncore functions, e.g., grass cutting, that lend themselves to being centralized in order to improve overall efficiency. On the other hand, the study has shown that certain functions should be decentralized, so that workers at facilities can be empowered to work as an independent team and control their work. Functions that benefit from decentralization include purchasing and procurement, and personnel hiring and firing.

Case Study 2.4

Reengineering the Water Department
Tampa, Fla.

- In order to increase efficiency, reduce costs, and become competitive with the private sector, the City of Tampa conducted a competitive assessment of the Production Division.

- The assessment recommendations include increasing automation, replacing maintenance-intensive equipment, reducing and retraining the workforce, and outsourcing noncore functions.

BACKGROUND

The Tampa Water Department is responsible for delivering water to a service population of more than 430,000, and manages two water treatment plants (80 mgd surface water and 24 mgd groundwater). The Production Division has 111 funded positions that include operations, maintenance, and laboratory functions.

In order to avoid a takeover of the water system, cut costs, and capitalize on the efficiencies of the private sector, the Tampa Water Department conducted a competitive assessment of its operations in March 1996. The goal of the assessment was to establish ways to reengineer the system in order to increase efficiency, lower the cost of operating the system, and improve services.

PARTIES INVOLVED

Public agency

Tampa Water Department

Consultants

CH2M Hill (performed competitive assessment)

APPROACH

The assessment team included a private company water official, a contract operations manager, a management services specialist, and an automation specialist. Six major recommendations emerged:

- Invest to implement extensive automation at all facilities.

- Replace maintenance-intensive equipment.

- Develop a system with skill based pay, that supports workforce flexibility.

- Use remote control at one water treatment plant.

- Reduce staff by 27 percent (30 positions) and reduce the number of classifications by creating cross-functional staff capable of performing maintenance and operations tasks.

- Outsource noncore functions such as janitorial, grounds maintenance, and major corrective maintenance.

The anticipated savings are:

- Reduced chemical costs (10 percent) $400,000/year

- Reduced corrective maintenance costs $100,000/year

- Staff reduction (30 positions) $1,200,000/year

The implementation of the reengineering program is expected to take place over two years, and be completed at the end of 1999. This completion date coincides with the scheduled completion of the automation and equipment replacement construction projects and the two-year technical training program to develop a cross-functional staff.

The financial break even point of the 1997 program investment is 2001. The first steps have been to create a new organization table and job classifications, and to initiate employee training programs.

LESSONS LEARNED

- A collaborative effort among the Water Department management, personnel department, employee relations, union(s), and employees is critical.

- Employee involvement is a must.

- A two-year time period for implementation is reasonable.

CONTACT

Mike Bennett
Production Manager
Tampa Water Department
7125 North 30th Street
Tampa, FL 33610
813-231-5254
813-231-5283 FAX

RESULTS

The reengineering program is underway and is expected to be completed by the end of 1999. The current organization chart reduces the number of job classifications from 44 to 24. Of the 30 positions to be eliminated in 1999, nine vacancies have occurred to date. A hiring freeze has been implemented, and as needed personnel assistance is provided by temporary staffing agencies.

Technical training curriculums have been developed through a local technical training institute. The personnel department is creating a new pay structure for the new flexible workforce positions.

Case Study 2.5

Continuous Improvement at BHC Company
Bridgeport, Conn.

- From 1857 to 1986 BHC Company (BHC) had invested some $105 million in facilities. From 1986 to the present those investments have grown to $210 million, a 100 percent increase. Despite doubling in size, the increase in annual rates was held to the rate of inflation during the same time period.

- The BHC program focused on five integral component parts: a long-term strategic plan, the development of an integrated operational management information system, a comprehensive customer satisfaction program, implementation of new technology accompanied by appropriate training and development, and a quality driven employee involvement process. As a result, BHC won the 1997 Baldridge based Connecticut Small Organization Quality Award.

BACKGROUND

BHC, the principal subsidiary of Aquarion Company, collects, treats, and distributes water to approximately 130,000 residential, commercial, and industrial customers in 24 cities and towns in Connecticut's Fairfield, New Haven, and Litchfield counties. In 1986, with the Amendments to the Safe Drinking Water Act and the Surface Water Treatment Rule, it became apparent to company leadership that filtration was going to be required for its surface water supplies, and that this would represent an additional $105 million investment in four additional water filtration and treatment facilities.

Senior leaders recognized the need to transform the company from the traditional monopoly to focus on customer service and continuous improvement in order to diminish the pressure for rising rates.

433

PARTIES INVOLVED

Public Agency

BHC Company

APPROACH

Through restructuring, management was reduced by 40 percent and operational layers went from seven to four without any lay-offs. In 1989 BHC launched "Excellence" as its quality improvement and problem solving process and "Excellence through Participation" (ETP), a team based approach to improving customer service and reducing cost. The company also developed several information technology resource options that are used to capture meaningful data and to assist in running the business as an efficient enterprise. These new systems include an automated mapping and facility management system, a supervisory control and data acquisition system, and a company information system that is enhanced with a facility management and maintenance system. These systems individually help departments manage their work process efficiently and share information throughout the company. A comprehensive customer satisfaction program was designed and implemented based upon customer feedback and experiences. Product and service features are introduced as a result of customer participation and feedback.

A long-term BHC strategic plan was developed and implemented. BHC revised its mission to be the service provider, employer, and investment of choice through a relentless commitment to excellence. The company also adopted core values: respect our customers, each other, and the environment; be responsive to our customers, each other, and our communities; take responsibility; and, reach for excellence. To achieve the mission the company identified the factors considered most important to its success—critical success factors. Departments also created business plans that include goals, strategies, action plans, and measures/metrics.

LESSONS LEARNED

- Keep the mission, values, goals, critical success factors, and strategies simple and repeat them often.

- Identify SMART goals (Specific, Measurable, Agreeable, Realistic, and Time-bounded) and tie all goals to critical success factors. Identify measures for product and service quality, productivity and asset management, and employee growth, development, and satisfaction and promote them throughout the company.

- Involve employees in a variety of activities ranging from improvement teams and committees to becoming in-house training facilitators.

- Determine how to best measure performance and do it. Align mission, critical success factors, and measures at all levels of the company. And most importantly, communicate, communicate, communicate.

- Build in a feedback loop for all aspects of the organization (meetings, programs, and processes) and act on the input. This demonstrates a true commitment to continuous improvement.

CONTACT

Jennifer Paul
Manager of Total Quality
BHC Company
600 Lindley Street
Bridgeport, CT 06606
203-337-5933
203-337-5938 FAX

RESULTS

- The ETP process has improved the working environment and is credited with $1,750,000 cost savings in the last five years. Customer service complaints have been reduced from above 20 to below 15 per 1,000 customers per year.

• Based on investor owned utilities with revenues of more than $30 million, BHC's O&M/Revenue Ratio has become 37.1 percent, the best of its class.

• The overall employee satisfaction rating has risen, 78 percent of BHC employees rate themselves as extremely/very satisfied.

• The number of customers per employee has risen from 400 in 1986 to 490 in 1996.

• BHC is the proud recipient of the 1997 Baldridge-based Connecticut Small Organization Quality Award.

Case Study 3.1

Contract Operation and Maintenance of Water and Wastewater Treatment Plants
City of Evansville, Ind.

- The City successfully contracted operations for its wastewater treatment plant and sewer system.

- The City pursued a similar contract for its Water Utility, including capital improvements.

- This was one of the first long-term water treatment plant operations contracts in the United States under the new IRS Revenue Procedure 97-13.

BACKGROUND

In the past, the City of Evansville has entered into public-private partnerships for many City services. From 1992 to 1995 some of these partnerships included the City's wastewater treatment system and sewer collection system.

After several years of success with short-term contract operations of its wastewater treatment plant, Evansville became one of the first cities in the nation to take advantage of new tax laws by entering into a 10-year agreement to manage its water system. The City discovered that public-private partnerships are beneficial by enhancing government services, reducing costs, and minimizing or stabilizing utility rates. The approach described here explains how, with the right timing, contract, partner, and support, public-private partnerships can provide guaranteed savings and service to a municipality.

PARTIES INVOLVED

Public Agency

City of Evansville, Ind.
Evansville Utility Board
Evansville Water and Sewer Utility

Contract Operator

> Environmental Management Corporation
> EA2/Systems

APPROACH

In the early 1990s, the City of Evansville began searching for alternative ways to control its costs and improve its services. At that time public-private partnerships were becoming a new trend in government. In 1991 Evansville's Utility Board subcommittee took a year-long look at the possibility of forming a public-private partnership contract to manage the City's wastewater treatment system as a means of increasing efficiency while decreasing costs.

After reviewing and comparing candidates for contract operations, in 1992 the City entered into a five-year, $2.3 million per year contract to operate the wastewater treatment plants. Three years later the City created a new contact with the existing operator, combining the wastewater treatment plant and sewer collection system management. Under the new contract, the City engaged the existing operator for a guaranteed cost of $2.6 million per year for operation of the wastewater plant, and $1.5 per year for operation and maintenance of the sewers and collection system. The City's combined wastewater and sewer contract expires in August 2000.

In 1996 the Evansville Water and Sewer Utility needed to fund a $13 million bond issue for future water capital projects. The City solicited statements of qualifications from several firms to form a public-private partnership to operate the water utility. In February 1997 the Evansville Utility Board approved a 10-year management contract with EA2/Systems, establishing a partnership between the private company and the Evansville Water and Sewer Utility. This contract was one of the first of its kind to take advantage of the new IRS regulation changes that allow utilities to enter into long-term private contracts to gain greater savings.

LESSONS LEARNED

- Implementation of successful public-private partnerships requires three critical elements: timing, contract partner, and support. The Evansville contract is truly a partnership effort, not the complete privatization of a government service.

- Public-private partnerships require close ties between the City and the contract operator so that there are no surprises.

- Public-private partnerships can offer municipalities the use of a wide range of management talent and technical experts, which in turn enable the public sector to operate more efficiently.

Contact

Jack Danks
Evansville Water and Sewer Utility
P.O. Box 19
Evansville, Indiana 47740
812-426-5771
812-426-5721 FAX

RESULTS

The City has saved more than $500,000 annually in wastewater treatment plant management, while increasing cost savings by $200,000 annually in sewer management. The City expects to implement $13 million in future capital projects, with a rate increase of 11.6 percent, almost 6 percent below what was projected prior to the partnership.

Case Study 3.2

Contract Operations of Wastewater Treatment Plants
City of Indianapolis, Ind.

- The City contracted the operation and maintenance of two wastewater treatment plants.

- Savings of 44 percent are projected over the five-year contract.

- City employees competed for work and bid 10 percent lower than existing costs, but did not win.

BACKGROUND

The City of Indianapolis operates two 125 mgd advanced wastewater treatment plants that serve the City's one million residents. As part of a movement to reduce costs, increase efficiency, and "reinvent" government, the City decided to contract operation and maintenance of the two plants to a private operator for a five-year period starting in 1994.

In 1993, the City conducted a study to determine whether there would be savings seen through privatization. The study concluded that the plant was well operated, and private operation could generate perhaps a 5 percent savings.

PARTIES INVOLVED

Public Agency

Department of Capital Assessment Management
City of Indianapolis Department of Public Works

Private Party

White River Environmental Partnership

Consultants

Ernst & Young
Camp, Dresser & McKee

APPROACH

The City then prepared an RFP for a five-year operations and maintenance contract for both plants. The RFP contained performance standards for the operation and maintenance of the two plants, but left some flexibility as to materials and procedures used. The RFP also provided flexibility regarding employee issues, citing equal or better benefits, wages, pension vesting, and recognition of the existing union for bargaining purposes. Five firms responded to the RFP.

LESSONS LEARNED

- Privatization can reduce costs and improve efficiency and operations.

- Equipment reliability and availability improved as total maintenance costs decreased.

- Employee training and safety programs are more effective, as evidenced by a reduction in the number of accidents.

- Privatization has led to additional employee compensation and career enhancement.

- Under current management there is wiser use of capital investment.

CONTACT

Skip Stitt
Deputy Mayor
City of Indianapolis
Mayor's Office
200 E. Washington St.
Suite 2501
Indianapolis, IN 46204
317-327-3601
317-327-3980 FAX
www.ci.indianapolis.in.us (web)

RESULTS

In order to compare actual costs to projected costs submitted by bidders, the City established that in 1993 it spent $30.1 million to operate the two plants. This figure includes operating costs but does not include capital expenditures and some city and departmental overhead costs. The shortlisted proposals were as follows for the operating cost component: White River Environmental Partnership at about $18.1 million, PSG at about $ 18.1 million, and City management employees at about $27.7 million.

The lowest bidder and winner of the contract was the White River Environmental Partnership, at 44 percent below existing costs for five years of operation. The City management team bid was approximately 10 percent below existing costs for five years of operation.

Of the 325 employees originally employed by the City, 206 were offered positions by the White River Environmental Partnership, and the others have retired, moved, or were shifted to other City departments. Since the privatization wages and benefits have increased, employee training has expanded, and the number of job-related accidents has decreased. The improvement in the accident rate is attributed to increased training and private sector opportunities.

The City reports that violations of water quality standards have decreased by 50 percent under private management.

The current five-year contract will expire in 1999, at which time the City plans to enter into a longer-term contract in conformance with IRS Rule 97-13.

Case Study 3.3

Contract Operations of Wastewater Treatment Plant
City of New Orleans, La.

- The City contracted the operation and maintenance of its wastewater treatment plants, the largest wastewater operations contract at the time.

- With a savings of more than $1 million annually, the City saves nearly 15 percent of its budgeted costs.

BACKGROUND

The City of New Orleans operated two wastewater plants with a combined average capacity of 132 mgd, and two water treatment plants with a combined capacity of 198 mgd, serving more than one million urban residents. In 1990, after investigating privatization alternatives, a decision was made to privatize only the wastewater plant. As part of a movement to reduce costs and increase efficiency, the City decided to contract operation and maintenance of the plant to a private operator for a five-year period, starting in 1992. With an average flow of approximately 122 mgd and peak flow capacity of 239 mgd, the East Bank wastewater plant was the largest facility of its kind to be contract operated.

PARTIES INVOLVED

Public Agency

City of New Orleans
Department of Public Works
Public Utility Board

Contract Operator

Professional Services Group (PSG)

APPROACH

The City began by reviewing its operations of the wastewater plant. A major area of concern for the City was that hiring restrictions and high employee turnover were resulting in noncompetitive salaries and poor performance of the plant staff. Operational problems were correlated with the staffing difficulties.

The utility board and plant staff embarked on an inspection tour of 13 different wastewater facilities across the country in order to identify and understand the contracting process. Following this learning experience, an RFP for a five-year operations and maintenance contract was prepared. Of the 14 firms initially interested in the project, 11 responded to the RFP. A shortlist of three firms was selected for best and final offers. Because this procurement was a service contract and not a traditional construction contract, price was not the only consideration for awarding the contract. A single firm, Professional Services Group, was selected and offered the contract to operate the plant.

Since this contracting project was the largest of its kind at the time, it was critical that the utility have assurances that concerns about facility maintenance and upkeep would be at a level consistent with long-term operation of the plant. Therefore a two-part maintenance plan was developed in the RFP, in which the contract operator identified routine and major maintenance separately. Major maintenance items and capital improvements required the board's approval, while routine maintenance did not. This allowed some flexibility to the contract operator, while providing a check and balance review for the utility.

The utility board has maintained constant involvement with the project. Most importantly, a public utility cannot "divorce itself the project once the privatization is complete." The board requires daily monitoring reports, and holds monthly meetings to communicate progress and concerns on a regular basis. The services have been contracted for, but since the utility retained ownership it maintains legal responsibility for performance of the plant within the regulations.

LESSONS LEARNED

- Contracting O&M services can establish competitive employee salaries and improve morale through training, safety programs, and clearly defined career paths and goals.

- Operational costs were reduced while efficiency improved.

- Under current management, ongoing monitoring must be maintained through regular reports and meetings between the utility and the operator.

CONTACT

Harold G. Gorman, P.E.
Executive Director
Sewerage and Water Board
of New Orleans
625 St. Joseph Street
New Orleans, LA 70165
504-585-2190
504-585-2540 FAX

RESULTS

The City currently spends approximately $1 million less per year than it did under the previously budgeted $7 million annual operation budget. Responsibility for capital improvements and long-range planning have been retained by the City.

While PSG had the lowest cost proposal, and was the selected contractor, saving the City nearly 15 percent on the operation and maintenance of the plant, the proposals were not evaluated solely on the basis of cost. The current five-year contract expired in 1996. PSG has been granted a one year extension to the contract with second extension anticipated shortly.

Case Study 3.4

Contract Operations of Wastewater Treatment Plant
City of New Haven, Conn.

- The New Haven Water Pollution Control Authority procured an operation and maintenance contract for its wastewater treatment plant, collection system, and pump stations.

- This was one of the first procurements for a long-term operations wastewater plant contract in the United States under the new IRS Revenue Procedure 97-13.

BACKGROUND

The New Haven Water Pollution Control Authority was established in 1979, with jurisdiction over the operation, maintenance, capital assets, and administration of the City's sanitary sewer facilities. It currently operates 15 pump stations, one 40 mgd secondary treatment plant, and other collection facilities that include 250 miles of pipeline. In 1993, the Authority expanded its service to perform capital planning, accounting, and budgeting functions that were previously handled by the City's Department of Engineering.

The New Haven Water Pollution Control Authority (NHW-PCA) sought to improve efficiency through procurement of a contract operator for the City's 40 mgd wastewater treatment plant, pump stations, and the collection system.

PARTIES INVOLVED

Public Agency

The City of New Haven/New Haven Water Pollution
Control Authority

Contract Operator

Currently under evaluation

446

Consultants

> Public Financial Management
> Malcolm Pirnie, Inc.
> Verner, Liipfert, Bernfard, McPherson, and Hand
> Susman, Duffy and Segaloff, P.C.

APPROACH

The transition to contract operations required an integrated technical, legal, and financial team of advisers to assist with the following steps:

- Preparing RFQ/RFP solicitation documents

- Reviewing capital projects

- Analysis of IRS Rev. Proc. 97-13

- Evaluating selection proposals

- Contract negotiations

As one of the nation's first procurement processes for long-term contract operation under the new IRS Revenue Procedure 97-13, this project included numerous technical, financial, economical, and performance-related challenges that were addressed as part of the procurement process.

Prior to the promulgation of IRS Revenue Procedure 97-13 in January 1977, most operation and maintenance contracts were for a maximum of five years. Because the procedure was promulgated during the early stages of the procurement process, the WPCA sought proposals for five, 10, and 15 years. It was envisioned that this approach would provide the necessary cost information to make an informed decision between short- and long-term contracts.

Capital projects included as part of the procurement process consisted of all repair and replacement activities and upgrading of all pump stations. Proposals were received in July 1997 and are confidential until final selection is made and the final contract is negotiated. It is anticipated that the contract will be executed by October 1997.

The procurement documents required the contract operator to provide employment for all existing employees for at least one year. Thereafter the workforce could only be reduced via attrition, buyouts, and transfers.

LESSONS LEARNED

- There is a noticeable difference in contracted costs between a five-year and 10- and 15-year contracts. Longer-term contracts exhibited reduced costs.

- Creative and innovative operating approaches from the private sector appear to have contributed to significant cost savings.

CONTACT

Raymond C. Smedberg, P.E.
General Manager
Water Pollution Control Authority of the City of New Haven
345 East Shore Parkway
New Haven, CT 06512
203-466-5280
203-466-5286 FAX

RESULTS

The City expects to save at least 20 percent when compared to the existing operation budget.

Results are anticipated to be available by the end of 1997.

Case Study 3.5

Benchmarking, Contract Negotiation, and Contract
Operations of Wastewater Treatment Plant
City of Schenectady, N.Y.

- The City of Schenectady contracted the operations and maintenance of its Wastewater Treatment Plant.

- The City prepared detailed five-year plans for operations and capital expenditures, and performs long-term capital planning.

BACKGROUND

The City of Schenectady provides wastewater collection and treatment for its 65,000 residents, contributing industries, and residential customers in surrounding communities. The 1996 average wastewater treatment flow was nearly 15 mgd.

In 1996, as it neared the end of its first five-year contract with Professional Services Group (PSG) for contract operations of the wastewater treatment plant and appurtenant facilities, the City of Schenectady was concerned about three issues: (1) how the City could reform the contractor relationship so that PSG would share in the risks (additional costs) of operating the plant, (2) how it could be assured that PSG was offering a good price on the upcoming five-year contract renewal, and (3) how it could encourage PSG to operate the plant in an economically efficient manner, while still ensuring that equipment and property were being properly maintained.

PARTIES INVOLVED

Public Agency

The City of Schenectady, N.Y.

Contract Operator

Professional Services Group

449

Consultants

Public Financial Management
Malcolm Pirnie, Inc.

APPROACH

Two primary tools enabled the City to establish a framework for achieving assurances that it was receiving a fair deal from the existing contract operator.

1. Benchmarking. The City hired a consultant to undertake a comprehensive facility audit of the wastewater facilities to compare the probable cost of operation of the facility by PSG to the cost of operation by the City, and to study how the current cost of operation compared to other facilities of similar size, type, and effluent permit requirements.

2. Contract Revision. In consultation with the City, a new form of contract, designed to encourage efficiency and cost reduction in operations was developed for negotiations with PSG. The new contract contained an annual operating fee provision that comprised a fixed fee, a variable fee, and a maintenance and repair fee. The annual fee is based on the probable cost of operation previously established.

The earlier contract required the City to pay for repairs and to make capital investments without any opportunity to benefit from the expenditure. As an example, if a piece of equipment failed, the City would pay to replace it. If the equipment was more efficient, the contractor would receive a reduced electrical cost benefit.

In the negotiated contract, either party may undertake capital improvements that will improve the operations at the plant. The party that implements the improvement is entitled to recover its costs and the residual benefit is shared. In addition, a fixed budget was established for maintenance and repair, in recognizing that proper maintenance is required for continued successful operation. This budget is subject to a separate accounting to discourage deferments intended to increase profits at the expense of the equipment.

The negotiated contract also recognizes that many of the costs to operate the treatment plant are directly or indirectly influenced by the quantity and strength of wastewater received at the plant. The variable fee provisions of the contract establish unit costs for labor, electricity, fuel oil, natural gas, chemicals, residual disposal, and materials for the biosolids composting operation. In addition, the contract establishes a variable cost budget for influent wastewater flow, total suspended solids, and biochemical oxygen demand within a fixed operating range and provides for adjustment when the influent characteristics vary outside the operating range.

LESSONS LEARNED

- Contract operations costs were reduced while the plant continued to provide high quality treatment.

CONTACT

Milton G. Mitchell, P.E.
Commissioner of Public Works
City of Schenectady
105 Jay Street
Schenectady, NY 12305
518-382-5082

RESULTS

After several months of contract negotiation, the City and PSG executed a contract that will save the City an estimated $1 million when compared to the previous PSG contract. The new contract is valued at approximately $15 million, over five years. Thus through benchmarking and informed contract negotiations the City achieved its goals.

Case Study 4.1

Managed Competition for Operations and Maintenance
City of Houston, Texas

- Managed competition was conducted by the City of Houston, managing partner on behalf of the 13 owners of the Southeast Water Purification Plant for a five-year operations and maintenance contract.

- Detailed performance specifications gave clear direction to bidders, made bid evaluation relatively simple, and allowed for creativity in approaches to operations and maintenance.

- The City of Houston formed a "company" composed of employees of the City to submit a proposal in competition with private operators.

BACKGROUND

The Southeast Water Purification Plant (80 mgd) serves 700,000 people in the Houston area. The facility is owned by 13 partners, made up of cities and utility districts or water authorities, each with varying percentages of production capacity ownership. The City of Houston is the managing partner of the facility and owns 23.02 percent. Under a five-year contract (1991 to 1996), PSG (owned by Compagnie Générale Des Eaux of France) operated and maintained the plant.

When it came time to rebid the work, the City of Houston decided to seek contract proposals for a private firm to operate and maintain the plant. Further, it decided to conduct the process as a managed competition with sealed bids, as opposed to negotiations, in order to simplify the evaluation of bids and get the best contractor at the lowest cost. The bid documents included a detailed set of performance specifications that allowed bidders to clearly understand the project's level of service and performance standard requirements, allowed equal comparison of bidders, and left room for innovative methods and approaches. A consultant was hired to work with auditors, accountants, and engineers to assist with qualifying the proposers and evaluating proposals.

PARTIES INVOLVED

Public Agency

City of Houston
Department of Public Works and Engineering

Private Party

JMM-OSI

Consultants

HNTB
Public Works and Engineering

APPROACH

Five private firms and the City of Houston employees were qualified. All bids were lower than the projected cost of continuing the original contract with PSG. The City of Houston employees came in fourth lowest. Every qualified bid was under the projected cost of continuing with the original contractor.

The contract was awarded to the lowest bidder, JMM-OSI, owned by Suez-Lyonnaise des Eaux, of France. JMM-OSI's cost is a substantial 25 percent lower than the next bidder, and their projected savings is due to innovative changes in the types of chemicals used and to process changes.

LESSONS LEARNED

- Sealed bids with detailed performance specifications eased the evaluation process and allowed innovative and cost saving methods.

- Managed competition forced bidders to be cost-competitive.

CONTACT

Peter W. Dobrolski, P.E.
Chief Engineer
City of Houston
Dept. of Public Works and Engineering
1801 Main St., 13th Flr.
Houston, TX 77087
713-754-0807
713-754-0525 FAX

RESULTS

- $12.7 million will be saved over the duration of the five-year contract, which represents a 43 percent savings over the previous contract.

- Savings to each city, utility district, and water authority will allow them to keep utility rates in check.

Case Study 4.2

Managed Competition of a Water Treatment Facility
San Diego, Calif.

- The City of San Diego embarked on a city-wide managed competition process to demonstrate how the public provision of services compares to the provision of the these services by private companies.

- The City issued an RFQ and RFP for the operation of a 40 mgd water treatment facility.

BACKGROUND

The City of San Diego operates a municipal water system that supplies domestic water to about 1.2 million people. The city established a city-wide Managed Competition Program within the Financial and Technical Services Business Center to identify the optimal provider of city services. The impetus for the creation of the program was to respond to the general perception on the part of tax payers and rate payers that the government was ineffective in providing services. Additionally, the City wanted to proactively address impending pressure from private companies that wanted to assume the operation of large public systems.

The program assists city departments in conducting self-assessments and self-optimization of specific services, in addition to delineating the costs and level of performance of public and private provision of these services. Two committees have been established that determine whether services identified by the program should be put out for an RFP. The first committee is composed of employees and union representatives, while the second committee is composed of citizens, members of the business community, and union representatives. Both committees forward their recommendations to the city manager, who then takes these into consideration in determining whether to issue an RFP. If an RFP is to be issued, a portion of the city's staff works on the procurement side while another portion handles the responses to RFPs. A well-established boundary separates the two.

Parties Involved

San Diego Water Department
San Diego Competition Program
Citizen Advisory Board
Labor Organizations

Consultants

Malcolm Pirnie, Inc.
Hawkins, Delafield & Wood
HDR Engineering, Inc.

APPROACH

The San Diego Water Department's 40 mgd water treatment facility was identified as a candidate for managed competition. The RFQ was issued first, for which the employee team was prequalified. Following the RFQ, the Managed Competition Program issued an RFP. The issuance of the RFP included a six week comment period to allow for questions and comments from responders to the RFP, and to clearly outline how public and private responders would be treated during the process. Four companies and the employee team were shortlisted and the comments received during the six week period, as well as any questions and responses were shared with the shortlisted teams. Proposals were due August 22 and the notice to proceed is expected by December, 1997.

The self-assessment continues. Recommendations about whether or not to pursue an RFP for water distribution are expected by March of 1998. The City is also looking extensively at water operations such as preventative maintenance and SCADA systems, as well as wastewater treatment.

LESSONS LEARNED

- The Managed Competition Program had difficulty estimating and comparing costs within the water utility industry.

- Especially in managed competition, where the private sector may be hesitant to participate, a fair, open, and objective procurement process is critical.

- Developing both a service agreement, if a private proposer is selected, and a memorandum of understanding, if the current city workforce is selected, is necessary for accountability and to ensure the City's desired results for the performance of the contract.

CONTACT

Lisa Irvine
San Diego Department of Finance and Administration
619-236-6892
619-236-5584 FAX

RESULTS

An RFP was issued for the operation of a 40 mgd water treatment facility. Five respondents have been shortlisted: one employee team and four private companies, one of which purchases water wholesale from the Department.

Case Study 4.3

Managed Competition
Martin County Utilities, Fla.

- Martin County Utilities successfully competed against five shortlisted private service providers for a five-year contract for operation of water and wastewater facilities.

- The key to success was embracing new organizational concepts and the appropriate use of technology.

BACKGROUND

Martin County Utilities provides water and wastewater services to a population of approximately 45,000 near Stuart, Fla. In late 1995, a private firm began courting the Board of County Commissioners, which then issued a Request for Qualifications and subsequent Request for Proposals for contract operations and maintenance. The qualifications process shortlisted three respondents in addition to Martin County Utilities staff. The RFP solicited sealed technical and price proposals from the shortlisted firms for a five-year contract. Martin County Utilities retained a consultant to help them prepare their competitive proposal, which was the successful proposal, offering savings of more than $10 million in five years.

The utility operates four advanced secondary wastewater plants ranging in size from 1.0 to 1.2 mgd. It also operates five water treatment plants ranging from 1.0 to 3.5 mgd. These nine treatment plants are located at five separate sites and were being operated with staff at each site. Therefore, redefining staffing and organizational structure was one of the high leverage components of the new way of operating.

PARTIES INVOLVED

Public Agency

Martin County Utilities

Consultant

EMA Services, Inc.

APPROACH

One of the first major initiatives put in motion was the concept called Total Productive Operations (TPO). First, plant and field maintenance functions for both water and wastewater were centralized. The maintenance division reorganization included designating team leaders for operations and maintenance, broadening job classifications, and creating a new job classifications. In taking another step toward performing more preventive maintenance, a computerized maintenance management system (CMMS) is being implemented.

Closely related to implementing the other cost-saving strategies is a utility-wide Supervisory Control and Data Acquisition (SCADA) system. Selective use of appropriate technology supports the new work practices and opens the possibility of partially unattended operation at some of the plants. The SCADA system's network provides for remote operation of any of the treatment plants from any other plant or via modem and laptop computer from any location.

Early in the process management recognized the critical role of employees, not only in the managed competition but in remaining successful in the long term. Management communicated frequently with union representatives and employees during the managed competition process. Today communication channels are kept open using an employee newsletter and routinely scheduled meetings.

LESSONS LEARNED

- Public utilities can successfully compete with the private sector if they embrace new organizational concepts, implement new practices throughout the organization, and employ technology appropriately.

- Support of the employees must be obtained; improve communication channels with employees and clarify roles and expectations.

CONTACT

Mark Wehmeyer, P.E.
Vice President
EMA Services, Inc.
2180 West S.R. 434
Suite 6100
Orlando, FL 32779
407-865-6601
407-865-6615 FAX

RESULTS

- Half way through the first year of operation, approximately $550,000 in savings had been achieved. When the SCADA system is operational, additional labor and overtime savings will result from making final staff adjustments. Many other expenses, including energy and chemicals, will be reduced.

- The relationship between management and the union is more cooperative with focus on joint goals for the organization. Employees are assuming more responsibility and accountability in the reorganized utilities.

- Operations and maintenance employees are broadening their skills through training and are becoming a more flexible, productive workforce.

Case Study 4.4

Managed Competition
Charlotte, N.C.

- The Charlotte-Mecklenburg Utility Department conducted a managed competition to select a private company to operate and maintain two water treatment plants.

- Utility employees formed a private company that successfully competed with other bidders for the contract.

BACKGROUND

The Charlotte-Mecklenburg Utility Department (CMUD) is a municipal agency that manages three water treatment plants (with a total capacity of 138 mgd) and five wastewater treatment plants (with a total capacity of 92 mgd). The regional system serves 550,000 people, and is the largest public utility in the Carolinas. The City of Charlotte has had a long-term policy to encourage competition in order to provide the best services at the lowest possible cost. Further, the policy states that services that are better performed by private companies should be outsourced, and other functions should be filled through competition.

In 1995 the CMUD received an inquiry from a private operator regarding the purchase of one of the CMUD's wastewater treatment plants. The CMUD concluded that selling the plant would not be in its best interest, and that it seemed like an ideal time to open a managed competition to award operations and maintenance contracts. The operations and maintenance of one wastewater treatment plant and one water treatment plant were selected for managed competition.

PARTIES INVOLVED

Public Agency

Charlotte-Mecklenburg Utility Department

461

Consultants

Camp Dresser & McKee
HDR Engineering
Raftelis Environmental Consulting Group

APPROACH

The managed competition process began with a Request for Qualifications, with 10 firms responding. Respondents included CMUD's own bid team, formed by public employees for the purpose of competing with private firms for the contract. With the same agency both conducting procurement and bidding for the contract, it was important to separate the members of each team to avoid conflicts of interest. The procurement and bidding teams were separated administratively, and separate staff and consultants were used for each team.

Nine firms qualified to receive a Request for Proposals, based on personnel qualifications, financial strength, performance history, regulatory experience, and firm qualifications, among other factors. A draft Request for Proposals was then issued to all qualified firms for review. The CMUD bid team reviewed the document carefully, so that technical errors, omissions, or misleading statements would be corrected in the final Request for Proposals. The CMUD bid team review was conducted because of concern that the integrity of the process would be jeopardized if there was a perception of advantage for their team.

A final Request for Proposals was issued, and all prequalified firms submitted proposals. After an evaluation of the proposals and approval processes were completed, the contract was awarded to the CMUD bid team, who submitted the lowest bid.

The CMUD bid team approached the process as a private firm, and had to learn to view the bid process and the project as a private firm. As such, they toured other facilities, used an engineering consultant to test their approach and ideas, and worked to change their perspective from utility employees to that of a private company.

LESSONS LEARNED

- Managed competition results in significant cost savings.

- Employees can successfully compete for contracts with the private sector, and must approach the project with a private mentality and with a goal of providing efficient service.

Since the managed competition is partly a political process, attention must be given to the process selected for the competition.

CONTACT

Charlotte-Mecklenburg
Utility Department
Douglas O. Bean, Director
5100 Brookshire Blvd.
Charlotte, NC 28216
704-391-5073
704-393-2219 FAX

RESULTS

The contract was awarded to the CMUD bid team, the lowest bidder. Bids ranged from $7.5 million (net present value of annual fees for both plants, five years) to $15 million. The bid submitted by the CMUD bid team was $1.6 million less than that of the next highest bidder, and $3.6 million less than the projected CMUD budget. Further, internal organizational changes resulting from the managed competition improved the utility's efficiency and organization, and ultimately the operation of the facilities.

Case Study 4.5

Managed Competition
Jefferson Parish, La.

- Two private operators and a public water department competed for a five-year operations and maintenance contract for a large water system.

- The public water department submitted the lowest bids for a variety of options.

- Significant cost savings and staff reductions have been realized.

BACKGROUND

Jefferson Parish is part of the New Orleans metropolitan area, and has a population of 450,000. The Water Department system includes six water treatment plants at two facilities constructed between 1930 and 1970, and 1,600 miles of water main. The East facility has a design maximum of 87 mgd, and the West facility has a design maximum of 44 mgd. Between 100 and 200 water main repairs are performed each year.

In 1992, the water rate in Jefferson Parish rose from $0.73 to $0.92 per 1,000 gallons. At the time, the new rate was the lowest in the country for water systems serving populations of 100,000 or more. Despite the relatively low rate, the Jefferson Parish Council decided to pursue contract operations for the entire water system, based in part on claims from contract operations firms that they could save costs and avoid future rate increases. The council had previously contracted out operation of a new 50 mgd wastewater treatment plant in 1988. The council anticipated savings from contract operations of $2 to $4 million.

The operations that the council wanted to contract out included administrative management and operation of the water treatment plants, distribution system maintenance, a water/wastewater laboratory, utility billing (later removed due to the lack of experience of proposers), meter reading, and the water warehouse. The last four components were historically partially subsidized by other departments, mainly the wastewater department.

PARTIES INVOLVED

Public Agency

Jefferson Parish Water Department

Consultants

Howard, Needles, Tammen and Bergendoff (technical)
Verner, Liipfert, Bernhard, McPherson and Hand (legal)

APPROACH

In 1993, technical and legal consultants prepared an RFQ, and six firms and the Water Department submitted qualifications. Two private firms (PSG and Montgomery Watson) and the Water Department were shortlisted.

The RFP described the scope of services and required responding firms to describe how the cost savings would be achieved. The term of the contract was five years, with an optional five-year extension.

Various options were included in the RFP, as follows:

1. Operation and maintenance of the East and West Water Treatment Plants, and laboratory with staff reductions through attrition.

2. Operation and maintenance of the East and West Water Treatment Plants, laboratory, with staff reductions through layoffs.

3. Operation and maintenance of the East and West Water Treatment Plants and laboratory distribution system maintenance, and water warehouse with staff reduction through attrition.

4. Operation and maintenance of the East and West Water Treatment Plants, laboratory, distribution system maintenance, and water warehouse with staff reduction through layoffs.

The key components of the RFP were organization and staffing (education, training, and experience of lead personnel), management strategy (normal operations, emergency response, equipment maintenance, and quality assessment) and cost reduction strategy (e.g., staffing reductions).

The proposal submitted by the Water Department, the ultimate winner of the contract, included cost savings in personnel reduction and reduction in the use of consumables. Personnel reduction was to be achieved through reorganization and increased automation, requiring capital improvements as well as the implementation of private sector contracts. The reduction of the use of consumables was to be achieved through close monitoring of ammonia and chlorine residuals, corrosion inhibitor, powdered activated carbon, and electricity usage.

LESSONS LEARNED

- When a public entity competes for an operations contract, the process forces the management and staff to rethink operations and management with increased efficiency and lower costs as a goal.

- When a public entity wins a contract, savings that are achieved become community resources.

- Keeping community associations and the media involved in the selection process encourages the parties involved to operate in the public's best interest.

CONTACT

Randy P. Schuler, Director
Jefferson Parish Water Dept.
1221 Elmwood Park Boulevard
Harahan, LA 80123
P.O. Box 10007
Jefferson, LA 70181
504-736-6050
504-736-6835 FAX

RESULTS

Technical and legal consultants evaluated the technical proposals and concluded that the two private operators and the Water Department were qualified to operate and manage the water system. The Water Department, however, was the lowest bidder under all options. High and low five-year contract operations bid results for options 1,2,3, and 4 were as follows: (1) $40.4 million versus $33.9 million, (2) $36.4 million versus $32.1 million, (3) $71.9 million versus $52.9 million, and (4) $65.3 million versus $49 million.

The Water Department was selected through a Memorandum of Understanding negotiated among the Parish Council, the administration, and the Water Department, covering the period from 1994 to 1999.

As of 1996, staff had been reduced from 388 to 219, which is a 44 percent reduction, just under the Memorandum of Understanding goal of 46 percent. Most of the decrease is attributable to abolishing vacant positions and to attrition. Some other employees were transferred or failed a drug test and were terminated.

By the end of 1996, savings due to contract operations totaled $5.6 million, and the savings are expected to exceed $10 million by the end of 1999. In 1994, 1995, and 1996, actual expenditures were less than those agreed on in the Memorandum of Understanding. Based on current and projected Water Department expenditures, no rate increase is expected to be required until after the year 2005.

Case Study 5.1

Privatization and Competition
City of Scottsdale Water Resources Department

- Scottsdale experienced privatization of a different sort. The Water Resources Department competed with the private sector for the operation of two wastewater reclamation plants that were operated by private sector firms and bought out the managing partner of a water treatment plant.

- Operations of both wastewater reclamation facilities were awarded to the City and operation of the water treatment plant was assumed by the City after it demonstrated savings compared to the private sector.

BACKGROUND

The City of Scottsdale, Ariz. owned two small wastewater reclamation facilities (1.7 mgd and 350 kgd) that were both operated by private companies. During a renegotiation of the contract for the operation of the 1.7 mgd facility, a staff member of the Water Resources Department felt that the Department could compete with the private sector for operation of both facilities.

In addition, a private venture was operating a water treatment plant rated for 22 mgd that had been built through a DBO process. The City was the guarantor of the bonds and paid debt service and operation costs along with a 7.5 percent management fee. Staff from the Water Resources and Finance Departments examined these costs borne by the City and determined that the City could benefit from buying out the managing partner and assuming management and operational responsibilities of the plant.

Finally, the City had had trouble with the premature failure of water service lines between the main and house meters, experiencing massive failures during summer peak demand. The City put out a request for bids to provide emergency assistance in waterline replacement. After two years of this arrangement, the

City felt that it could provide a higher quality service than what it was receiving from its contractor.

Parties Involved

> Scottsdale Water Resources Department
> Scottsdale City Council

APPROACH

The City issued an RFP for operation of the wastewater reclamation facilities. A finance Subcommittee of the Scottsdale City Council conducted an independent review of the bid submittals. The Water Resources Department submitted the lowest bid and was awarded the contract for operation of the facilities.

In the case of the water treatment plant, the City reexamined its contract with the private venture and determined that there were significant opportunities to save money. It bought out the managing partner in the venture, retaining one silent partner, with the Water Resources Department assuming operations of the plant. The City realized savings of $400,000–$500,000 per year, which can be attributed partially to the removal of the management fee, the elimination of the profit embedded in the private sector operation of the plant, and to the exertion of the city purchasing power.

With respect to the emergency maintenance assistance, the City was finding that its crews typically did a better repair job than that of its contractor, with quality complaints on 75 percent of the repairs done by the contractor, versus 5 percent for repairs conducted by city crews. The Water Resources Department put together a proposal to provide the emergency maintenance assistance. It was reviewed by and concurrence received from Quality Resource Management (QRM), which is a financial and performance auditing division of the City of Scottsdale. The proposal was included and approved in the next budget, and the city realized operational savings of $100,000 per year over the contractual costs.

There are several influencing factors in Scottsdale's experience: its workforce is not unionized, it utilizes a work management system with performance standards for most tasks, it has an exceptional benefits package, its compensation program is at the top level for the geographical area, and it has a bonus program in place for each employee that awards up to 2.5 percent of an employee's salary.

LESSONS LEARNED

- Facilities operated by the private sector are subject to competitive pressures similar to publicly run facilities.

CONTACT

Bob Berlease
Scottsdale Water Resources Department
9388 East San Salvador Drive
Scottsdale, AZ 85258
602-391-5655
602-391-5663 FAX
E-mail: berlease@ci.scottsdale.az.us

RESULTS

The City of Scottsdale realized significant savings through a process that subjected the operations of facilities to competition. The Water Resources Department continues to implement cost savings measures such as extensive cross-training of employees to minimize staff requirements for the operation of an expanded water treatment plant, a new advanced water treatment plant, and a new 8 mgd wastewater reclamation facility. The Department is presently participating in a city-wide council called the Managed Competition Board of Directors, which is taking a proactive approach to make city departments as competitive as possible with respect to the private sector.

Case Study 5.2

Privatization of Water Reclamation Facility
City of Chandler, Ariz. Department of Public Works

- In 1984 Chandler was experiencing population growth rates of 15 to 20 percent per year.

- The City determined it needed a 5 mgd wastewater reclamation facility but it was already at its bonding capacity due to extensive borrowing for other infrastructure needs.

- At the time, the tax structure was favorable to private construction and operation of such facilities, so the City pursued this option.

BACKGROUND

In 1984 Chandler was experiencing significant population growth. Due to the current growth and anticipated future growth, the City decided to build a 5 mgd wastewater reclamation facility. However, existing investments in infrastructure had used all of the City's bonding capacity. The limits of the City's bonding capacity, coupled with tax advantages for the DBO of such facilities, led the City to decide to privatize the construction and operation of the facility.

Parties Involved

> Chandler Public Works Department
> Malcolm Pirnie, Inc.
> Parsons Municipal Group
> ES Environmental Services

APPROACH

The City had originally planned to construct and operate the facility itself and had already created the design for the facility. It then issued an RFP for the construction and operation and awarded the contract to Parsons Municipal Group, which is operated by ES2, a subsidiary of Parsons.

The contract was for 20 years, at the end of which the City has the option to purchase the plant for fair market value. The City also pays for the capitalization of the plant, which presently is $2 million per year. The operation and maintenance costs are estimated on an annual basis and divided into 12 installments. ES2 passes this cost on to the City, along with its profit, on a monthly basis. The City is responsible for maintaining a major maintenance and repair fund at $1.1 million, which is accessed only for aforementioned activities.

The City recently completed what was essentially a "paper" upgrade (only minor investments of new equipment) of the plant from 5 mgd to 7.5 mgd. It followed the paper upgrade with the issuance of an RFP to increase the capacity from 7.5 to 10 mgd.

At this time the City has determined that it requires another 5 mgd wastewater reclamation plant. It has issued an RFP for the construction of the plant and will issue a separate RFP for plant operation. A city team is interested in responding to the operations RFP. This would not be without precedent, because the City presently operates a 10 mgd wastewater reclamation facility.

CONTACT

Paul Bishop
Chandler Public Works Dept.
Water and Wastewater Administration
602-786-2857
602-786-2780 FAX
pbishop344@aol.com

RESULTS

By privatizing the construction and operation of a wastewater reclamation facility, the City was able to build on existing tax advantages to creatively address bonding capacity limitations while still meeting infrastructure needs. As the City looks to expand its wastewater reclamation system through the addition of a new facility, it is likely that both a public and a private team will respond to the RFP when it is issued.

Case Study 5.3

Outsourcing Sludge Management
Passaic Valley Sewerage Commission, N.J.

- Under a Consent Decree with the New Jersey Department of Environmental Protection (NJDEP), the Passaic Valley Sewerage Commission (PVSC) was required to provide for beneficial use of its sludge.

- PVSC used the RFQ and RFP processes in a unique way to let the market determine the most suitable cost effective system.

- PVSC placed the responsibility for compliance with the Consent Decree on a private vendor under a 10-year contract.

BACKGROUND

The Passaic Valley Sewerage Commission (PVSC) operates a 330 mgd secondary wastewater treatment plant that serves 1.5 million people and 4,500 businesses in four New Jersey counties. Prior to the project, combined primary and secondary thickened digested sludge was disposed of at a landfill as an interim solution to the ban on ocean dumping. As a result of the Ocean Dumping Elimination Act and the Federal Ocean Dumping Ban Act (1988), Consent Decree was issued between PVSC and the NJDEP requiring that it beneficially use its sludge.

The Consent Decree allowed for privatization of sludge management services via a full service (off-site) design-build-operate and privately financed arrangement. The PVSC chose this option. The Consent Decree required the PVSC to procure a full-service vendor using a preestablished milestone schedule for issuing procurement documents and executing a contract.

PARTIES INVOLVED

Public Agency

Passaic Valley Sewerage Commission

Private Party

Wheelabrator Clear Water New Jersey

Consultant

Malcolm Pirnie, Inc.

APPROACH

The PVSC used the RFQ and RFP process, in effect, to let the market determine the most suitable system. Specifically, the RFQ invited vendors to submit qualifications for any sludge composting technology; an evaluation of the responses dictated the treatment methods the PVSC could realistically consider. Based on the Statements of Qualifications of 33 vendors, four categories emerged: composting, land application, sludge-derived product for agricultural use, and innovative technologies.

The RFP process proved to be highly effective in allowing the PVSC to select both the most qualified vendor and the most cost effective contract arrangement. Based on the results of the RFQ process, PVSC first issued four distinct RFPs (one for each sludge management category), leaving open the technology selection. Second, the RFPs allowed responders to submit proposals that varied with respect to the following options: (1) base sludge commitment as a percent of sludge generated; (2) service agreement duration; (3) sludge preparation requirements; (4) existence of put-or-pay commitment from PVSC; and(5) percent standby sludge commitments. While this led to the proposal of more than 600 options, requiring an extensive and methical evaluation process, it allowed PVSC to comprehensively evaluate all options.

The RFPs contained performance specifications and a service contract, both of which required only minor adjustment during contract negotiations. In order to optimize final contract conditions, several proposers were selected to undergo simultaneous negotiations withthe PVSC.

LESSONS LEARNED

- The solicitation process worked effectively to define a cost effective solution.

- A simultaneous negotiation process maximized the competitive process and reduced cost of original proposals.

CONTACTS

Passaic Valley Sewerage Commission
Robert J. Davenport
Executive Director
600 Wilson Avenue
Newark, NJ 07105
201-344-1800
201-817-5738 FAX

RESULTS

The procurement process was conducted in compliance with the New Jersey Wastewater Treatment Privatization Act. This act requires a multiple stage approval process, including public hearings on the proposed contract and approval from the New Jersey Department of Community Affairs and the New Jersey Department of Environmental Protection. Resulting disposal fees were $65 per wet ton compared to the previous $95 per wet ton for landfill disposal. This translates into an annual savings of more than $2 million per year.

Case Study 5.4

Design-Build-Operate
Seattle Public Utilities, Wash.

- The City of Seattle is pursuing a design-build-operate approach to develop its first filtration plant.

- DBO is a nontraditional approach to project implementation wherein proposers compete for a contract that includes design, construction, and long-term facility operation.

BACKGROUND

Through a Design-Build-Operate (DBO) approach, the City of Seattle is developing a filtration plant to treat water from the Tolt River, which provides about a third of the City's current supply for a service population of 1.2 million. The $100 million plant with a capacity of 120 mgd will permit continuous operations through periods of high turbidity, increasing yield and reliability and allowing long-run conformity with anticipated USEPA water quality standards.

PARTIES INVOLVED

Public Agency

Seattle Public Utilities

Consultants

RW Beck
Malcolm Pirnie, Inc.
J. Gilbert, Inc.
A. Yarmuth, Wilson, Inc.
A. Hawkin, Dealfield and Wood

APPROACH

The City initially planned to follow the traditional practice in building water system facilities by awarding a competitively bid construction contract based on an independent design engineer's specifications. However, at the time the City was considering entering into final design for the project, the opportunity existed through state legislation to utilize a DBO procurement process.

In the fall of 1995 the Department determined that substantial benefit to rate payers could potentially be achieved through adoption of a DBO contracting method for implementation of the Tolt Treatment Facilities.

Consultants were selected to provide support services to the DBO process and augmented a high level City project team comprised of water quality, engineering, and policy and administrative leadership in SPU. Additional specialty skills were brought in with other independent technical, legal, and financial advisers.

In the summer of 1996, a city council ordinance authorized the issuance of the Request for Proposals (RFP) and continuation of this process subject to the condition that the DBO proposals meet the City's technical and business specifications and produce savings of at least 15 percent over the estimated cost of pursuing the project using a conventional procurement process (that is, the net present value of the proposals was required to be 15 percent below the benchmark of $156 million). In seeking proposals, the City also required proposers to comply with the City's minority and business contracting standards to ensure that workers were paid prevailing wages and to pursue environmentally sensitive approaches to plant design and operations. The RFP also included a draft service agreement, setting forth SPU's risk posture, and 25 draft schedules for the treatment process, facilities, and operations.

The RFP required proposers to submit two separate proposals: Proposal A and Proposal B. In Proposal A, facility specifications were intended to result in a facility that meets currently enacted regulations and those that could be reasonably anticipated as of the projected facility on-line date. Selection of a Proposal A facility would most likely result in treatment processes similar to those envisioned during the predesign (the benchmark of $156 million is based on a Proposal A facility). Proposal B specifications were intended to result in a facility that meets current regulations, takes a more conservative estimate of potential Enhanced Surface Water Treatment Rule requirements, and addresses other enhancements that could be desired in a long-term operational period for the facility. Proposal B would likely require additional treatment processes such as ozonation. Proposers were also asked to include costs and methods for upgrading a Proposal A facility to Proposal B performance levels in the future.

Proposals were received on November 27, 1996. SPU's evaluation committee conducted a series of internal work sessions to better understand various aspects of the proposals and concluded that issuing requests for best and final offers would serve the City's interests. Best and final proposals from the four shortlisted proposers were received on February 7, 1997.

LESSONS LEARNED

- The project's diverse and multidisciplinary team was critical in bringing in diverse perspectives and better solutions to complex issues.

- The project philosophy, developed early in the process, was to key to balancing performance versus prescriptive interests, choices between innovation and traditional approaches, and tradeoffs between control and costs.

- The project partnership with Seattle elected officials ensured that stakeholders were on board throughout the process.

CONTACT

Liz Kelly, Project Coordinator
Resources Development Division
Seattle Public Utilities
720 Third Ave., 10th Floor
Seattle, WA 98104
206-386-9779
206-684-0206 FAX
E-mail: liz.kelly@ci.seattle.wa.us
http://www.ci.seattle.wa.us/seattle/util/DW/TOLT/
projorg.htm (web)

RESULTS

On March 3, 1997, the Seattle City Council formally authorized SPU to proceed with the DBO process. CDM Philip and the Proposal B alternative were selected. A contract was signed on May 21, 1997, and the plant is expected to be on line by the end of 2000. This innovative procurement process produced a savings of $70 million when compared to the estimated cost of building and operating a comparable facility using the conventional process of consultant design, low bid construction, and City operation. The SPU internal and consultant costs for the procurement process was approximately $2 million.

Case Study 5.5

Wastewater Treatment Plant Asset Sale
Miami Conservancy District Franklin, Ohio

- As a demonstration project, this was the first successful asset sale under Executive Order 12803.

- Part of the funds generated by the sale retired the original construction loan debt, with the rest going to participating communities for sewer upgrades.

- The contract between the communities and the owner/operator of the plant includes provisions for changing the arrangement after 20 years.

BACKGROUND

Until 1995 the Miami Conservancy District (MCD) was responsible for providing wastewater services in three communities (Germantown, Carlisle, and Franklin) and unincorporated areas in Warren and Montgomery Counties. About 25,000 people reside in this area. The Franklin Area Wastewater Treatment Plant, built in the 1970s and owned by MCD, is one of the plants servicing this area.

In 1993, the MCD made a policy decision to discontinue their direct involvement in wastewater treatment, and to sell the wastewater treatment facilities. At about the same time, the contractor, who had been operating the Franklin Area Wastewater Treatment Plant under contract since 1987, expressed an interest in buying it from the MCD.

The plant was selected by USEPA to be a first demonstration project under Executive Order 12803, which defines the conditions under which a public facility may be transferred to a private entity. USEPA selected the plant because its characteristics fit the intent of the order and MCD wanted to sell it.

The communities selected Wheelabrator to operate the plant on a sole source basis.

PARTIES INVOLVED

Public Agencies

Franklin Area Wastewater Treatment Corporation
USEPA
Ohio Water Development Authority
Ohio EPA
Federal Office of Management and Budget
Internal Revenue Service

Private Party

Wheelabrator EOS

In mid-1997 Wheelabrator EOS was purchased by U.S. Filter, which operates the plant using the same employee team that was used by Wheelabrator.

Consultants

Sierra Consulting Associates
    Consultant to Wheelabrator; EOS
Jack Dowd, Bond Counsel
Squire, Sanders & Dempsey Defeasance Counsel
Raftelis Environmental Services, Adviser to Communities
Applied Technologies, Inc., Consultant to Franklin Area
    Wastewater Treatment Corp.

APPROACH

Several critical hurdles had to be overcome in order for the sale to take place. First, the plant retained the domestic sewage exclusion, considered a necessary condition for the sale. Second, the Franklin Area Wastewater Treatment Corporation and Wheelabrator became copermittees of the plant.

Third, participants had to determine the value of the plant. Fourth, the construction grants under which the facility was built were amortized so that the transfer could take place. Fifth, Wheelabrator supplied the money to pay back outstanding tax exempt debt associated with bond defeasance. Sixth, Public Utility Commission of Ohio (PUCO) regulations were avoided by establishing that the communities themselves, not the individuals in them, be considered Wheelabrator's customers.

The Franklin Area Wastewater Treatment Plant Corporation was formed as the agency representing the communities served by the plant. They in turn hired Raftelis Environmental Services to be their consultant during the negotiations with Wheelabrator.

The contract between Wheelabrator and the Franklin Area Wastewater Treatment Corporation includes the following:

- Determination of future rates.

- Stipulation that communities could require Wheelabrator to provide additional capacity as needed.

- Requirement that if Wheelabrator could save money through capital improvements, both Wheelabrator and the communities benefit.

- Requirement that communities have the right to review operations in order to protect their assets.

- Determination of potential impact in the case of performance failure.

At the end of the 20-year contract term, one of three options is possible: the communities buy back the plant for the remaining book value of capital; the communities decide not to participate in the plant any longer; or Wheelabrator continues to own the plant under a contract extension or new contract.

During the implementation of the agreement, numerous meetings and public forums were held in order to inform the public and industry of the details of the project.

An independent firm was hired by the Franklin Area Wastewater Treatment Corporation to inspect the plant on their behalf each month, and to consult with and represent the Corporation in settling disputes with the operator.

LESSONS LEARNED

- Project implementation goes more smoothly if all related staff people are involved from the start.

- Involvement of the public during the planning process eases the implementation of the project, particularly when the process is unique.

CONTACT

Paul Hillard, President
Franklin Area Wastewater Treatment Authority
City of Franklin
35 East 4th Street
Franklin, OH 45005
513-746-9921
513-746-1136 FAX

RESULTS

Wheelabrator paid $7 million for the plant, most of which was used to retire the debt of the original construction loan. The remaining portion of the $7 million (about 20 percent) was used by the communities (proportionally, based on percentage of use) for sewer upgrades and extensions.

The Franklin Area Wastewater Treatment Corporation is pleased with the operation of the plant and with its working relationship with the operator.

Case Study 5.6

Design, Build, Own and Operate - Water Treatment Plant
Franklin, Ohio

- The City of Franklin selected Earth Tech Operation Services to design, build, own, and operate a 5 mgd water treatment plant for 20 years.

- Earth Tech is responsible for design and operational risks and for regulatory approval.

- Bidders were required to submit a nonrefundable $20,000 concession fee to the City.

- The City enjoys one of the lowest water rates in Ohio.

BACKGROUND

The City of Franklin, Ohio, has a population of about 212,000, and since the early 1960s has owned and operated its 5 mgd water treatment facility, consisting of a well field and disinfection and pumping equipment. Franklin is the home of several boxboard and fiberboard plants for whom the price of water is critical.

In 1995 the Ohio EPA informed the City that because of excessive levels of iron and magnesium in the water, additional treatment would be necessary. With the help of Sierra Consulting, the City examined options for the additional treatment required:

- The City builds and staffs the plant.

- The City builds the plant and it is operated by others.

- The City selects a private company to design, build, and operate the plant.

- The City selects a private company to design, build, own, and operate the plant.

The City elected the last option because it provided the most comprehensive service package, minimized City staff involvement and expertise requirements, and provided off balance sheet financing at lower net costs than the alternative, tax exempt financing.

APPROACH

In early 1996 the City solicited expressions of interest and 27 firms responded. The field was narrowed to four qualified firms, based on responses to the RFQ.

The RFP included the requirement that the proposal submittals be on a performance output basis, leaving room for variation in equipment and process choices by each submitter. The restriction on processes and equipment technology focused on the requirement for demonstrated, proven technology implemented in operating facilities.

The City required those responding to the RFP to submit $20,000 as a nonrefundable cost to participate. This fee was collected to cover the City's cost for the RFQ/RFP process and to offset the City's costs for outside consultants.

In August 1996, Earth Tech was selected to be the contractor. Several provisions are included in the contract:

- Price and schedule guarantees during construction.

- Annual rate adjustments are constrained to published indices for the term of the contract.

- Earth Tech is responsible for design and operation risks, including those associated with the throughput rate exceeding the 10-state regulatory standard.

- Regulatory approval and capital improvements are the responsibility of Earth Tech.

- Earth Tech is required to comply with surface water standards, not groundwater standards, as an added assurance of treatment and costs, if the well field becomes influenced by the surface water in the Miami River.

- In the event that the agreement is terminated, Earth Tech is required to provide a bridge loan to the City to allow the City to assume plant operations and ownership while the City obtains permanent financing.

- If government programs that provide advantageous financing are created, the City can ask Earth Tech to match the financing terms or the City can become the facility owner in order to take advantage of such government programs.

At the end of the 20-year contract term, the City has three options: It can buy the facility from Earth Tech, it can end its relationship with the facility and with Earth Tech, or it can have Earth Tech continue as contractor. Earth Tech, on the other hand, cannot require the City to continue the relationship.

LESSONS LEARNED

- Significant guarantees and risk assumption can be obtained from the private sector.

- Costs and compliance risk can be transferred to the private sector.

- Private sector financing, under these circumstances, can be less than tax exempt debt issued by the City.

- The RFQ/RFP process can be developed and completed in about six months and a contract awarded within eight months.

CONTACTS

Jim Mears, Mayor
City of Franklin
313 S. Main Street
Franklin, OH 45005
513-743-7755
513-746-2166 FAX

Paul Hillard
City Councilman and Selection Committee Head
City Hall
35 E. Fourth Street
Franklin, OH 45005
513-746-8152
513-746-1145 FAX

RESULTS

Franklin is enjoying the second lowest water rates in Ohio, and is guaranteed rate stability and the availability of water. Further, the boxboard and fiberboard plants in Franklin have been able to expand, attributable partly to lower water and wastewater rates.

Case Study 5.7

Lease of a Public Water System
Hawthorne, Calif.

- The City of Hawthorne, responding to a financial crisis, evaluated options for its water system.

- Leasing to a private water management company proved to be the simplest and quickest option, and has resulted in an infusion of funds to the City, and relief of most of the administrative responsibility for the system.

BACKGROUND

The City of Hawthorne, population 76,000, owns a water system consisting of 51 miles of pipeline, 9.2 million gallons of storage, and a 5 mgd treatment plant. Seventy-five percent of its water originates from the Metropolitan Water District of Southern California, and 25 percent originates from the City's groundwater wells.

In early 1995, financial difficulties that the City had been experiencing evolved into a crisis, and the city council began to search for ways to reduce costs quickly. The water system was self-funding and even returning some "profit" to the City, but the City as a whole needed an infusion of funds.

Several changes to the structure of the water system were considered, including creating a utility district, selling bonds, and selling or leasing the system. Creating a water utility district would have required more time than the City desired to spend to make the change. Selling bonds was ruled out because the City did not want to enter into debt. Selling the system would have required a special referendum involving only the people living in the area served by the system, and the City was not certain that such a vote would pass.

The best option for the City was to lease the system, since it would bring in funds relatively quickly, but still allow the City to retain control over the system.

PARTIES INVOLVED

Public Agency

City of Hawthorne,
Calif. Department of Public Works

Consultants

Marshall & Stevens (Valuation and Financial Consultants)
McDonald, Holland and Allen (Legal Consultants)

Contractor

California Water Service Co.

APPROACH

A consultant prepared an evaluation of the water system to determine its lease value.

An RFQ was prepared by the City in late 1995, and two firms, Southern California Water Company and California Water Service Company, responded and were considered qualified. A lease was prepared and reviewed by both qualified firms. The lease stipulated that City water system employees must be hired by the selected firm with an "equal or better" total employment package.

California law requires that competition for leases be conducted by auction. Once the bidding began, each firm was given three days to better the bid of the other firm. The process took several months, and was finally completed when Southern California Water Company was unwilling to offer a higher price for the lease of the system.

LESSONS LEARNED

- In a lease situation, it is advisable to establish a minimum bid increase amount, to avoid a long process in which there is little increase in the bid price.

- A lease agreement provided a quick infusion of funds and allowed the City entity to maintain ownership and control of the system.

- A lease agreement relieved administrative pressure on the City and reduced its responsibility for the system to oversight of the leasing company.

CONTACT

Charles Herbertson, Director
Department of Public Works
4455 West 126th Street
Hawthorne, CA 90250
310-970-7955
310-970-7033 FAX

RESULTS

California Water Service Company was awarded the lease for a period of 15 years. The City achieved the financial aid that it needed, and was relieved of most of the administrative responsibilities for the system at a time when it was overtaxed. The City does continue to oversee the system.

Six employees who worked for the City were hired by California Water Services when they took over the lease for the system, as required by the contract.

Case Study 5.8

Reverse Privatization
Huber Heights, Ohio

- Huber Heights decided to purchase an existing water and wastewater system from a private operator in order to allow industrial growth, which was restricted by boundaries imposed by the public utilities commission.

- After a valuation of the system, a due diligence study, and a public referendum, the City condemned the system and selected Earth Tech Operation Services for management of the system.

- As a result of the reverse privatization, water rates have been reduced and no City staff has been added.

BACKGROUND

Huber Heights is a community of 40,000 that was created in the 1950s. Within the City is a water treatment plant and wastewater treatment plant that were built with the community itself. In 1963 the plants and associated collection and distribution lines were sold to a private company, which was regulated by the Public Utilities Commission of Ohio (PUCO).

In the mid-1980s a regional wastewater treatment plant replaced the private plant, using USEPA's Construction Grants Program. The plant served several communities, and was owned by a public authority, the Miami Conservancy District.

In 1992 the City prepared a Water Feasibility Study to examine the feasibility of the City's acquisition of the water and wastewater systems. The study was prompted by a need to increase the tax base and to develop industrial sites, many of which were outside of the private utility's existing service boundary, as identified by PUCO. Additionally, the local water treatment plant was for sale. The study concluded that the City could operate the system effectively and reduce customer rates.

The City then attempted to purchase the private utility, but the system was sold to another private company. After a valuation of the system and a special referendum election confirming voter support, the City decided to condemn and take over the utility system.

As part of the referendum, the City committed to retain the services of a private firm to manage and operate the system and to lower and maintain reduced water rates for five years.

PARTIES INVOLVED

Public Agency

City of Huber Heights
Citizens Water and Sewer Advisory Board

Private Party

Earth Tech Operation Services

Consultants

Raftelis Environmental
Shaw, Weiss and DeNaples

APPROACH

The City issued an RFQ for a professional utility management firm in 1994, and eight firms responded. Four firms were invited to submit detailed proposals for system management: Earth Tech, Avatar Utility Services, Inc./Wheelabrator EOS, Citizen's Utilities, and Ohio Suburban Water Company/American Commonwealth Management Services Co.

Because of the condemnation court case, there was little information about day-to-day operations and costs available. This made putting proposals together more difficult for vendors than would be the case in usual circumstances.

Three proposals were received by the City, and the City selected Earth Tech Operation Services as the management firm. The City completed the acquisition of the system and Earth Tech began operation in 1995.

Funding for the City acquisition was through issuance of Water Revenue Bonds and through the Water Pollution Control Loan Fund.

In 1996 the City and neighboring communities formed the Tri-Cities Wastewater Authority, which purchased the 12.8 mgd wastewater treatment plant that served these communities.

LESSONS LEARNED

- Reverse privatization should be considered where restrictions placed on private operators conflict with government and public objectives.

CONTACT

Deborah Swan, P.E.
Project Manager
City of Huber Heights
6131 Taylorsville Road
Huber Heights, OH 45424
837-233-1423
937-233-1272 FAX

RESULTS

As the City promised customers, water rates were reduced when Earth Tech started operations. The City's Finance Department and Engineering Division have additional duties as a result of the reverse privatization, but no additional staff has been hired.

It is expected that the contract with Earth Tech will be extended through 1999.

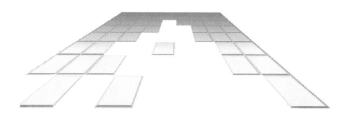

Outline Solicitation Documents

INTRODUCTION

Detailed descriptions of the RFQ and RFP process are provided in chapter 15. These two official procurement documents, the Request for Qualifications (RFQ) and the Request for Proposals (REP), can be used to obtain proposals from potential contract service providers. The documents can be combined into a single request and are similar whether the solicitation is for contract operations, managed competition, or capital facilities.

The following are outlines of RFQ and RFP documents that are also applicable to DBO and BOOT projects with certain additions.

Solicitation documents can range from a few pages to extensive volumes with detailed specifications. Neither approach is appropriate. With little detail scope and responsibilities remain unclear, leading to misinterpretation and disputes as to what is or is not included in the proposed costs. On the other hand, detailed requirements and specifications can hamper the creativity of the respondents in providing innovative solutions. It is important to strike a balance between brief requirements and detailed specifications. The following RFP document in this appendix assumes that a proposed form of a service agreement along with technical performance specifications would be included as an attachment to the RFP.

OUTLINE BASIC REQUEST FOR QUALIFICATIONS (RFQ)

1. Introduction

2. Project Information

   2.1 Background Information

   2.2 Overview of Service Provider Activities

   2.3 Information Requests and Addenda

3. Procurement Process Overview

   3.1 Procurement and/or Competition Process

   3.2 Competition Objectives

   3.3 Procurement Schedule

   3.4 Shortlist Selection Process

   3.5 Special Conditions

   3.6 Information Disclosure to Third Parties

4. Submittal of Qualifications And Comments

   4.1 General Instructions—Rules for Submission

   4.2 Information Requirements of Statement of Qualifications

       4.2.1 Cover Letter

       4.2.2 Executive Summary

       4.2.3 Project Understanding

       4.2.4 General Company/Team Information (Company Organization)

       4.2.5 Technical Qualifications

       4.2.6 Relevant Project Experience

       4.2.7 Financial Information

       4.2.8 Project Guarantor

MODIFICATIONS TO BASIC RFQ

For Managed Competition

For managed competition processes it is uncommon for municipal employees to submit a statement of qualifications. Most managed competitions include the assumption that municipal employees are prequalified.

For Inclusion of Capital Facilities

While the structure of the RFQ for inclusion of capital facilities is essentially the same, the submission requirements for the proposer will be more rigorous. Because of the permitting, design, and construction components of such projects, the RFQ should require qualifications for design and permitting elements, along with the qualifications for construction. In addition, qualifications should be sought on the ability to secure financing for BOOT projects where financing is required.

OUTLINE OF BASIC REQUEST FOR PROPOSALS (RFP)

LIST OF APPENDICES

Schedule 11:   Equipment and Chemicals Inventory
Schedule 12:   Pass Through Costs
Schedule 13:   Personnel
Schedule 14:   Historic Electricity Consumption Data
Schedule 15:   Performance Testing
Schedule 16:   Reporting Requirements
Schedule 17:   Liquidated Damages

B.   Historic Water Quality Characteristics
C.   Equal Opportunity Program Requirements
D.   Plant Permits and Related Correspondence

MODIFICATIONS TO BASIC RFP

For Managed Competition

While the same basic RFP can be used by municipal employees for managed competition processes, several distinctions are required:

- Proposal submission requirements for municipal employees for items such as performance bonds and insurance requirements will not be required.

- Municipal employees must include an audited proposal.

- Evaluation criteria must address the nuances of comparing municipal proposals with other contractor proposals.

- In lieu of a service agreement, a memorandum of understanding will be required with the municipal employees.

For Inclusion of Capital Facilities

Where capital facilities are required, submission requirements should include proposed conceptual designs and implementation approaches. In addition, technical specifications that address the minimum level of quality for equipment and systems should be included.

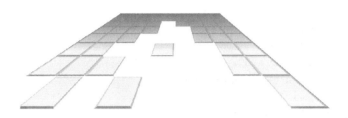

Outline Service Agreement

INTRODUCTION

A draft of the service agreement or contract can be included in the RFP, but once the solicitation process is completed the contractual provisions of the engagement of service provider by the utility are detailed in a final service agreement. The service agreement addresses all terms and conditions, ranging from contract term to performance requirements.

The service agreement includes technical, legal, business, and financial elements within either the text or appendices and exhibits. This document serves as the basis for performance measurement by the contractor. The agreement must be carefully developed to address foreseeable conditions during the project term.

OUTLINE OF BASIC SERVICE AGREEMENT

RECITALS

ARTICLE I   DOCUMENTS

Section 1.01   Agreement Documents

ARTICLE VI   INDEPENDENT ENGINEER

ARTICLE VII   INSURANCE AND BOND

ARTICLE VIII   EVENTS OF DEFAULT

ARTICLE IX   TERMINATION

ARTICLE X   REPRESENTATIONS

ARTICLE XI   MISCELLANEOUS

SERVICE AGREEMENT SCHEDULES:

Schedule 12:   Personnel
Schedule 13:   Capital Improvements

MODIFICATIONS TO BASIC SERVICE AGREEMENT

For Managed Competition

For a managed competition process the service agreement would be replaced with a Memorandum of Understanding (MOU) if municipal employees are selected. The MOU would incorporate the intent of the service agreement.

For Inclusion of Capital Facilities

If the project includes capital facilities, the service agreement should expand to include:

- design and permitting review procedures

- construction requirements and monitoring activities

- acceptance testing provisions

- construction drawdown (payment) schedules

- implementation schedules

- technical specifications and conceptual drawings of the proposed facilities

Additional Reading

A literature search on the topics discussed in this book would produce literally thousands of references. Given the availability of so many easily accessible resources, the authors have not attempted to compile a comprehensive bibliography. Readers are encouraged to do their own literature searches on the topics of particular interest to them, using databases such as American Water Works Association's WATERNET. The bibliography that follows is meant to be helpful to the reader seeking a few references for further reading. This listing also serves to illustrate the wide range of information available to utilities seeking to explore more fully the promising opportunities of organizational improvement.

Additional major online databases containing items relevant to the subject of this book include:

Commercial Databases—available via database service providers such as Knight Ridder (DIALOG).

NTIS_64-1997/Sep W3—National Technical Information Service, government reports.

Ei Compendex(R)_1970-1997/Sep W2—Engineering Index, general engineering.

Wilson Appl. Science & Technology Abstracts_1983-1997/ Jul—H.W. Wilson, general engineering.

Water Resources Abstracts_1967-1997/Jul—Cambridge Scientific, all aspects of water resources.

IAC PROMT(R)_1972-1997/Aug 22—Information Access, business perspectives.

Also available on the Internet:

ASCE Publications Database—searchable database of all ASCE works 1975–present.

Austin, Sarah, and Shapiro, Gillian. 1996. "Equality-Driven Employee Involvement." *Journal of General Management* 21(4):62(16).

Basu, Ron, and Wright, Nevan. "Measuring Performance Against World Class Standards." *IIE Solutions* 28(12):32(4).

Cairo, Patrick R., Smith, Donald L., and Ballard, Ronald J. "Delegation of Municipal Water and Wastewater Services to Private Companies." *WEF/AWWA Joint Management Conference: Facing the Management Challenge.* Tulsa, Okla. Water Environment Federation and American Water Works Association, 1995.

Campbell, John Dixon. *Uptime: Strategies for Excellence in Maintenance Management.* Productivity Press. Portland, Maine: 1995.

Crossley, Ian A., Brailey, Donald; Melamud, Susan. *What Can the U.S. Utilities Learn from the English Experience?* Annual Conference Proceedings; American Water Works Association Management and Regulations. Toronto, Ontario, Canada. American Water Works Association, 1996.

Deb, Arun, K., Hasit, Yakir J., and Grablutz, Frank M. *Distribution System Performance Evaluation.* AWWA Publication: 1995.

"Developing Comprehensive Performance Indicators." 1997. CMA–*The Management Accounting Magazine.* 71(2):39(1).

Edwards, Keith. "Benchmarking in Customer Services: A new Competitive Posture." *WEF/AWWA Joint Management Conference: Facing the Management Challenge.* Tulsa, Okla. Water Environment Federation and American Water Works Association, 1995.

Giardina, Richard D., Ambrose, Robert D., and Olstein, Myron. "Private Sector Financing of Public Facilities—When and Why It May be Appropriate." 1994 *Annual Conference*

Proceedings; American Water Works Association; Engineering and Operations. New York, N.Y., 1994.

Gouillart, Francis J., and Kelly, James N. *Transforming the Organization: Reframing Corporate Direction, Restructuring the Company, Revitalizing the Enterprise and Renewing People.* New York: McGraw Hill, 1995.

Groom, J. 1995. "Maintenance Management in Australia." *Australian Water and Wastewater Association.* 22(4):29.

Gullet, Barry M., and Bean, Douglas O. "The Charlotte Model for Competition." 1996 *Annual Conference Proceedings; American Water Works Association; Management and Regulations.* Toronto, Ontario, Canada, 1996.

Hammer, Michael, and Champy, James. *Reengineering the Corporation: A Manifesto for Business Revolution.* New York: Harper Business, 1993.

Harrington, H. J. *Business Process Improvement: The Breakthrough Strategy for Total Quality, Productivity, and Competitiveness.* New York: McGraw Hill, 1991.

Hayes, Bob E. *Measuring Customer Satisfaction: Development and Use of Questionnaires.* Milwaukee: ASQC, 1992.

Kotter, John P. *Leading Change.* Boston: Harvard Business School Press, 1996.

LaChance, Peter A. "Practical Application of Performance Benchmarking." 1995 *Annual Conference Proceedings; American Water Works Association Management and Regulations.* Anaheim, Calif., 1995.

Light, Paul C. 1997. "Organizational Performance and Measurement in the Public Sector: Toward Service, Effort, and Accomplishment Reporting." *Journal of Policy Analysis and Management* 16(2):328(6).

Manion, Patrick W. 1997. "Accountability for Performance: Measurement and Monitoring in Local Government." *Public Productivity and Management Review* 20(3): 346(1).

Mohrman, Allan M. *Large-Scale Organizational Change.* San Francisco: Jossey-Bass Publishers, 1991.

Moylan, Robert L. Jr. "Performance Based Budgeting: An Essential Management Tool for the 90's." 1995 *Annual Conference Proceedings; American Water Works Association; Management and Regulations.* Anaheim, Calif., 1995.

Moylan, Robert L. Jr., Ashe, Carol R., and Ridge, Joseph T. "Evaluating Contract Operations for a New Water Treatment Facility." *1996 Annual Conference Proceedings; American Water Works Association; Management and Regulations.* Toronto, Ontario, Canada, 1996.

Okun, Daniel, A. "Privatization and the Consumer." *1995 Annual Conference Proceedings; American Water Works Association; Management and Regulations.* Anaheim, Calif., 1995.

Osborne, David. *Reinventing Government.* Redding: Addison-Wesley Publishers, 1992.

Rothstein, Eric P. "Modeling Privatization Decisions." *Annual Conference Proceedings; American Water Works Association.* Toronto, Ontario, Canada, 1996.

Russell, John C. 1997. "Driving Change Through Performance Measurement." *Strategy and Leadership* 25(2):40(2).

Shanker, Amy, and Len Rodman. 1996. "Public-Private Partnerships." *Journal American Water Works Association,* 88(4): 102-107.

Siu, Stanley. "Improving Productivity in Water Maintenance Operations." *Proceedings of the 45th Annual Conference of the Western Canada Water and Wastewater Association.* Saskatoon, Saskatchewan, Canada, 1993.

Tuck, Nancy, and Zaleski, Gary. 1996. "Criteria for Developing Performance Measurement Systems in the Public Sector." *International Journal of Public Administration.* 19(11-12):1945(34).

Warren, Lloyd. "Linking Customer Satisfaction to Performance Measures." *1996 Annual Conference Proceedings; American Water Works Association; Management and Regulations.* Toronto, Ontario, Canada, 1996.

Westerhoff, Garret P., Lane, Thomas J. 1996. "Competitive Ways to Run Water Utilities." *Journal American Water Works Association* 88(4): 96-101.

Zigon, Jack. 1997. "Team Performance Measurement: A Process For Creating Team Performance Standards." *Compensation and Benefits Review* 29(1):38(10).

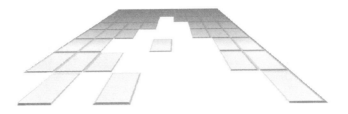

Index

NOTE: *f.* indicates a figure, *t.* indicates a table, *s.* indicates a sidebar.

Occupational Safety and Health
Administration
(OSHA), 389
Ocean Dumping Elimination
Act, 473
Office of Management and
Budget (OMB), 359, 481
Office of Water Regulation
(OFWAT), 213
Ohio Suburban Water
Company/American
Commonwealth Man-
agement Services Co.,
492
Ohio Water Development
Authority, 481
Oklahoma City, Okla., 11, 271
Oldham, Michael, 402
O'Neal and Associates, 416
One stop service, 112
Online operations and mainte-
nance information, 138t.
Operations and maintenance
(O&M)
best practices evaluation for,
76–77t.
building, 81, 82t.
contracting of. See
Contracting
Houston, 452–454
integration of, 73–74, 76t.
opportunities for enhance-
ment in, 70–77
optimization of, 67–68
references and resources for,
74, 76t.
Operations contracts, 24–25
Organizational area best practices evaluation, 86–87t.
Organizational constraints, 69,
83–87
Organizational Horizons, Inc.,
416
Organizational performance
measures, 212–213t.

Organizational structure, 83,
86t.
Organizational support teams,
244
OSHA. See Occupational
Safety and Health
Administration
Outsourcing, 123, 201–207. See
also Contracting
capability for in marketplace,
206–207
defined, 25
good candidates for, 203–205,
207f.
legal issues in, 207
nine categories of questions
for, 201–203
Passaic Valley Sewerage
Commission, 473–
475
political and organizational
support for, 205–206
Oversight meetings, 54
Oversight systems, 178
Ownership, 354

P

Parking facilities, 116
Parsons Municipal Group, 471
Passaic Valley Sewerage Com-
mission (PVSC), N.J.,
473–475
Pass-through costs, 300–301
Paul, Jennifer, 435
Pennsylvania, 153
Pennsylvania American Water
Company, 44
Perceived loss of control, 358
Perceptions
about customer service, 114
of water quality, 9
Performance, 15–26
core responsibility of utilities
in, 20–21

customer's central place and,
16–20
obligation for competitive, 22
path to competitive, 23–24
public participation and, 22–
23
Performance assessment, 35–48,
197–220
against others, 211–219
building list of strategies in,
45–46
choosing key areas for, 41–42
of control issues, 207–208
of customer alignment, 198–
200
of energy efficiency, 156–158
framework for, 39–41
gap analysis in. See Gap
analysis
identifying improvement
strategies in, 45
of information technology,
136
initiation of steering and process teams in, 36–37
internal, 38, 200–207
setting priorities in, 46
skills, 227–228
in strategic planning, 189–190
third party, 218–219
of treatment works, 63–64
Performance audits, 303–304
Performance evaluations, 57
Performance goals
for customer service, 121
for treatment works, 64
Performance improvements
internal. See Internal performance improvement
in maintenance and rehabilitation, 101–102
methodology in, 64–65
regulatory compliance and,
383–385
reorganization in, 383